DESIGNS 2002

Mathematics and Its Applications

Managing Editor:

M. Hazewinkel
Centre for Mathematics and Computer Science, Amsterdam, The Netherlands

Volume 563

DESIGNS 2002
Further Computational and
Constructive Design Theory

Edited by

W. D. WALLIS
Southern Illinois University

Kluwer Academic Publishers
Boston/Dordrecht/London

Distributors for North, Central and South America:
Kluwer Academic Publishers
101 Philip Drive
Assinippi Park
Norwell, Massachusetts 02061 USA
Telephone (781) 871-6600
Fax (781) 871-6528
E-Mail <kluwer@wkap.com>

Distributors for all other countries:
Kluwer Academic Publishers Group
Post Office Box 322
3300 AH Dordrecht, THE NETHERLANDS
Telephone 31 78 6576 000
Fax 31 78 6576 474
E-Mail <orderdept@wkap.nl>

 Electronic Services <http://www.wkap.nl>

Library of Congress Cataloging-in-Publication

Wallis, Walter D.
Designs 2002: Further Computational and Constructive Design Theory
ISBN 1-4020-7599-5

Printed in United Kingdom by Biddles/IBT Global

Contents

Contents

Preface

This volume is a sequel to our 1996 compilation, *Computational and Constructive Design Theory*. Again we concentrate on two closely related aspects of the study of combinatorial designs: design construction and computer-aided study of designs.

There are at least three classes of constructive problems in design theory.

The first type of problem is the construction of a specific design. This might arise because that one particular case is an exception to a general rule, the last remaining case of a problem, or the smallest unknown case. A good example is the proof that there is no projective plane of parameter 10. In that case the computations involved were not different in kind from those which have been done by human brains without electronic assistance; they were merely longer.

Computers have also been useful in the study of combinatorial spectrum problems: if a class of design has certain parameters, what is the set of values that the parameters can realize? In many cases, there is a recursive construction, so that the existence of a small number of "starter" designs leads to the construction of infinite classes of designs, and computers have proven very useful in finding "starter" designs.

A third major use is the exhaustive construction of designs. For example, the (hand) construction of all 80 Steiner triple systems of order 15 is regularly used by researchers in design theory. Far larger computerized complete listings have been constructed (for example, the Dinitz-Garnick-McKay compilation of all 526,915,620 non-isomorphic one-factorizations of K_{12}), and these tabulations give some feeling of the richness of the designs involved. They are obviously impossible without the computer.

Computers can arise in other ways in the study of combinatorial designs such as searching large lists for the existence of designs with desired properties. As an example, Beaman's discovery of a skew Room square of side 10 was made by first constructing a complete list of Room squares of side 10 (by computer), then searching the list (by computer).

Sometimes the computer is used as an aid to theoretical work. One can interpret the work on the projective plane of parameter 10 in this way, and another combinatorial example is the proof of the Four Color Theorem.

When *Computational and Constructive Design Theory* appeared in 1996, I said in the preface

> There is very little mathematical literature on computers in design theory. This book is a step toward remedying this deficiency. I have aimed at two audiences, the established researcher and the graduate student in combinatorics. It is hoped that both will gain new insights into these fascinating techniques.

The situation has changed a little, but there is still a need for more examples.

The earlier book included two tutorial papers and some general survey material. I saw no point in repeating that sort of material, as the new reader or researcher can refer to that volume. The papers included here contain some rather specific surveys and some papers on current research, written by many of the leading researchers in this field.

Chapter 1

THE EXISTENCE OF 2-SOLSSOMS

R. Julian R. Abel

School of Mathematics,
University of New South Wales,
Sydney, 2052, Australia

F. E. Bennett *

Department of Mathematics
Mount Saint Vincent University
Halifax, Nova Scotia B3M 2J6,
Canada

Abstract   In this paper, we investigate the existence of two self-orthogonal Latin squares with a symmetric orthogonal mate (2-SOLSSOMs). It is found that a 2-SOLSSOM(v) exists for all $v \geq 701$ with at most 183 possible exceptions below this value. Also given are six MOLS of orders 45, 55, five of which give 2-SOLSSOM(45) and 2-SOLSSOM(55).

1.   Introduction

A Latin square is called self-orthogonal if it is orthogonal to its transpose. An α-SOLSSOM of order v (α-SOLSSOM(v)) is a set of $2\alpha + 1$ Latin squares $A_1, A_2, \ldots, A_\alpha,\ B_1, B_2, \ldots, B_\alpha,\ C$ such that $A_i = B_i^T$ and $C = C^T$. Here SOLSSOM stands for self-orthogonal Latin squares with

* Researcher supported by NSERC Grant OGP 0005320

a symmetric orthogonal mate. When $\alpha = 1$, the term SOLSSOM (rather than 1-SOLSSOM) is more commonly used.

The existence problem for SOLSSOM(v) has been investigated for many years and the solution now is almost complete. More specifically we have the following result [1]:

Theorem 1.1 *If v is a positive integer, then a SOLSSOM(v) exists, except for $v \in \{2, 3, 6\}$ and possibly for $v \in \{10, 14\}$.*

On the other hand, much less is known about the existence of α-SOLSSOMs for $\alpha \geq 2$. In [9], Lee gave the following result:

Theorem 1.2 *A 2-SOLSSOM(v) exists for all $v \geq 4309$, with at most 328 possible exceptions below this value. Further, if $v \equiv 0 \mod 8$, then a 2-SOLSSOM(v) exists except possibly for $v \in \{24, 40, 48\}$, and if $v \equiv 1 \mod 2$, then a 2-SOLSSOM(v) exists except for $v = 3, 5$ and possibly for $v \in \{15, 21, 33, 35, 39, 45, 51, 55, 65, 69, 87, 123, 129, 135\}$.*

We also make some use of α-SOLSSOMs with an orthogonal mate which is symmetric in pairs. We say a Latin square is *symmetric in pairs* if one point occupies all cells on the main diagonal and the others can be put into pairs (x, y) such that if x occupies any cell then y occupies its symmetric cell. A Latin square of order v can only be symmetric in pairs if v is odd. We have the following known result (which will be confirmed later for odd q after Theorem 2.1) for existence of α-SOLSSOMs of prime power order:

Theorem 1.3   *1 If q is an even prime power, there exists a $(q-2)/2$-SOLSSOM(q).*

  2 If q is an odd prime power, there exists a $(q-3)/2$-SOLSSOM(q) with an orthogonal mate that is symmetric in pairs.

For our constructions of α-SOLSSOMs we also frequently need to use incomplete α-SOLSSOMs (α-ISOLSSOMs) and *holey* α-SOLSSOMs (α-HSOLSSOMs). Let S be a set and $\mathcal{H} = \{S_1, S_2, \cdots, S_n\}$ be a set of disjoint subsets of S. A *holey Latin square* having *hole set* $\mathcal{H}$ is an $|S| \times |S|$ array L, indexed by S, satisfying the following properties:

(1) every cell of L either contains an element of S or is empty,

(2) every element of S occurs at most once in any row or column of L,

(3) the subarrays indexed by $S_i \times S_i$ are empty for $1 \le i \le n$ (these subarrays are all referred to as *holes*),

(4) element $s \in S$ occurs in row or column t if and only if $(s, t) \in (S \times S) \backslash \bigcup_{1 \le i \le n} (S_i \times S_i)$.

The *order* of L is $|S|$. Two holey Latin squares, say L_1 and L_2, on symbol set S and hole set $\mathcal{H}$, are said to be *orthogonal* if their superposition yields every ordered pair in $(S \times S) \backslash \bigcup_{1 \le i \le n} (S_i \times S_i)$. We shall use the notation IMOLS$(s; s_1, \cdots, s_n)$ to denote a pair of orthogonal holey Latin squares on symbol set S and hole set $\mathcal{H} = \{S_1, S_2, \cdots, S_n\}$, where $s = |S|$ and $s_i = |S_i|$ for $1 \le i \le n$. If $\mathcal{H} = \emptyset$, we obtain a MOLS(s). If $\mathcal{H} = \{S_1\}$, we simply write IMOLS(s, s_1) for the orthogonal pair of holey Latin squares.

If $\mathcal{H} = \{S_1, S_2, \cdots, S_n\}$ is a partition of S, then a holey Latin square is called a *partitioned incomplete Latin square*, denoted by PILS. The *type* of the PILS is defined to be the multiset $\{|S_i| : 1 \le i \le n\}$. Two orthogonal PILS of type T will be denoted by HMOLS(T).

A holey Latin square is called *self-orthogonal* if it is orthogonal to its transpose. For self-orthogonal holey Latin squares we use the notations SOLS(s), ISOLS(s, s_1) and HSOLS(T) for the cases of $\mathcal{H} = \emptyset$, $\{S_1\}$ and a partition $\{S_1, S_2, \cdots, S_n\}$, respectively.

If any two PILS in a set of t PILS of type T are orthogonal, then we denote the set by t HMOLS(T). Similarly, we may define t MOLS(s) and t IMOLS(s, s_1). For later use we mention that the existence of t IMOLS(v, n) is equivalent to that of an incomplete TD, i.e. a TD$(t + 2, v) - TD(t + 2, n)$.

A *holey α-SOLSSOM* having partition $\mathcal{H}$ consists of $2\alpha + 1$ HMOLS (having partition $\mathcal{H}$), say $A_1, \cdots, A_\alpha$, $B_1, \cdots, B_\alpha$, C, where $B_i = A_i^T$ and $C = C^T$. A holey α-SOLSSOM of type T will be denoted by α-HSOLSSOM(T). Similarly, we may define α-ISOLSSOM(s, s_1). Also for any α-HSOLSSOM, it is known that all hole sizes must be identical mod 2.

2.   Direct Constructions

Existence of an α-SOLSSOM(v) is equivalent to that of a transversal design TD($2\alpha + 3, v$) with an order 2 automorphism that permutes the first $2\alpha + 2$ groups in pairs while leaving the last group unaltered. However, if this TD is obtained from a difference matrix, then a stronger result can sometimes be obtained, as in the next theorem. A $(v, k, 1)$ *difference matrix* over an abelian group G is a $k \times v$ array D with entries from G such that for any 2 distinct rows i, j, the v differences $D(i, t) - D(j, t)$ are all different.

Theorem 2.1 *Suppose v is odd, and there exists a $(v, 2\alpha + 3, 1)$ difference matrix D over an abelian group G. Suppose also one can define an automorphism T of order 2 on the rows of D an automorphism T in such a way that (1) T leaves one row unaltered and permutes the others in pairs (2) applying T also leaves one column unaltered and permutes the others in pairs. Then there exists an α-SOLSSOM(v) with an orthogonal mate which is symmetric in pairs.*

Proof: Label the rows of D as $0, 1, 2, \ldots, 2\alpha + 2$, and the columns using the elements of G; this labelling should be such that the ordering of the columns is of the form $0, x_1, x_2, x_3, \ldots, x_n, -x_n, \ldots, -x_3, -x_2, -x_1$ for $n = (q - 1)/2$. Without loss of generality we can assume that (1) T permutes rows $2t$ and $2t + 1$ for $0 \le t \le \alpha$, (2) T leaves the last row (row $2\alpha + 2$) unaltered and (3) this last row consists entirely of zeros. We can also assume without loss of generality that the columns of D are labelled so that $D(2\alpha, j) - D(2\alpha + 1, j) = j$ for any j; note that this condition implies that the pairs of columns interchanged by T are j and $-j$ for all j. The required Latin squares are A, B, and C_t, $(t = 0, \ldots, 2\alpha - 1)$ (whose rows and columns should be indexed using the same ordering as for the columns of D) where:

(i) $A(i, j) = j - i$,
(ii) $B(0, j) = D(2\alpha, j)$, $C_t(0, j) = D(2\alpha, j) - D(t, j)$,
(iii) For any $E \in \{B, C_0, C_1, \ldots, C_{2\alpha-1}\}$, $E(i, j) = E(0, j - i) + i$ for all i, j.

It is easily verified that A is symmetric in pairs (the symmetric pairs are of the form $(x, -x)$ for $x \in G \backslash \{0\}$), and that, since D is a difference matrix, the resulting Latin squares are all orthogonal. It remain to establish that B is symmetric and C_{2t} is the transpose of C_{2t+1}. Because of relation (iii) above, it is sufficient to establish this for entries in row zero and column zero. We have:

$$B(i,0) = B(0,-i) + i = D(2\alpha,-i) + i = D(2\alpha+1,i) + i$$
$$= D(2\alpha,i) = B(0,i)$$

so B is symmetric. Similarly:

$$C_{2t}(j,0) = C_{2t}(0,-j) + j = D(2\alpha,-j) - D(2t,-j) + j$$
$$= D(2\alpha+1,j) - D(2t+1,j)) + j \quad \text{(applying the automorphism } T\text{)}$$
$$= D(2\alpha,j) - D(2t+1,j) = C_{2t+1}(0,j),$$

and similarly, $C_{2t}(0,j) = C_{2t+1}(j,0)$ so C_{2t} is in fact the transpose of C_{2t+1}. This completes the proof. $\qquad\qquad\square$

If q is an odd prime power, then there exists a $(q,q,1)$ difference matrix D over $GF(q)$ satisfying the conditions of Theorem 2.1; thus in this case for $\alpha = (q-3)/2$, there exists an α-SOLSSOM(q) with an orthogonal mate symmetric in pairs. To each row i of the difference matrix D we associated a different element s of G; the entries in row i of D are then given by $D(i,j) = s \cdot j$ for all $j \in GF(q)$. In addition if $s = m$ is used for row $2x$ then we should take $s = -m$ for row $2x+1$. This condition ensures that the automorphism T which permutes rows $2x$ and $2x+1$ for any x also permutes columns j and $-j$ for any j. For the last row (row $q-1 = 2\alpha+2$), s should be 0, and it's convenient to take $s = (2)^{-1}$ and $-(2)^{-1}$ for rows $q-3 = 2\alpha$ and $q-2 = 2\alpha+1$ respectively, so that the condition $D(2\alpha,j) - D(2\alpha+1,j) = j$ for all j is satisfied without having to permute any columns. Here is an example of such a difference matrix when $q = 7$. The rows and columns of D are labelled as $0,1,2,\ldots,6$:

0	1	2	3	4	5	6
0	6	5	4	3	2	1
0	2	4	6	1	3	5
0	5	3	1	6	4	2
0	4	1	5	2	6	3
0	3	6	2	5	1	4
0	0	0	0	0	0	0

The relevant Latin squares A, B, C_0, C_2 are given below ($C_1 = C_0^T$, $C_3 = C_2^T$). For $t = 0,\ldots,3$, the first row of C_t is obtained by subtracting row t of D from the third last row.

$$
A = \begin{matrix}
0 & 1 & 2 & 3 & 4 & 5 & 6 \\
6 & 0 & 1 & 2 & 3 & 2 & 1 \\
5 & 6 & 0 & 1 & 2 & 3 & 5 \\
4 & 5 & 6 & 0 & 1 & 2 & 3 \\
3 & 4 & 5 & 6 & 0 & 1 & 2 \\
2 & 3 & 4 & 5 & 6 & 0 & 1 \\
1 & 2 & 3 & 4 & 5 & 6 & 0
\end{matrix}
\qquad
B = \begin{matrix}
0 & 4 & 1 & 5 & 2 & 6 & 3 \\
4 & 1 & 5 & 2 & 6 & 3 & 0 \\
1 & 5 & 2 & 6 & 3 & 0 & 4 \\
5 & 2 & 6 & 3 & 0 & 4 & 1 \\
2 & 6 & 3 & 0 & 4 & 1 & 5 \\
6 & 3 & 0 & 4 & 1 & 5 & 2 \\
3 & 0 & 4 & 1 & 5 & 2 & 6
\end{matrix}
$$

$$
C_0 = \begin{matrix}
0 & 3 & 6 & 2 & 5 & 1 & 4 \\
5 & 1 & 4 & 0 & 3 & 6 & 2 \\
3 & 6 & 2 & 5 & 1 & 4 & 0 \\
1 & 4 & 0 & 3 & 6 & 2 & 5 \\
6 & 2 & 5 & 1 & 4 & 0 & 3 \\
4 & 0 & 3 & 6 & 2 & 5 & 1 \\
2 & 5 & 1 & 4 & 0 & 3 & 6
\end{matrix}
\qquad
C_2 = \begin{matrix}
0 & 2 & 4 & 6 & 1 & 3 & 5 \\
6 & 1 & 3 & 5 & 0 & 2 & 4 \\
5 & 0 & 2 & 4 & 6 & 1 & 3 \\
4 & 6 & 1 & 3 & 5 & 0 & 2 \\
3 & 5 & 0 & 2 & 4 & 6 & 1 \\
2 & 4 & 6 & 1 & 3 & 5 & 0 \\
1 & 3 & 5 & 0 & 2 & 4 & 6
\end{matrix}
$$

We now give two more difference matrices satisfying the conditions of Theorem 2.1:

Theorem 2.2 *There exist 6 MOLS(45) and a 2-SOLSSOM(45) with an orthogonal mate symmetric in pairs.*

Proof: We construct a $(45, 7, 1)$ difference matrix. Let A_1, A_2 be the following arrays over $Z_5 \times Z_3 \times Z_3$:

$$
A_1 = \begin{matrix}
000 \\
000 \\
000 \\
000 \\
000 \\
000 \\
000
\end{matrix}
\qquad
A_2 = \begin{matrix}
221 & 311 & 412 & 401 & 011 & 021 & 322 \\
121 & 422 & 120 & 410 & 311 & 300 & 212 \\
411 & 221 & 320 & 120 & 210 & 100 & 321 \\
010 & 211 & 400 & 002 & 422 & 322 & 122 \\
312 & 210 & 022 & 421 & 021 & 201 & 112 \\
211 & 122 & 301 & 201 & 100 & 421 & 110 \\
000 & 000 & 000 & 000 & 000 & 000 & 000
\end{matrix}
$$

Now let A_3 be the array obtained by permuting rows i and $7-i$ in A_2 (for $1 \leq i \leq 3$) while leaving row 7 unaltered. A $(45, 7, 1)$ difference matrix is then obtained by adding the following 3 vectors to each of the 15 columns of $[A_1|A_2|A_3]$: ($(0,0,x),(0,2x,0),(0,x,2x),(0,2x,x),(0,x,0),$ $(0,0,2x),(0,0,0)$)^T for $x = 0, 1, 2$. This difference matrix gives 6 MOLS of order 45, and because of the above permutation which interchanges rows i and $7-i$, while leaving row 7 unaltered, it also gives a 2-SOLSSOM(45) with an orthogonal mate symmetric in pairs. $\qquad\square$

The next theorem makes use of a $(55, 7, 1)$ difference matrix from [12]:

Theorem 2.3 *There exist 6 MOLS(55) and a 2-SOLSSOM(55) with an orthogonal mate symmetric in pairs.*

Proof: Let A be the 6×9 matrix below. A $(55, 7, 1)$ difference matrix over Z_{55} is then obtained by cyclically permuting the rows of A, and then adding a row and column of zeros. The set of columns in this difference matrix clearly remains invariant under the permutation which interchanges rows 1 and 4, rows 2 and 5, plus rows 3 and 6 while leaving the last row (consisting of zeros) unaltered; thus this difference matrix gives a 2-SOLSSOM(55) with an orthogonal mate symmetric in pairs.

$$A = \begin{array}{ccccccccc}
1 & 7 & 14 & 19 & 28 & 33 & 40 & 46 & 50 \\
2 & 13 & 25 & 38 & 52 & 12 & 20 & 32 & 45 \\
39 & 6 & 8 & 26 & 24 & 51 & 11 & 34 & 37 \\
54 & 48 & 41 & 36 & 27 & 22 & 15 & 9 & 5 \\
53 & 42 & 30 & 17 & 3 & 43 & 35 & 23 & 10 \\
16 & 49 & 47 & 29 & 31 & 4 & 44 & 21 & 18
\end{array}$$

$\square$

Also related to α-SOLSSOMs are Steiner k-cycle systems. A Steiner k-cycle system of order v, or SkCS(v), is a pair (X, C), where C is a collection of k-cycles of K_v such that for any integer r, $1 \le r \le k/2$, and for any two distinct vertices x and y of X there exists in C a unique k-cycle along which the distance between x and y is r. (Here if $(a_1, a_2, \ldots, a_k)$ is a k-cycle, then the distance between a_i, a_j is defined to be the minimum of $\mathrm{mod}(i - j + k, \ k)$ and $\mathrm{mod}(j - i + k, \ k)$.) If instead, X can be partitioned into holes T_i of sizes t_i $(i = 1, \ldots, n)$ and there is a unique k-cycle along which the distance between x and y is r (for $1 \le r \le k/2$) only when x, y do not belong to the same T_i, then we have a holey Steiner k-cycle system (or HSkCS) of type $(t_1, t_2, \ldots, t_n)$. Existence of an HSkCS of any type implies existence of an α-HSOLSSOM of the same type where $\alpha = (k - 3)/2$ [2]. Also, as for α-SOLSSOMs, exponential notation is commonly used for hole type of HSkCS's when several holes have the same size.

Theorem 2.4 *There exists a HS7CS of type $8^9 10^1$ and hence also a 2-SOLSSOM of this type.*

Proof: Let $X = (Z_2 \times Z_2 \times Z_{18}) \cup \{\infty_1, \infty_2, \ldots, \infty_{10}\}$; the size 10 hole consists of the infinite points, and other holes are of the form $Z_2 \times Z_2 \times \{x, x + 9\}$ for $0 \le x \le 8$. The required 7-cycles are obtained by developing $C_1, C_2, \ldots, C_6$ below mod $(2, 2, 18)$. Also, replace ∞_i by

∞_{i+5} when adding to C_i any value (x, y, z) satisfying either (1) $x = 1$ and $1 \leq i \leq 4$ or (2) $y = 1$ and $i = 5$.

$$C_1 = \{(0,0,0),(0,0,4),(1,0,14),(0,0,1),(1,0,7),(1,1,15),\infty_1\}$$
$$C_2 = \{(0,0,0),(0,0,7),(1,0,11),(0,0,13),(1,1,5),(1,0,3),\infty_2\}$$
$$C_3 = \{(0,0,0),(0,1,6),(1,0,13),(0,1,10),(1,0,12),(1,0,2),\infty_3\}$$
$$C_4 = \{(0,0,0),(0,0,2),(1,1,7),(0,1,6),(1,1,17),(1,0,12),\infty_4\}$$
$$C_5 = \{(0,0,0),(0,0,5),(0,1,2),(0,0,3),(0,1,7),(0,1,8),\infty_5\}$$
$$C_6 = \{(0,0,0),(0,0,15),(1,1,1),(0,1,4),(1,0,16),(1,1,5),(1,1,17)\}$$

$\square$

3.　　Recursive Constructions

The remaining constructions used in this paper are mainly recursive. We start by summarizing a number of commonly used known constructions. The first of these was given in [10] for 1-SOLSSOMs and is easily generalized to α-SOLSSOMs for $\alpha > 1$:

Construction 3.1 [10] (Singular Indirect Product) *If there exist an α-HSOLSSOM of type t^u, $2\alpha + 1$ IMOLS$((q - a)/t, (w - a)/t)$, and an α-ISOLSSOM(q, w), then an α-ISOLSSOM$(u(q - a) + a, u(w - a) + a)$. exists. If further there exists an α-SOLSSOM(s) for $s = u(w - a) + a$ then an α-SOLSSOM(v) also exists for $v = u(q - a) + a$.*

We also have the simpler direct product construction:

Construction 3.2 *If an α-SOLSSOM(m) and an α-SOLSSOM(n) both exist, then for $v = m \cdot n$, there exist an α-ISOLSSOM(v, m), an α-ISOLSSOM(v, n) and an α-SOLSSOM(v).*

Construction 3.3 (Filling in the holes) *Suppose an α-HSOLSSOM of type $T = (t_1, \ldots, t_n)$ exists and $w \geq 0$. For $i = 1, 2, \ldots, n - 1$, suppose there exists an α-ISOMSSOM$(t_i + w, w)$. Then, for $t = \sum_{i=1}^{n} t_i$, there exists an α-ISOLSSOM$(t+w, t_n+w)$. If further, an α-SOLSSOM(t_n+w) exists, then an α-SOLSSOM$(t + w)$ also exists.*

The next construction uses group divisible designs (GDDs). A GDD of index 1, or K-GDD of type $(t_1, t_2, \ldots, t_m)$ is a triple $(\mathbf{X}, \mathbf{G}, \mathbf{B})$ such that:

1 **G** is a partition of **X** (a set of *points*) into subsets called *groups* of sizes $t_1, t_2, \ldots, t_m$,

2 **B** is a collection of subsets of **X** (called *blocks*) whose sizes all belong to K,

3 Any two points in **X** appear in exactly one block or one group but not both.

When repeated group sizes occur, exponential notation is more commonly used for group type. For instance, type $x^a y^b \ldots$ means there are a groups of size x, b groups of size y and so on. Also, if $K = \{k\}$ and the group type is g^k, then the GDD is a *transversal design*, denoted as $TD(k, g)$.

Construction 3.4 (Weighting) *Suppose there exists a K-GDD with groups $G_1, G_2, \ldots, G_n$. Suppose also we can assign each point x a weight $w(x)$ so that for any block $B = \{x_1, x_2, x_3, \ldots, x_s\}$,, an α-HSOLSSOM of type $(w(x_1), w(x_2), \ldots, w(x_s))$ exists. Then an α-HSOLSSOM of type $(w(G_1), w(G_2), \ldots, w(G_n))$ (where $w(G_i) = \sum_{x \in G_i} w(x)$) exists.*

When k is odd, all values appear exactly once on the main diagonal of all Latin squares in an α-SOLSSOM(k); hence deleting all entries on the main diagonal gives an α-HSOLSSOM(1^k). Using the previous construction, if we give all points weight 1, we have the following simpler result:

Construction 3.5 *Suppose a K-GDD of type $(t_1, t_2, \ldots, t_n)$ exists, all block sizes in K are odd, and there exists an α-SOLSSOM(k) for all $k \in K$. Then there exists an α-HSOLSSOM of type $(t_1, t_2, \ldots, t_n)$.*

The next lemma gives three types of GDD which are useful for the previous construction, since they frequently only have odd block sizes:

Lemma 3.6   *1 (Brouwer, [4]) If q is a prime power, then for*
$$0 < t < q^2 - q + 1, \text{ a } \{t, q + t\}\text{-GDD of type } t^{q^2 - q + 1} \text{ exists.}$$

2 (Greig, [8]) If q is a prime power, then for $0 \le x \le q - 1$, a $\{q - x - 1, q - x + 1\}$-GDD of type $(q - 1)^{q - x}(x + 1)^1$ exists.

3 (Greig, [8]) If q is a prime power, then for $0 \le x \le q - 1$, a $\{q - x - 1, q - x + 1, q\}$-GDD of type $(q - x)^{q - 1}(x + 1)^1$ exists.

The next construction is well known and was given in [11]:

Construction 3.7 *If $2\alpha + 1$ MOLS(m) plus an α-HSOLSSOM of type $(t_1, t_2, \ldots, t_n)$ both exist, then there exists an α-HSOLSSOM of type $(mt_1, mt_2, \ldots, mt_n)$.*

The next two constructions were given in [3] for 1-SOLSSOMs and are easily generalized to α-SOLSSOMs. Before giving these, we need the new concept of 3 compatible IMOLS. Denote by ILS(s, s_1) a holey Latin square of order s with just one hole of size s_1. An element in the hole of an ILS is said to be *evenly distributed* if (1) it does not appear on the main diagonal and (2) whenever it appears in any cell, it also appears in the symmetric cell. If every element in the hole is evenly distributed, then we say that the ILS is *balanced*. Given 3 IMOLS, one of which is balanced, then for each holey point x, there will be a total of $s - s_1$ cells above the main diagonal in the other two squares that are occupied by x in the balanced square. If the first square is balanced, and if for every holey point x, each non-holey point lies in exactly one of these the $s - s_1$ cells then we say that the 3 IMOLS are *compatible*. Note that if the hole size s_1 is zero, then three IMOLS (or MOLS) are always compatible from this definition.

For $k \geq 2$, $2k + 1$ IMOLS(s, s_1), are said to be compatible if one of the squares, C is balanced, and the others can be put into pairs (A_i, B_i) such that for all i, C, A_i, B_i are 3 compatible IMOLS.

Remark 3.8 *For 3 or more compatible IMOLS(v, h) to exist, $v - h$ must be even. An α-ISOLSSOM(v, h) always gives $2\alpha + 1$ compatible IMOLS(v, h) if none of its holey points lies on the main diagonal; hence any α-ISOLSSOM(v, h) with v and h both odd gives $2\alpha + 1$ compatible IMOLS(v, h). However, this may not be the case when v and h are even. One way of constructing $2\alpha + 1$ compatible IMOLS(v, h) is to obtain an α-HSOLSSOM on v points, and fill in all but one hole of size h; in fact most of the known sets of compatible IMOLS(v, h) with v, h even are obtained this way. An example of 3 compatible IMOLS($10, 2$) is given in Figure 4 of [3]. Filling in the holes of an α-HSOLSSOM with odd hole sizes by using an infinite point doesn't give compatible IMOLS, since one holey point (the infinite point) will appear on the main diagonal of the SOM. For instance filling in the holes of a 2-HSOLSSOM(7^7) with 2-SOLSSOM(8) does not give 5 compatible IMOLS($50, 8$).*

Construction 3.9 *Suppose q is a product of odd prime powers $\geq 2\alpha + 3$ (or more generally, suppose there exists an α-SOLSSOM(q) with an orthogonal mate which is symmetric in pairs). Suppose also, m is even, $2\alpha+1$ MOLS(m) exist, $t \leq (q-1)/2$ and $2\alpha+1$ compatible IMOLS($m + e_i, e_i$) exist for $1 \leq i \leq t$. Then there exists an α-HSOLSSOM of type $(m)^q(2e)^1$ for $e = \sum_{i=1}^t e_i$. If also, an α-ISOLSSOM($m + w, w$) exists, then for $v = mq + 2e + w$, an α-ISOLSSOM($v, 2e + w$) exists. If further, an α-SOLSSOM($2e + w$) exists, then an α-SOLSSOM(v) exists.*

Construction 3.10 *Suppose q is an odd prime power such that $\alpha + t \leq (q-3)/2$ (or more generally, suppose there exists an $(\alpha+t)$-SOLSSOM(q) with an orthogonal mate which is symmetric in pairs). Suppose also m is even and there exist $2\alpha + 1$ MOLS(m) plus $2\alpha + 1$ compatible IMOLS($m + e_i, e_i$) for $1 \leq i \leq t$. Then there exists an α-HSOLSSOM of type $m^{q-1}(m+2e)^1$ for $e = \sum_{i=1}^t e_i$. If also, an α-ISOLSSOM($m+w, w$) exists, then for $v = mq + 2e + w$, an α-ISOLSSOM($v, m+2e+w$) exists. If further, an α-SOLSSOM($m + 2e + w$) exists, then an α-SOLSSOM(v) exists.*

Remark 3.11 *In Constructions 3.9 and 3.10, if for every t there is cell on the main diagonal of the $2\alpha + 1$ compatible IMOLS($m + e_t, e_t$) that is not occupied by a holey point in any of the IMOLS, then an α-ISOLSSOM(v, q) will also exist. This condition will always be satisfied if the $2\alpha+1$ compatible IMOLS($m+e_t, e_t$) are an α-ISOLSSOM($m+e_t, e_t$) for each t.*

An important ingredient in Construction 3.9 is the existence of an α-SOLSSOM(q) with q disjoint transversals, of which one is the main diagonal, and the others are symmetric in pairs. Similarly, in Construction 3.10 a necessary ingredient is an α-SOLSSOM(q) with $2t + 1$ transversals (the main diagonal and $2t$ others, in symmetric pairs) all intersecting in a common cell. If q is an even prime power $\geq 2\alpha + 2t + 2$ then there exists an α-SOLSSOM(q) with $2t$ transversals in symmetric pairs, all of which intersect in one common cell on the main diagonal but which are otherwise disjoint. Using this, we have the following result which is similar to Construction 3.10 and an extension of Lemma 2.1 in [5]:

Construction 3.12 *Suppose m is even, q is an even prime power $\geq 2\alpha + 2t + 2$, and there exists an α-SOLSSOM(m). Suppose also, $2\alpha + 1$ compatible IMOLS($m + e_i, e_i$) exist for $1 \leq i \leq t$. Then there exists* an α-ISOLSSOM$(mq + 2e, m + 2e)$ for $e = \sum_{i=1}^{t} e_i$. If further, an α-SOLSSOM$(m + 2e)$ exists, then an α-SOLSSOM$(mq + 2e)$ also exists.

We are now ready to investigate the existence of 2-SOLSSOM(v). We start with the easiest cases, where v is odd or $v \equiv 0$ mod 8:

Lemma 3.13   *1 If $v \equiv 0$ mod 8, then a 2-SOLSSOM(v) exists, except possibly for $v \in \{24, 40, 48\}$.*

2 If $v \geq 7$ is an odd integer, then a 2-SOLSSOM(v) exists, except possibly for $v \in \{15, 21, 33, 35, 39, 51, 65, 87, 123, 135\}$.

Proof: Lee [9] proved these two results for all but four values of v namely, $45, 55, 69, 129$. For $v = 45$, and 55, see Theorems 2.2 and 2.3, and for $v = 69$, see [2]. For $v = 129$, existence was mentioned in [2] but the construction was inadvertently omitted; a 2-HSOLSSOM(16^8) is obtainable by Construction 3.7, inflating the 2-HSOLSSOM(2^8) from [2] with 5 MOLS(8); we then add an infinite point and use 2-SOLSSOM(17) to fill the holes. The proof is complete.   □

The next lemma gives some 2-HSOLSSOMs which are very important for our main recursive construction for SOLSSOM(v) with v even:

Lemma 3.14 *There exist 2-HSOLSSOMs of the following types:* $8^8 2^1$, $8^8 10^1$, 8^9, $8^9 2^1$, $8^9 10^1$, 8^{10}.

Proof: For types 8^{10} and $8^8 2^1$, Construction 3.5 can be applied. Deleting one point from an affine plane of order 9 gives a 9-GDD of type (8^{10}), while a $\{7, 9\}$-GDD of type $(8^8 2^1)$ is obtainable from Lemma 3.6.2 with $q = 9$, $x = 1$. Construction 3.5 now gives 2-SOLSSOMs of these types. For type $8^9 10^1$, a 2-HSOLSSOM is obtainable from the HS7CS of the same type in Theorem 2.4. 2-HSOLSSOMs of types $8^9 u^1$, $u = 0, 2$ are obtainable by Construction 3.9 with $m = 8$, $q = 9$, and $e_1 = e = 0$ or 1. Finally for type $8^8 10^1$, apply Construction 3.10 with $m = 8$, $q = 9$, and $e_1 = e = 1$.   □

The next theorem provides our main construction for v even and $v \not\equiv 0$ mod 8:

Theorem 3.15 *Suppose a TD$(10, m)$ exists, and (1) $0 \leq u, x, y, z \leq m$, (2) $x + y + z = m$. Then there exists a 2-HSOLSSOM of type $(8m)^8 (8u)^1 (2x + 8y + 10z)^1$.*

Proof: Start with a TD(10, m), and apply Construction 3.4. Here, we give weight 8 to all points in the first 8 groups; in group 9 give weight 8 to u points and zero weight to the rest. In group 10 give weight 2 to x points weight 8 to y points and weight 10 to z points. All the required input 2-HSOLSSOMs are given in Lemma 3.14. □

Remark 3.16 *If $m \geq 3$, then any even integer in the range $[2m + 6, 10m]$ (except $2m + 10$) can be written as $2x + 8y + 10z$ where x, y, z are non-negative integers and $x + y + z = m$. Therefore, when using the above theorem, we normally don't give the values of x, y, z explicitly.*

Theorem 3.17 *If $v \in \{106, 148, 158, 574\}$, a 2-SOLSSOM($v$) exists.*

Proof: For $v = 106$, an HS7CS of type 7^{15} (and hence also a 2-HSOLSSOM(7^{15})) is given in [2]; fill in each group with the aid of an infinite point using 2-SOLSSOM(8). For $v = 148, 158, 574$ respectively, we can obtain a $\{7, 11\}$-GDD(7^{21}) by Lemma 3.6.1 with $q = 4, t = 7$, a $\{9, 11\}$-GDD($15^{10}7^1$) by Lemma 3.6.2 with $q = 16, x = 6$ and a $\{17, 19\}$-GDD($31^{18}15^1$) by Lemma 3.6.2 with $q = 32, x = 14$. Apply Lemma 3.5 to these GDDs to obtain 2-HSOLSSOMs of these types; then add an infinite point and use 2-SOLSSOMs of orders 8, 16 and 32 to fill in the holes. □

Theorem 3.18 *If $v \in \{50, 78, 92, 134, 162, 190, 204, 218, 260, 302, 316, 330, 342, 372, 386, 404, 414, 428, 442, 470, 484, 498, 526, 540, 554, 582, 590, 638, 652, 666, 694, 708, 722, 750, 764, 778, 806, 820, 890, 932, 1030, 1044\}, then a 2-SOLSSOM(v) exists.*

Proof: For $v = 342, 404, 590$, start with a 2-HSOLSSOM of type 1^n for $n = 11, 13, 19$ and apply Construction 3.7 with $m = 31$ to obtain a 2-HSOLSSOMs of type 31^n. Then add an infinite point and form a 2-SOLSSOM(32) on each hole plus the infinite point. Similarly, the other values are of the form $7q + 1$ where a 2-HSOLSSOM(1^q) exists; inflate this by 7 to obtain a 2-HSOLSSOM(7^q); again we add an infinite point, and form a 2-SOLSSOM(8) on each hole plus the infinite point. □

The next theorem, like the previous one gives one of our main construction methods for 2-SOLSSOM(v) where $v < 650$:

Theorem 3.19 *If $v \in \{356, 362, 374, 380, 394, 468, 486, 524, 542, 556, 562, 566, 580, 596, 598, 612, 636, 642\}, then a 2-SOLSSOM(v) exists.*

Proof: These are obtained by the singular indirect product construction (Construction 3.1) with $t = 1$. Values of the relevant parameters u, q, w, a and s are indicated below. In all cases, the required sets of 5 IMOLS$(q - a, w - a)$ exist by [6]. Most of the required 2-ISOLSSOM(q, w)'s are obtainable by Construction 3.2; for $(q, w) = (50, 8)$ it is obtained by adding an infinite point to the 2-HSOLSSOM(7^7) obtained in Theorem 3.18, and for $(q, w) = (80, 9)$, $(96, 11)$, we can use Remark 3.11 and Construction 3.9 since $80 = 8 \cdot 9 + 8$, $96 = 8 \cdot 11 + 8$.

v	u	q	a	w	s		v	u	q	a	w	s
356	7	56	6	7	13		362	7	56	5	7	19
374	7	56	3	7	31		380	7	56	2	7	37
394	9	50	7	8	16		468	7	72	6	9	27
468	7	72	3	9	45		524	7	80	6	9	27
542	7	80	3	9	45		556	7	88	10	11	17
562	7	88	9	11	23		566	11	56	5	7	27
580	7	88	6	11	41		596	11	56	2	7	57
598	7	88	3	11	59		612	7	96	10	11	41
636	7	96	6	11	41		642	7	96	5	11	47

Theorem 3.20 *A 2-SOLSSOM(v) exists for all $v \in \{350, 450, 546, 550, 644, 650, 828, 1036\}$.*

Proof: These 2-SOLSSOMs can be obtained by using the direct product construction (Construction 3.2) since $350 = 7 \cdot 50$, $450 = 9 \cdot 50$, $546 = 7 \cdot 78$, $550 = 11 \cdot 50$, $644 = 7 \cdot 92$, $650 = 13 \cdot 50$, $828 = 9 \cdot 92$, $1036 = 7 \cdot 148$. □

Theorem 3.21 *Let:*

$S_2 = \{466, 474, 490, 514, 522, 530, 538, 578, 586, 610, 618, 674, 706, 714, 754, 762, 850, 858, 906, 914, 922, 1050, 1058, 1066, 1074, 1098, 1106, 1114, 1122\}$,

$S_4 = \{892, 900, 908, 916, 940, 948, 956, 964, 1052, 1060, 1084, 1092, 1100, 1108, 1132, 1140, 1524, 1532, 1540, 1572, 1580, 1588, 1596\}$,

$S_6 = \{702, 734, 742, 878, 886, 894, 902, 926, 934, 942, 950, 1558, 1566, 1574, 1582\}$.

Then a 2-SOLSSOM(v) exists for all $v \in S_2 \cup S_4 \cup S_6$.

Proof: The values in S_2, S_4, S_6 are all equivalent to 2, 4 and 6 mod 8 respectively. All the given values of v (except 490) can be written

as $8q + 2e$ where q is an odd prime power, and $2e \in \{42, 50, 98, 106,$
$84, 92, 140, 148, 70, 78, 126, 134\}$. As a result, we can apply either Construction 3.9 or Construction 3.10 with $m = 8$ and $e_t = 1$ for all t. If
$2e \in \{42, 98, 154, 84, 140, 70, 126, 150\}$, a 2-SOLSSOM$(2e + 8)$ exists and
we can use Construction 3.10, while if $2e \in \{50, 106, 162, 92, 148, 78, 134\}$
a 2-SOLSSOM$(2e)$ exists and we use Construction 3.9. The table below
indicates suitable values of $2e$ for each v. Note that when we use Construction 3.10, the condition $\alpha + t \leq (q - 3)/2$ implies $e \leq (q - 7)/2$,
and when using Construction 3.9, the condition $t \leq (q - 1)/2$ gives
$e \leq (q - 1)/2$. Finally, for $v = 490$ we apply Construction 3.9 with
$m = 8$, $q = 55$, $2e = 50$, using the 2-SOLSSOM(55) with an orthogonal
mate symmetric in pairs from Theorem 2.3.

$2e$	v
42	$466, 514, 530, 578, 610, 674, 706, 754, 850, 914, 1058$
50	$474, 522, 538, 586, 618, 714, 762, 858, 906, 922, 1050$
98	$1066, 1098, 1114$
106	$1074, 1106, 1122$
84	$892, 908, 940, 956, 1052, 1084, 1100, 1132, 1532$
92	$900, 916, 948, 964, 1060, 1092, 1108, 1140, 1524, 1540$
140	$1572, 1588$
148	$1580, 1596$
70	$702, 734, 878, 894, 926, 942$
78	$742, 886, 902, 934, 950$
126	$1558, 1574$
134	$1566, 1582$

$\square$

Theorem 3.22 *For all* $v \in \{564, 572, 692, 732, 738, 746, 748, 756, 814,$
$822, 1076\}$, *a 2-SOLSSOM(v) exists.*

Proof: For these 2-SOLSSOMs, we first use either Construction 3.9
(for $v = 572, 738, 1076$) to obtain a 2-HSOLSSOM of type $m^{q-1}(2e)^1$
or Construction 3.10 (for other values of v) to obtain a 2-HSOLSSOM
of type $m^{q-1}(m + 2e)^1$, where in both cases, $e = \sum_{i=1}^{t} e_i$. Then, for
some w we add w infinite points and use Construction 3.3 to fill in the
holes. The following table indicates the values of m, q, $w_,$, e_i, used in
Constructions 3.9, 3.10, plus values of x and (y, z) for which we use a
2-SOLSSOM(x) and 2-ISOLSSOM(y, z) to fill in the holes.

With just 2 exceptions $((y, z) = (66, 10)$ or $(102, 6))$, all the required
2-ISOLSSOM(y, z)'s (where (y, z) is either $(m + e_i, e_i)$ or one of the fill

in pairs) are obtainable by Construction 3.9 or Construction 3.10 with $m = 8$, sometimes in conjunction with Remark 3.11. For instance, for $v = 814$, we have $76 = 8 \cdot 9 + 4$, hence Construction 3.9 gives us a 2-ISOLSSOM$(76, 4)$ and a 2-HSOLSSOM of type $8^9 4^1$; by Remark 3.8, it also gives 5 compatible IMOLS$(76, 4)$. For $v = 738$, $106 = 8 \cdot 13 + 2$, so Construction 3.10 gives us a 2-HSOLSSOM$(8^{12} 10^1)$ and hence a 2-ISOLSSOM$(106, 10)$ and 5 compatible IMOLS$(106, 10)$. For $v = 822$, to obtain the required 2-ISOLSSOM$(102, 6)$, we first construct a $\{7, 9, 13\}$-GDD of type $8^{12} 6^1$ by Lemma 3.6.3 with $q = 13, x = 5$; Construction 3.5 now gives a 2-HSOLSSOM of this type. By filling in the size 8 holes with 2-SOLSSOM(8), we obtain the required 2-ISOLSSOM$(102, 6)$. For $v = 738$, a non-compatible 2-ISOLSSOM$(66, 10)$ is obtainable by Construction 3.12, since $66 = 8 \cdot 8 + 2$.

v	m	q	e_i	HSOLS-SOM type	w	Fill in parameters
564	50	11	$e_i = 7, i = 1$	$50^{10} 64^1$	0	50, 64
572	80	7	All $e_i = 0$	80^7	12	(92, 12), 92
692	50	13	$e_i = 7, i = 1, 2, 3$	$50^{12} 92^1$	0	50, 92
732	64	11	$e_i = 10, i = 1$	$64^{10} 84^1$	8	(72, 8), 92
738	96	7	$e_i = 10, i = 1, 2;$ $e_3 = 8$	$96^7 56^1$	10	(106, 10), (66, 10), 106
746	64	11	$e_i = 8, i = 1, 2$	$64^{10} 96^1$	10	(106, 10), (74, 10), 106
748	80	9	$e_i = 8, i = 1$	$80^8 96^1$	12	(92, 12), (108, 12), 92
756	80	9	$e_i = 12, i = 1$	$80^8 104^1$	12	(92, 12), (116, 12), 92
814	72	11	$e_i = 4, i = 1, 2$	$72^{10} 88^1$	6	(78, 6), (94, 6), 78
822	72	11	$e_i = 6, i = 1, 2$	$72^{10} 96^1$	6	(78, 6), (102, 6), 78
1076	88	11	$e_1 = 8; e_i = 11,$ $i = 2, \ldots, 5$	$88^{11} 104^1$	4	(92, 4), (108, 4), 92

$\square$

The table below gives several more 2-SOLSSOM(v)'s using Construction 3.9 in a more standard manner. First Construction 3.9 is used to obtain a 2-HSOLSSOM$(m^q 2e^1)$; next, for some w (usually 0 or 8) we add w infinite points, and finally, we apply Construction 3.3, using 2-ISOLSSOM$(m + w, w)$ plus 2-SOLSSOM$(2e + w)$ to fill in the holes. As an example,

$$1564 = 84 \cdot 17 + (10 \cdot 11 + 6 \cdot 3) + 8$$

means that we first use Construction 3.9 to obtain a 2-HSOLSSOM of type $(84^{17} u^1)$ for $u = 128 = 10 \cdot 11 + 6 \cdot 3$ by taking $m = 84$, $q = 17$, and $e_i \in \{3, 11\}$; or more precisely, $e_i = 11$ for $i = 1, \ldots, 5$, 3 for $i = 6, \ldots, 8$. We then apply Construction 3.3, with $w = 8$, adding 8 infinite points, and using 2-ISOLSSOM$(84 + 8, 8)$ plus 2-SOLSSOM$(128 + 8)$ to fill the holes. Note that 2-ISOLSSOM$(87, 3)$ and 2-ISOLSSOM$(95, 11)$

are obtainable by combining Remark 3.11 with Construction 3.9 (taking $m = 12$ or 8, $q = 7$ or 11, $w = 1$, $e_i = 0, 1$) since $87 = (12 \cdot 7 + 2) + 1$ and $95 = (8 \cdot 11 + 6) + 1$. These 2-ISOLSSOMs are compatible since $87, 3, 95, 7$ are all odd. Also, 2-ISOLSSOM(92, 8) is obtained by constructing a 2-HSOLSSOM(7^{13}) as in Theorem 3.18, and then using 2-SOLSSOM(8) to fill in all holes except one with the aid of an infinite point.

v

$506 = 50 \cdot 9 + (8 \cdot 7)$

$562 = 42 \cdot 13 + (8 \cdot 1) + 8$

$730 = 42 \cdot 17 + (8 \cdot 1) + 8$

$786 = 42 \cdot 17 + (8 \cdot 7 \ + \ 8 \cdot 1) + 8$

$802 = 70 \cdot 11 + (2 \cdot 9 \ + \ 6 \cdot 1) + 8$

$1538 = 78 \cdot 19 + (4 \cdot 13 \ + \ 4 \cdot 1)$

$1554 = 78 \cdot 19 + (10 \cdot 7 \ + \ 2 \cdot 1)$

$1570 = 78 \cdot 19 + (8 \cdot 11)$

$1586 = 78 \cdot 19 + (8 \cdot 13)$

$1602 = 78 \cdot 19 + (2 \cdot 11 \ + \ 14 \cdot 7)$

$1618 = 78 \cdot 19 + (10 \cdot 13 \ + \ 6 \cdot 1)$

v

$554 = 42 \cdot 13 + (0) + 8$

$602 = 42 \cdot 13 + (6 \cdot 7 + 6 \cdot 1) + 8$

$770 = 42 \cdot 17 + (6 \cdot 7 \ + \ 6 \cdot 1) + 8$

$794 = 42 \cdot 17 + (10 \cdot 7 \ + \ 2 \cdot 1) + 8$

$930 = 80 \cdot 11 + (6 \cdot 8 \ + \ 2 \cdot 1)$

$1546 = 78 \cdot 19 + (8 \cdot 7 \ + \ 8 \cdot 1)$

$1562 = 78 \cdot 19 + (6 \cdot 13 \ + \ 2 \cdot 1)$

$1578 = 78 \cdot 19 + (8 \cdot 11 \ + \ 8 \cdot 1)$

$1594 = 78 \cdot 19 + (8 \cdot 13 \ + \ 8 \cdot 1)$

$1610 = 78 \cdot 19 + (4 \cdot 11 \ + \ 12 \cdot 7)$

$1626 = 78 \cdot 19 + (10 \cdot 13 \ + \ 2 \cdot 7)$

$604 = 84 \cdot 7 + (2 \cdot 3 \ + \ 2 \cdot 1) + 8$

$668 = 64 \cdot 9 + (6 \cdot 11 \ + \ 2 \cdot 9) + 8$

$780 = 78 \cdot 9 + (6 \cdot 13)$

$796 = 64 \cdot 11 + (6 \cdot 8 + 4 \cdot 9) + 8$

$844 = 84 \cdot 9 + (6 \cdot 11 \ + 2 \cdot 7) + 8$

$972 = 80 \cdot 11 + (4 \cdot 9 \ + \ 6 \cdot 8) + 8$

$1148 = 96 \cdot 11 + (4 \cdot 9 \ + \ 6 \cdot 8) + 8$

$1164 = 84 \cdot 13 + (2 \cdot 11 \ + \ 6 \cdot 7) + 8$

$1556 = 84 \cdot 17 + (2 \cdot 11 \ + \ 14 \cdot 7) + 8$

$1604 = 84 \cdot 17 + (14 \cdot 11 \ + \ 2 \cdot 7) + 8$

$620 = 84 \cdot 7 + (2 \cdot 7 \ + \ 2 \cdot 5) + 8$

$772 = 84 \cdot 9 + (8 \cdot 1) + 8$

$788 = 84 \cdot 9 + (2 \cdot 11 \ + \ 2 \cdot 1) + 8$

$836 = 92 \cdot 9 + (8 \cdot 1)$

$924 = 64 \cdot 13 + (2 \cdot 10 \ + \ 8 \cdot 9)$

$1068 = 92 \cdot 11 + (4 \cdot 13 \ + \ 4 \cdot 1)$

$1156 = 84 \cdot 13 + (4 \cdot 11 \ + \ 4 \cdot 1) + 8$

$1548 = 84 \cdot 17 + (16 \cdot 7) + 8$

$1564 = 84 \cdot 17 + (10 \cdot 11 \ + \ 6 \cdot 3) + 8$

$1612 = 84 \cdot 17 + (16 \cdot 11) + 8$

$478 = 42 \cdot 11 + (8 \cdot 1) + 8$

$606 = 50 \cdot 11 + (8 \cdot 7)$

$758 = 78 \cdot 9 + (8 \cdot 7)$

$774 = 78 \cdot 9 + (4 \cdot 11 + \ 4 \cdot 7)$

$790 = 78 \cdot 9 + (8 \cdot 11)$

$958 = 80 \cdot 11 + (2 \cdot 11 \ + \ 6 \cdot 8) + 8$

$1086 = 78 \cdot 13 + (10 \cdot 7 \ + \ 2 \cdot 1)$

$1102 = 78 \cdot 13 + (8 \cdot 11)$

$1118 = 78 \cdot 13 + (8 \cdot 13)$

$1134 = 78 \cdot 13 + (6 \cdot 13 \ + \ 6 \cdot 7)$

$1150 = 78 \cdot 13 + (2 \cdot 13 \ + \ 10 \cdot 11)$

$1542 = 128 \cdot 11 + (4 \cdot 17 \ + \ 6 \cdot 11)$

$1590 = 84 \cdot 17 + (14 \cdot 11) + 8$

$510 = 48 \cdot 9 + (6 \cdot 9 \ + \ 2 \cdot 8) + 8$

$646 = 70 \cdot 9 + (8 \cdot 1) + 8$

$766 = 78 \cdot 9 + (2 \cdot 11 + \ 6 \cdot 7)$

$782 = 78 \cdot 9 + (6 \cdot 13 \ + \ 2 \cdot 1)$

$798 = 78 \cdot 9 + (4 \cdot 13 \ + \ 4 \cdot 11)$

$1078 = 78 \cdot 13 + (2 \cdot 11 \ + \ 6 \cdot 7)$

$1094 = 78 \cdot 13 + (6 \cdot 13 \ + \ 2 \cdot 1)$

$1110 = 78 \cdot 13 + (2 \cdot 13 \ + \ 10 \cdot 7)$

$1126 = 78 \cdot 13 + (10 \cdot 11 \ + \ 2 \cdot 1)$

$1142 = 84 \cdot 13 + (6 \cdot 7) + 8$

$1534 = 78 \cdot 17 + (16 \cdot 13)$

$1550 = 64 \cdot 23 + (6 \cdot 9 + 2 \cdot 8) + 8$

$1598 = 80 \cdot 19 + (6 \cdot 11 + 12 \cdot 1)$

Theorem 3.23 *If* $v \in \{626, 634, 642, 654, 658, 662, 670, 676, 678, 686,$ $804, 812, 910, 918, 1082, 1090, 1116, 1124\}$, *then a 2-SOLSSOM($v$) exists.*

Proof: Apply Theorem 3.15. For $v = 654, 662, 670, 686$, take $m = 9$, $u = 0, 1, 2$ or 4, and $x = 0$, $y = 6$, $z = 3$ to obtain 2-HSOLSSOMs of type $(72^8 s^1 78^1)$ for $s = 0, 8, 16, 32$; for $v = 626, 634, 642, 658$ take $m = 9$, $u = 0, 1, 2$ or 4, and $x = 4$, $y = 4$, $z = 1$ to obtain 2-HSOLSSOMs of type $(72^8 s^1 50^1)$ for $s = 0, 8, 16, 32$; for $v = 804, 812$, take $m = 11$, $u = 1$ or 2, $x = 1$, $y = z = 5$ to obtain 2-HSOLSSOMs of type $(88^8 s^1 92^1)$ for $s = 8, 16$; for $v = 910, 918$, take $m = 13$, $u = 0$ or 1, $x = 6$, $y = 2$, $z = 5$ to obtain 2-HSOLSSOMs of type $(104^8 s^1 78^1)$ for $s = 0, 8$; for $v = 1082, 1090$, take $m = 16$, $u = 1$ or 2, and $x = 13$, $y = 3$, $z = 0$ to obtain 2-HSOLSSOMs of type $(128^8 s^1 50^1)$ for $s = 8, 16$, and for $v = 1116, 1124$, take $m = 16$, $u = 0$ or 1, and $x = 6$, $y = 10$, $z = 0$ to obtain 2-HSOLSSOMs of type $(128^8 s^1 92^1)$ for $s = 0, 8$.

We can then fill in the holes by using 2-SOLSSOMs of orders $0, 8, 16$, $32, 50, 72, 78, 88, 92, 128$. Finally, for $v = 676$ take $m = 9$, $u = 1$, $x = 0, y = 3$, $z = 4$, and for $v = 678$ take $m = 9$, $u = 3$, and $x = y = 2$, $z = 5$. This gives 2-HSOLSSOMs of types $(72^8 8^1 84^1)$ and $(72^8 24^1 70^1)$; we then add 8 infinite points and use 2-SOLSSOM(16) or 2-SOLSSOM(32) plus 2-ISOLSSOM$(t, 8)$, $t = 80, 78, 92$ for the fill. 2-ISOLSSOM$(80, 8)$ is obtainable by Construction 3.9 since $80 = 8 \cdot 9 + 8$, while 2-ISOLSSOM$(78, 8)$ and 2-ISOLSSOM$(92, 8)$ are obtained by filling all holes of a 2-HSOLSSOM(7^{11}) or 2-HSOLSSOM(7^{13}) (except one) with the aid of an infinite point.   □

Theorem 3.15 also works for all v in a given residue class mod 8 within some quite large ranges. Below we indicate several of these. Recall a 2-SOLSSOM$(8u)$ exists for all $w \geq 7$; the appropriate range for $8u$ is $56 \leq 8u \leq 8m$ (or $48 \leq 8u \leq 8m$ when 8 infinite points are needed to fill in the holes, as is the case for the ranges $[716, 740]$, $[830, 870]$ and $[966, 1070]$).

v mod 8	Range for v	m	HSOLSSOM	type
2	$[682, 698]$	9	$72^8(8u)^1 s^1,$	$s = 50$
2	$[810, 842]$	11	$88^8(8u)^1 s^1,$	$s = 50$
2	$[866, 898]$	11	$88^8(8u)^1 s^1,$	$s = 106$
2	$[938, 1042]$	13	$104^8(8u)^1 s^1,$	$s = 50, 106$
2	$[1130, 1202]$	16	$128^8(8u)^1 s^1,$	$s = 50$
2	$[1194, 1330]$	17	$136^8(8u)^1 s^1,$	$s = 50, 106$
2	$[1322, 1530]$	19	$152^8(8u)^1 s^1,$	$s = 50, 106, 162$
2	$[1634, 1762]$	23	$184^8(8u)^1 s^1,$	$s = 106$
2	$[1762, 2018]$	25	$200^8(8u)^1 s^1,$	$s = 106, 218$
2	$[2018, 2306]$	29	$232^8(8u)^1 s^1,$	$s = 106, 218$
2	$[2210, 2522]$	32	$256^8(8u)^1 s^1,$	$s = 106, 218$
2	$[2530, 2882]$	37	$296^8(8u)^1 s^1,$	$s = 106, 218$
2	$[2842, 3282]$	41	$328^8(8u)^1 s^1,$	$s = 162, 330$
2	$[3226, 3714]$	47	$376^8(8u)^1 s^1,$	$s = 162, 330$

$v \bmod 8$	Range for v	m	HSOLSSOM	type
2	[3666, 4314]	53	$424^8(8u)^1 s^1$,	$s = 218, 498$
2	[4178, 5002]	61	$488^8(8u)^1 s^1$,	$s = 218, 610$
2	[4946, 5970]	73	$584^8(8u)^1 s^1$,	$s = 218, 714$
2	[5970, 7122]	89	$712^8(8u)^1 s^1$,	$s = 218, 714$
4	[716, 740]	9	$72^8(8u)^1 s^1 + 8$,	$s = 84$
4	[852, 884]	11	$88^8(8u)^1 s^1$,	$s = 92$
4	[980, 1028]	13	$104^8(8u)^1 s^1$,	$s = 92$
4	[1172, 1244]	16	$128^8(8u)^1 s^1$,	$s = 92$
4	[1236, 1372]	17	$136^8(8u)^1 s^1$,	$s = 92, 148$
4	[1364, 1516]	19	$152^8(8u)^1 s^1$,	$s = 92, 148$
4	[1620, 1804]	23	$184^8(8u)^1 s^1$,	$s = 92, 148$
4	[1804, 2004]	25	$200^8(8u)^1 s^1$,	$s = 148, 204$
4	[2004, 2292]	29	$232^8(8u)^1 s^1$,	$s = 92, 204$
4	[2196, 2564]	32	$256^8(8u)^1 s^1$,	$s = 92, 260$
4	[2516, 2924]	37	$296^8(8u)^1 s^1$,	$s = 92, 260$
4	[2772, 3308]	41	$328^8(8u)^1 s^1$,	$s = 92, 356$
4	[3212, 3700]	47	$376^8(8u)^1 s^1$,	$s = 148, 316$
4	[3652, 4340]	53	$424^8(8u)^1 s^1$,	$s = 204, 524$
4	[4164, 4996]	61	$488^8(8u)^1 s^1$,	$s = 204, 604$
4	[4932, 5964]	73	$584^8(8u)^1 s^1$,	$s = 204, 708$
4	[5956, 7116]	89	$712^8(8u)^1 s^1$,	$s = 204, 708$
6	[710, 726]	9	$72^8(8u)^1 s^1$,	$s = 78$
6	[830, 870]	11	$88^8(8u)^1 s^1 + 8$,	$s = 70$
6	[966, 1070]	13	$104^8(8u)^1 s^1$,	$s = 78, 134$
6	[1158, 1230]	16	$128^8(8u)^1 s^1$,	$s = 78$
6	[1222, 1382]	17	$136^8(8u)^1 s^1$,	$s = 78, 158$
6	[1350, 1526]	19	$152^8(8u)^1 s^1$,	$s = 78, 158$
6	[1606, 1814]	23	$184^8(8u)^1 s^1$,	$s = 78, 158$
6	[1814, 1990]	25	$200^8(8u)^1 s^1$,	$s = 158, 190$
6	[1990, 2278]	29	$232^8(8u)^1 s^1$,	$s = 78, 190$
6	[2238, 2606]	32	$256^8(8u)^1 s^1$,	$s = 134, 302$
6	[2558, 2966]	37	$296^8(8u)^1 s^1$,	$s = 134, 302$
6	[2814, 3326]	41	$328^8(8u)^1 s^1$,	$s = 134, 374$
6	[3198, 3758]	47	$376^8(8u)^1 s^1$,	$s = 134, 374$
6	[3638, 4342]	53	$424^8(8u)^1 s^1$,	$s = 190, 526$
6	[4150, 4998]	61	$488^8(8u)^1 s^1$,	$s = 190, 606$
6	[4918, 5958]	73	$584^8(8u)^1 s^1$,	$s = 190, 702$
6	[5942, 7110]	89	$712^8(8u)^1 s^1$,	$s = 190, 702$ □

It is also possible to use Theorem 3.15 to obtain 2-SOLSSOMs for larger v, but these are more easily obtained by Construction 3.9 with $m = 8$ and $e_i \in \{0, 1\}$ as was done by Lee [9]. Any $v > 6884 = 8 \cdot 765 + 764$ can be written as $8q + e$ where q is any one of four consecutive odd integers ≥ 765, and $702 \leq e \leq 764$. Further, in any set of four consecutive odd integers, at most two values are divisible by 3, and one by 5 so at least one will always satisfy the required conditions on q in Construction 3.9.

To conclude, the following table provides a list of v values for which a 2-SOLSSOM(v) is unknown. Values are grouped into five residue classes: 2 mod 8, 4 mod 8, 6 mod 8, 0 mod 8, and 1 mod 2. These five classes contain respectively, 51, 60, 57, 3 and 12 possible exceptions, giving a total of 183 possible exceptions.

2	10	18	26	34	42	58	66	74	82	90	98
114	122	130	138	146	154	170	178	186	194	202	210
226	234	242	250	258	266	274	282	290	298	306	314
322	338	346	354	370	378	402	410	418	426	434	458
482	570	594									

4	12	20	28	36	44	52	60	68	76	84	100
108	116	124	132	140	156	164	172	180	188	196	212
220	228	236	244	252	268	276	284	292	300	308	324
332	340	348	364	388	396	412	420	436	444	452	460
476	492	500	508	516	532	548	588	628	660	684	700

6	14	22	30	38	46	54	62	70	86	94	102
110	118	126	142	150	166	174	182	198	206	214	222
230	238	246	254	262	270	278	286	294	310	318	326
334	358	366	382	390	398	406	422	430	438	446	454
462	494	502	518	534	558	614	622	630			

24	40	48

3	5	15	21	33	35	39	51	65	87	123	135

Remark 3.24 *In* [9], 434, 548 *and a few other values which we solved (for instance* 782, 790*) were not included in the exception list, but no construction appears to have been given. We were unable to find any constructions for* 434 *and* 548*, so these values are included in our exception list.*

Acknowledgments

This research was supported in part by NSERC grant OGP 0005320 and a grant from the Mount Saint Vincent University Committee on Research and Publications for the second author. The second author also wishes to thank the University of New South Wales for their kind hospitality, services and financial assistance while a visiting professor in

July, 2001. The authors would also like to thank the referee for several helpful comments.

References

[1] R.J.R. Abel, F.E. Bennett, H.Zhang and L. Zhu, A few more self-orthogonal Latin squares and related designs, *Aust. J. Comb.* **21** (2000), 85-94.

[2] R.J.R. Abel, F.E. Bennett, G. Ge and L. Zhu, Existence of Steiner seven-cycle systems, *Discrete Math.* **252** (2002), 1-16.

[3] F.E. Bennett and L.Zhu, The spectrum of HSOLSSOM(h^n) when h is even, *Discrete Math.* **158**, (1996) 11-25.

[4] A.E. Brouwer, A series of separable designs with application to pairwise orthogonal Latin squares, *Europ. J. Comb.* **1** (1990), 39-41.

[5] B. Du, A few more resolvable spouse-avoiding mixed double round robin tournaments, *Ars Comb.* **36**, (1993), 309-314.

[6] C.J. Colbourn and J.H. Dinitz, *CRC Handbook of Combinatorial Designs*, CRC Press, Boca Raton FL., 1996. (New results are reported at http://www.emba.uvm.edu/~dinitz/newresults.html).

[7] G. Ge and R.J.R. Abel, Some new HSOLSSOMs of types h^n and $1^m u^1$, *J. Comb. Designs* **9** (2001), 435-444.

[8] M.Greig, Designs from projective planes and PBD bases, *J. Comb. Designs* **7** (1999), 341-374.

[9] T.C.Y. Lee, Tools for constructing RBIBDs, Frames, NRBs and SOLSSOMs, PhD Thesis, University of Waterloo, 1995.

[10] C.C.Lindner, R.C. Mullin, and D.R. Stinson, On the spectrum of resolvable orthogonal arrays invariant under the Klein group K_4, *Aequat. Math.* **26** (1983), 176-183.

[11] R. C. Mullin and D. R. Stinson, Holey SOLSSOMs, *Utilitas Math.* **25** (1984), 159-169.

[12] M. Wojtas, Three new constructions of mutually orthogonal Latin squares, *J. Comb. Designs* **8** (2000), 218-220.

Chapter 2

CONJUGATE ORTHOGONAL DIAGONAL LATIN SQUARES WITH MISSING SUBSQUARES

Frank E. Bennett

Department of Mathematics, Mount Saint Vincent University
Halifax, Nova Scotia B3M 2J6, Canada
frank.bennett@msvu.ca

Beiliang Du

Department of Mathematics, Suzhou University
Suzhou 215006, P.R.China
dubl@pub.sz.jsinfo.net

Hantao Zhang

Computer Science Department, The University of Iowa
Iowa City, IA 52242, U. S. A.
hzhang@cs.uiowa.edu

Abstract   We shall refer to a diagonal Latin square which is orthogonal to its $(3, 2, 1)$-conjugate, and the latter is also a diagonal Latin square, as a $(3, 2, 1)$-conjugate orthogonal diagonal Latin square, briefly CODLS. This article investigates the spectrum of CODLS with a missing subsquare. The main purpose of this paper is two-fold. First of all, we show that for any positive integers $n \geq 1$, a CODLS of order v with a missing subsquare of order n exists if $v \geq 13n/4 + 93$ and $v - n$ is even. Secondly, we show that for $2 \leq n \leq 6$, a CODLS of order v with a missing subsquare of order n exists if and only if $v \geq 3n + 2$ and $v - n$ is even, with one possible exception.

1.   Introduction

A *Latin square* of order v is a $v \times v$ array such that every row and every column is a permutation of a v-set, say $S = \{0, 1, \cdots, v - 1\}$. A *transversal* in a Latin square is a set of positions, one per row and one per column, among which the symbols occur precisely once each. A *transversal Latin square* is a Latin square whose main diagonal is a transversal. A *diagonal Latin square* is a transversal Latin square whose back diagonal is also a transversal. A Latin square is *idempotent* if the cell (i, i) has value i, $0 \leq i \leq v - 1$.

Two Latin squares are *orthogonal* if each symbol in the first square meets each symbol in the second square exactly once when they are superposed.

A *quasigroup* is an ordered pair $(Q, \otimes)$, where Q is a set and $\otimes$ is a binary operation on Q such that the equations $a \otimes x = b$ and $y \otimes a = b$ are uniquely solvable for every pair of elements $a, b \in Q$. It is well known that the multiplication table of a quasigroup defines a Latin square. That is, a Latin square can be viewed as the multiplication table of a quasigroup with the headline and sideline removed.

If $(Q, \otimes)$ is a quasigroup, we may define on the set Q six binary operations $\otimes_{(1,2,3)}, \otimes_{(1,3,2)}, \otimes_{(2,1,3)}, \otimes_{(2,3,1)}, \otimes_{(3,1,2)},$ and $\otimes_{(3,2,1)}$ as follows: $a \otimes b = c$ if and only if

$$a \otimes_{(1,2,3)} b = c, \quad a \otimes_{(1,3,2)} c = b, \quad b \otimes_{(2,1,3)} a = c,$$
$$b \otimes_{(2,3,1)} c = a, \quad c \otimes_{(3,1,2)} a = b, \quad c \otimes_{(3,2,1)} b = a.$$

These six (not necessarily distinct) quasigroups $(Q, \otimes_{(i,j,k)})$ are called the *conjugates* of $(Q, \otimes)$ (see Stein [13]). If the multiplication table of a quasigroup $(Q, \otimes)$ defines a Latin square L, then the six Latin squares defined by the multiplication table of its conjugates $(Q, \otimes_{(i,j,k)})$ are called the *conjugates* of L. For more information on Latin squares and quasigroups, the interested reader may refer to the book of Dénes and Keedwell [8].

A Latin square which is orthogonal to its (i, j, k)-conjugate will called an (i, j, k)-*conjugate orthogonal* Latin square. An (i, j, k)-conjugate orthogonal (idempotent, transversal, diagonal) Latin square is an (idempotent, transversal, diagonal) Latin square which is orthogonal to its (i, j, k)-conjugate, and the latter is also a (idempotent, transversal, diagonal) Latin square, where $\{i, j, k\} = \{1, 2, 3\}$. We denote by COLS (COILS, COTLS, CODLS) a $(3, 2, 1)$-conjugate orthogonal (idempotent, transversal, diagonal) Latin square.

For the spectra of COILS and CODLS, we have the following results.

Theorem 1.1 ([6, 16]) A COILS of order v exists for all positive integers v, except $v \in \{2, 3, 6\}$.

Theorem 1.2 ([1]) A CODLS of order v exists for all positive integers v, except $v \in \{2, 3, 6\}$ and except possibly $v = 10$.

The problem we study in this paper is the (i, j, k)-conjugate orthogonal diagonal Latin squares analogue of the Doyen-Wilson theorem [9]. We begin with some definitions. If an (i, j, k)-conjugate orthogonal diagonal Latin square has a subsquare occupying the central position, the subsquare itself must be an (i, j, k)-conjugate orthogonal diagonal Latin square. We refer to it as (i, j, k)-*conjugate orthogonal diagonal subsquare*. We denote by (i, j, k)-CODLS(v, n) an (i, j, k)-conjugate orthogonal diagonal Latin square of order v with (i, j, k)-conjugate orthogonal diagonal subsquare of order n. It is easy to see that the existence of an (i, j, k)-CODLS(v, n) requires that $v - n$ is even. In particular, any (i, j, k)-CODLS(v) is an (i, j, k)-CODLS$(v, 1)$ when v is odd. In view of Theorem 1.2, no $(3, 2, 1)$-conjugate orthogonal diagonal Latin square can contain $(3, 2, 1)$-conjugate orthogonal diagonal subsquare of order $2, 3$ or 6. However, we can construct $(3, 2, 1)$-conjugate orthogonal diagonal Latin square missing a subsquare of those orders. We also let CODLS(v, n) denote an (i, j, k)-conjugate orthogonal diagonal Latin square of order v with a missing subsquare of order n occupying the central position. It is not necessary for such a (i, j, k)-conjugate orthogonal diagonal Latin square of order n to exist. We refer to the subsquare as the *hole*.

Some simple computation shows the following property.

Theorem 1.3 If there exists an (i, j, k)-CODLS(v, n), then $v \geq 3n + 2$ and $v - n$ is even.

Proof The existence of an (i, j, k)-CODLS(v, n) implies the existence of a pair of orthogonal diagonal Latin squares of order v with missing subsquares of size n, hence $v \geq 3n + 2$. The other condition, $v - n$ is even, comes from the fact that the missing subsquare of order n must occupy the central position. $\qquad\square$

```
0 6 4 2 1 7 3 5
6 1 5 7 2 3 0 4
3 5 2 6 7 1 4 0
1 2 7     0 5 6
5 0 1     6 7 2
7 4 6 1 0 5 2 3
4 7 3 0 5 2 6 1
2 3 0 5 6 4 1 7
```

Figure 1.1.   CODLS(8, 2)

The (i, j, k)-CODLS(v, n) have been studied by several researchers. Du [10] and Bennett, Du and Zhang [2] investigated $(2, 1, 3)$-CODLS(v, n). It is proved that for any positive integers v and n, a $(2, 1, 3)$-CODLS(v, n) exists if and only if $v \geq 3n + 2$ and $v - n$ is even, except possibly for $(v, n) \in \{(3n + 2, n) : n = 6, 8, 10\}$. In this paper, we shall be restricting our attention to $(3, 2, 1)$-CODLS(v, n), a $(3, 2, 1)$-conjugate orthogonal diagonal Latin square of order v with a missing subsquare of order n, briefly CODLS(v, n). An example of a CODLS$(8, 2)$ is given in Figure 1.1.

This article investigates the spectrum of CODLS with a missing subsquare. The main purpose of this paper is two-fold. First of all, we show that for any positive integers $n \geq 1$, a CODLS of order v with a missing subsquare of order n exists if $v \geq 13n/4 + 93$ and $v - n$ is even. Secondly, we show that for $2 \leq n \leq 6$, a CODLS of order v with a missing subsquare of order n exists if and only if $v \geq 3n + 2$ and $v - n$ is even, with one possible exception. That is, our main objective is to establish the following results.

Theorem 1.4 For any positive integer $n \geq 1$, a CODLS(v, n) exists if $v \geq 13n/4 + 93$ and $v - n$ is even.

Theorem 1.5 For any positive integer $2 \leq n \leq 6$, a CODLS(v, n) exists if and only if $v \geq 3n + 2$ and $v - n$ is even, with the possible exception of $(v, n) = (11, 3)$.

We shall use both direct and recursive methods in our constructions. The direct construction method will be the "starter-adder" type where the plan is to use the first row to cyclically generate the entire Latin square. The recursive type of construction will include the pairwise balanced design construction and the "filling-in holes" construction.

2.   Preliminaries

In this section we shall define some of the auxiliary designs and some of the fundamental results which will be used later. The reader is referred to [4-7] for more information on designs, and, in particular, incomplete conjugate orthogonal idempotent Latin squares and pairwise balanced designs and group divisible designs.

A pair of *incomplete orthogonal Latin squares* based on the set $S \cup T$, missing a subsquare based on the set T, where $S = \{0, 1, 2, \cdots, v-n-1\}$ and $T = \{v - n, v - n + 1, \cdots, v - 1\}$, is a pair of orthogonal Latin squares of order v except for the common $n \times n$ subsquare (a_{ij}) based on the set T. Such Latin squares are idempotent if $a_{ii} = i$ for all i, $0 \le i \le v-n-1$. We shall denote by $a(i, j)$ the entry in the cell (i, j) of a Latin square L. An incomplete Latin square L based on the set $S \cup T$ will be called *diagonal* if both sets of cells $\{a(i, i) : 0 \le i \le v - n - 1\}$ and $\{a(i, v - n - i - 1) : 0 \le i \le v - n - 1\}$ represent a permutation of the $(v - n)$-set S. An incomplete (i, j, k)-conjugate orthogonal (idempotent, transversal) Latin square is an incomplete (idempotent, transversal) Latin square which is orthogonal to its (i, j, k)-conjugate and the latter is also an incomplete (idempotent, transversal) Latin square. We denote by ICOLS (ICOILS, ICOTLS) an incomplete $(3, 2, 1)$-conjugate orthogonal (idempotent, transversal) Latin square in this paper.

For ICOILS, we have the following existence result.

Lemma 2.1   For $2 \le n \le 6$, an ICOILS(v, n) exists if and only if $v \ge 3n + 1$, except $(v, n) = (11, 3)$.

Proof   It has been proved in [4-6,16] that, for $2 \le n \le 6$, an ICOILS (v, n) exists if and only if $v \ge 3n + 1$, except possibly $(v, n) = (11, 3)$. A computer search confirmed that an ICOILS$(11, 3)$ does not exist. $\square$

Let S be a set and $\mathcal{H} = \{S_1, S_2, \cdots, S_k\}$ be a set of subsets of S. A *holey Latin square* having hole set $\mathcal{H}$ is an $|S| \times |S|$ array L, indexed by S, satisfying the following properties:

1   every cell of L either contains a symbol of S or is empty,

2   every symbol of S occurs at most once in any row or column of L,

3   the subarrays indexed by $S_i \times S_i$ are empty for $1 \le i \le k$ (these subarrays are referred to as *holes*),

4   symbol $s \in S$ occurs in row or column t if and only if $(s, t) \in (S \times S) \setminus \cup_{1 \le i \le k} (S_i \times S_i)$.

The order of L is $|S|$. Two holey Latin squares on symbol set S and hole set $\mathcal{H}$ are said to be *orthogonal* if their superposition yields every ordered pair in $(S \times S) \setminus \cup_{1 \le i \le k} (S_i \times S_i)$.

We denote the holey $(3, 2, 1)$-conjugate orthogonal Latin square by $\text{HCOLS}(n; s_1, s_2, \cdots, s_k)$, where $n = |S|$ is the order and $s_i = |S_i|, 1 \le i \le k$.

If $\mathcal{H} = \{S_1, S_2, \cdots, S_k\}$ is a partition of S, then a holey COLS is called a *frame* COLS. The type of the frame COLS is defined to be the multiset $\{|S_i| : 1 \le i \le k\}$. We shall use an "exponential" notation to describe types: type $s_1^{n_1} s_2^{n_2} \cdots s_t^{n_t}$ denotes n_i occurrence of $s_i, 1 \le i \le t$, in the multiset. We briefly denote a frame COLS of type $s_1^{n_1} s_2^{n_2} \cdots s_t^{n_t}$ by $\text{FCOLS}(s_1^{n_1} s_2^{n_2} \cdots s_t^{n_t})$. We observe that the existence of a $\text{COILS}(n)$ is equivalent to the existence of an $\text{FCOLS}(1^n)$, and the existence of an $\text{ICOILS}(v, n)$ is equivalent to the existence of an $\text{FCOLS}(1^{v-n} n^1)$.

For FCOLS, we have the following existence result.

Lemma 2.2 ([3,6,16]) There exists an $\text{FCOLS}(h^n)$ if and only if $h \ge 1$ and $n \ge 4$, except $(n, h) = (6, 1)$.

Proof It has been shown in [3,6,16] that there exists an $\text{FCOLS}(h^n)$ if and only if $h \ge 1$ and $n \ge 4$, except $(n, h) = (6, 1)$, and except possibly $(n, h) = (6, 13)$. Using the starter-adder technique (see section 3), we construct an $\text{FCOLS}(13^6)$ from the vectors $\underline{e}$, $\underline{f}$, and $\underline{g}$ below, with $S = \{0, 1, ..., 64\} \cup X$, where $X = \{x_1, ..., x_9, y_1, ..., y_4\}$. □

$$
\begin{aligned}
\underline{e}: \quad & (\emptyset, y_4, 34, 36, 13, \emptyset, 52, 59, 32, 22, \emptyset, 49, 63, x_8, 61, \\
& \emptyset, 37, x_1, 62, 41, \emptyset, 64, x_3, 39, 23, \emptyset, 17, x_4, y_3, 58, \\
& \emptyset, 28, x_9, x_7, 46, \emptyset, 54, y_1, 57, 42, \emptyset, 24, x_5, y_2, 16, \\
& \emptyset, 44, x_2, 11, 18, \emptyset, 12, x_6, 27, 6, \emptyset, 14, 3, 29, 48, \\
& \emptyset, 4, 56, 2, 1) \\
\underline{f}: \quad & (7, 8, 9, 19, 21, 26, 31, 33, 38, 43, 47, 51, 53) \\
\underline{g}: \quad & (7, 41, 27, 42, 57, 61, 58, 6, 31, 14, 53, 49, 1)
\end{aligned}
$$

Lemma 2.3 For $2 \le n \le 6$ and $v \ge 3n + 1$, except for $(v, n) = (11, 3)$, there exists an $\text{FCOLS}(1^{v-n} n^1)$.

Proof This follows directly from Lemma 2.1. □

For our purpose, we also need the concept of pairwise balanced designs and group divisible designs.

Let K be a set of positive integers. A *pairwise balanced design* (PBD) of index unity $B(K, 1; v)$ is a pair $(X, \mathcal{B})$ where X is a v-set (of *points*) and $\mathcal{B}$ is a collection of subsets of X (called *blocks*) with sizes in K such that every pair of distinct points of X is contained in exactly one block of $\mathcal{B}$. The number $|X| = v$ is called the *order* of PBD.

Let K and M be sets of positive integers. A *group divisible design* (GDD) $GD[K, 1, M; v]$ is a triple $(X, \mathcal{G}, \mathcal{B})$ where

1. X is a v-set (of *points*),

2. $\mathcal{G}$ is a collection of nonempty subsets of X (called *groups*) with cardinality in M and which partition X,

3. $\mathcal{B}$ is a collection of subsets of X (called *blocks*) with cardinality at least two in K,

4. no block intersects any group in more than one point,

5. each pair set $\{x, y\}$ of points not contained in a group is contained in exactly one block.

The *group-type* (or *type*) of the GDD $(X, \mathcal{G}, \mathcal{B})$ is the multiset of sizes $|G|$ of the $G \in \mathcal{G}$ and we usually use the "exponential" notation for its description: group-type $1^i 2^j 3^k \cdots$ denotes i occurrences of groups of size 1, j occurrences of groups of size 2, and so on.

Let $(X, \mathcal{B})$ be a PBD $B(K, 1; v)$. A *parallel class* in $(X, \mathcal{B})$ is a collection of disjoint blocks of $\mathcal{B}$, the union of which equals X. The design $(X, \mathcal{B})$ is called *resolvable* if the blocks of $\mathcal{B}$ can be partitioned into parallel classes. A GDD $[K, 1, M; v]$ is resolvable if its associated PBD $B(K \cup M, 1; v)$ is resolvable with M as a parallel class of the resolution.

We need to establish some more notations. We shall denote by $B(k, 1; v)$ a $B(\{k\}, 1; v)$ and write $GD[k, 1, m; v]$ for a $GD[\{k\}, 1, \{m\}; v]$. We shall tacitly make use of the fact that in a $GD[k, 1, m; km]$, each block of size k intersects each group of size m in exactly one point, that is, each block is a transversal of the collection of groups. This GDD is usually called a *transversal design*, denoted as $TD[k, m]$. If $m \notin M$, the $GD[K, 1, M \cup \{m*\}; v]$ denotes a $GD[K, 1, M \cup \{m\}; v]$ which contains a unique group of size m and if $m \in M$, then a $GD[K, 1, M \cup \{m*\}; v]$ is a $GD[K, 1, M; v]$ containing at least one group of size m. We shall sometimes refer to a GDD $(X, \mathcal{G}, \mathcal{B})$ as a K-GDD if $|B| \in K$ for every block $B \in \mathcal{B}$.

3.   Direct Construction

First we state a "starter-adder" type construction for ICOLS(v, n). The main idea is to generate the square under a cyclic group of order $v - n$, from its first row and from the last n elements of the first column. Let $S = \{0, 1, \cdots, v - n - 1\} \cup X$, where $X = \{x_1, x_2, \cdots, x_n\}$. Suppose L is a Latin square based on S with a hole indexed by X. As before, we denote by $a(i, j)$ the entry in the cell (i, j) of the array L. The first row is given by the vectors $\underline{e}$ and $\underline{f}$, and the last n elements of the first column are given by the vector $\underline{g}$, where

$$\begin{aligned}
\underline{e} &= (a(0, 0), \cdots, a(0, v - n - 1)), \\
\underline{f} &= (a(0, v - n), \cdots, a(0, v - 1)), \\
\underline{g} &= (a(v - n, 0), \cdots, a(v - 1, 0)).
\end{aligned}$$

The ICOLS(v, n) is constructed modulo $v - n$ in the range $\{0, 1, \cdots, v - n - 1\}$, where the x_is act as "infinity" elements as follows:

1. $a(s + 1, t + 1) = a(s, t)$ if $a(s, t) = x_i$, and $a(s + 1, t + 1) \equiv a(s, t) + 1 \ (mod \ v - n)$ otherwise, where $0 \le s, t \le v - n - 1$,

2. $a(s + 1, v - n - 1 + t) \equiv a(s, v - n - 1 + t) + 1 \ (mod \ v - n)$, where $1 \le t \le n, 0 \le s < v - n - 1$,

3. $a(v - n - 1 + t, s + 1) \equiv a(v - n - 1 + t, s) + 1 \ (mod \ v - n)$, where $1 \le t \le n, 0 \le s \le v - n - 1$.

It is evident that there are necessary conditions which the vectors $\underline{e}$, $\underline{f}$ and $\underline{g}$ must satisfy to produce an ICOLS(v, n) (see, for example [6]), and this is the major task in our construction. However, we shall confine ourselves to presenting the vectors; the actual verification that they work will be omitted here.

For convenience, in what follows we define

$$\begin{aligned}
F \ = \ & \{(s, 2) : s = 8, 10, 12, 14, 16, 20, 24\} \ \cup \\
& \{(t, 3) : t = 13, 15, 17, 21, 25, 29, 37\} \ \cup \\
& \{(u, 4) : u = 14, 18, 26\} \ \cup \\
& \{(v, 5) : v = 19, 23, 27\} \ \cup \\
& \{(w, 6) : w = 20, 24, 28\}.
\end{aligned}$$

Lemma 3.1 For $(v, n) \in F$, there exists an ICOTLS(v, n) constructed by the starter-adder method, where $v - n$ is even and the first and the

Table 3.1.   Construction of ICOTLS(v, n)

(v, n)	$\underline{e}$	$\underline{f}$	$\underline{g}$
$(14, 2)$	$(0, 6, 1, 7, 10, 8, 4, x, 3, 5, 11, y)$	$(2, 9)$	$(2, 9)$
$(24, 2)$	$(0, 3, 10, 1, 20, 9, 12, 4, 19, 21, 15,$ $18, 13, 16, 6, 11, 7, 5, x, 14, y, 8)$	$(2, 17)$	$(21, 15)$
$(13, 3)$	$(0, 2, 8, 5, y, 3, x, 1, z, 4)$	$(6, 7, 9)$	$(3, 9, 7)$
$(37, 3)$	$(0, 8, 22, 15, 25, 7, 21, 16, 27, 31, 2,$ $9, 26, 30, 18, 6, 13, 20, 14, 24, 10, 29,$ $11, 5, 3, 1, 19, 33, x, 23, y, 32, z, 17)$	$(4, 12, 28)$	$(33, 29, 11)$
$(26, 4)$	$(0, 14, 6, 5, 13, 8, 16, 12, 1, 3, 18,$ $10, 7, 19, x, 4, y, 15, z, 20, u, 11)$	$(2, 9, 17, 21)$	$(19, 14, 7, 18)$
$(19, 5)$	$(0, 3, 12, 6, x, 9, y,$ $5, z, 1, u, 4, v, 7)$	$(2, 8, 10$ $11, 13)$	$(9, 1, 11$ $5, 13)$
$(23, 5)$	$(15, u, v, 14, 4, 3, 9, 17, 16,$ $5, 1, 0, 13, 12, x, y, 2, z)$	$(10, 11, 6,$ $7, 8)$	$(12, 13, 6,$ $2, 5)$
$(27, 5)$	$(0, 19, 6, 5, 13, 18, 16, 1, 3, 17, 15,$ $4, x1, 2, x2, 21, x3, 14, x4, 20, x5, 11)$	$(7, 8, 9,$ $10, 12)$	$(21, 20, 14$ $7, 3)$
$(24, 6)$	$(0, 12, 7, 5, 3, 17, x, 11, y,$ $15, z, 1, u, 4, v, 16, w, 9)$	$(2, 6, 8,$ $10, 13, 14)$	$(7, 13, 14$ $15, 3, 16)$
$(28, 6)$	$(13, 0, 16, 11, 15, 9, 2, 1, x, 10, 12,$ $6, y, z, u, v, 3, 5, 18, 4, w, 19)$	$(14, 17, 7,$ $8, 20, 21)$	$(3, 12, 19,$ $5, 15, 6)$

$(1+(v-n)/2)$th elements in the starter set $\underline{e}$ is not an infinity element, and the starter set $\underline{g}$ does not contain the elements 0 and $(v-n)/2$.

Proof   We can obtain each design by using the starter-adder technique. For the case $(v, n) = (10, 2)$ see [5]. For the case $(v, n) \in \{(14, 4), (18, 4)\}$ see [1]. For the case $(v, n) = (12, 2)$ see [6]. For the case $(v, n) \in \{(8, 2), (15, 13), (17, 3), (20, 6)\}$ see [4]. For the case $(v, n) \in \{(14, 2), (24, 2), (13, 3), (37, 3), (26, 4), (19, 5), (23, 5), (27, 5), (24, 6), (28, 6)\}$ see Table 1.1. For the other cases (v, n), see [16].   □

Using an ICOTLS(v, n) constructed by the starter-adder method, we have the following method to produce a CODLS(v, n).

Lemma 3.2([1]) Suppose there exists an ICOTLS(v, n) constructed by the starter-adder method, where $v - n$ is even and the first and the $(1+(v-n)/2)$th elements in the starter set $\underline{e}$ is not an infinity element, and the starter set $\underline{g}$ does not contain the elements 0 and $(v-n)/2$. Then there exists a $\check{\text{C}}$ODLS(v, n).

From Lemmas 3.1 and 3.2, we have the following result.

Lemma 3.3 There exists a CODLS(v, n) for $(v, n) \in F$.

For our result we need the following special COTLS. We denote by COTLS*$(v, 2)$ a COTLS(v) in which the cells $(v-2, v-1)$ and $(v-1, v-2)$ have entries $v - 2$ and $v - 1$, respectively. For COTLS*$(v, 2)$, we have the following result.

Lemma 3.4 There exists a COTLS*$(q, 2)$ with entry $(q - 3)/2$ in the cell $((q - 3)/2, (q - 3)/2)$ for every odd prime power $q \geq 7$.

Proof Let GF(q) be a finite field of q elements. Let us define the following Latin squares:

$$L_\lambda = (a_{xy}), \text{ where } a_{xy} = \lambda x + (1 - \lambda)y \text{ and } \lambda, x, y \in GF(q).$$

If $\lambda \neq 0, 1$, then it is easy to see that L_λ and $L_{1/\lambda}$ are a pair of COILS(q).

Note that q is an odd prime power, there exists a $\lambda' \in GF(q) \backslash \{0, 1\}$ such that $\lambda' + \lambda' = 1$. We look at $L_{\lambda'}$. Let GF$(q) = \{a_0 = 0, a_1, \cdots, a_{q-1}\}$ such that $a_0 + a_{q-1} = -1, a_1 + a_{q-2} = -1, \cdots, a_{(q-3)/2} + a_{(q+1)/2} = -1$. Then this Latin square's back diagonal is constant. Note that the Latin square is orthogonal to $L_\lambda, \lambda \in GF(q) \backslash \{0, 1, \lambda'\}$, and $q \geq 7$, we can choose λ such that $\lambda, 1/\lambda \in GF(q) \backslash \{0, 1, \lambda'\}$ and $\lambda \neq 1/\lambda$. L_λ is the design idempotent CODLS(q).

We rename a_i by i and permute rows and columns and entries with the permutation σ given by

$$\sigma = \begin{pmatrix} 0 & 1 & 2 & & q-2 & q-1 \\ & & & \cdots & \\ q-1 & 0 & 1 & & q-3 & q-2 \end{pmatrix}$$

We then obtain the design by permuting columns with the permutation μ given by

$$\mu = \begin{pmatrix} 0 & 1 & & q-4 & q-3 & q-2 & q-1 \\ & & \cdots & \\ q-3 & q-4 & & 1 & 0 & q-1 & q-2 \end{pmatrix}$$

Note that the first Latin square is the idempotent CODLS with a transversal which consists of the positions in $\{(q - 2, q - 1), (q - 1, q -$

2), $(i, q-3-i) : 0 \leq i \leq q-3\}$, so it is easy to check that the final square is the desired design, where the main diagonal consists of the transversal and the cells $(v-2, v-1)$ and $(v-1, v-2)$ have entries $v-2$ and $v-1$, respectively. □

We illustrate the construction of COTLS*(7, 2) in Figure 3.2.

$$
\begin{array}{ccccccc}
6 & 1 & 3 & 5 & 0 & 2 & 4 \\
2 & 4 & 6 & 1 & 3 & 5 & 0 \\
5 & 0 & 2 & 4 & 6 & 1 & 3 \\
1 & 3 & 5 & 0 & 2 & 4 & 6 \\
4 & 6 & 1 & 3 & 5 & 0 & 2 \\
0 & 2 & 4 & 6 & 1 & 3 & 5 \\
3 & 5 & 0 & 2 & 4 & 6 & 1
\end{array}
$$

Figure 3.2. COTLS*(7, 2)

4. Recursive Construction

In this section, we shall state some recursive constructions. For our first set of constructions, we employ the "filling-in holes" technique and its generalization.

Construction 4.1 Suppose there exists an FCOLS(h^n), where h is even. If a CODLS(h) exists, then there exists a CODLS(hn, h).

Proof We begin with the FCOLS and fill in all holes but the right lower hole with CODLS(h), and then we permute rows and columns and entries with the permutation σ given by

$$
\sigma(i) = \begin{cases} i - kh/2 & \text{if } kh \leq i < (k+1/2)h \\ i + (n - 3k/2 - 1)h & \text{if } (k+1/2)h \leq i < (k+1)h \end{cases} \quad 0 \leq k < n
$$

Note that the main diagonal and the back diagonal of the resulting square consist of the main diagonal and the back diagonal of each filling CODLS, respectively, we obtain the desired CODLS(hn, h). □

Observing that the condition requiring a CODLS(h) in Construction 4.1 can be replaced by a CODLS($h + a, a$), we have the following construction.

Construction 4.2 Suppose there exists an FCOLS(h^n), where h is even. If a CODLS($h + a, a$) exists, then there exist a CODLS($hn + a, a$) and a CODLS($hn + a, h + a$).

Our next set of constructions are the usual singular direct product constructions.

Construction 4.3 Suppose there exists an idempotent $CODLS(m)$, where m is even. Further suppose that a $COILS(n)$ and a $CODLS(n + a, a)$ all exist. Then there exists a $CODLS(mn + a, a)$. Moreover there exists a $CODLS(mn + 2, m)$, if n is odd and there exists a $COTLS^*(n + 2, 2)$ with entry $(n - 1)/2$ in the cell $((n - 1)/2, (n - 1)/2)$.

Proof We begin with the $CODLS(m)$ and replace each of its cells with an $n \times n$ array labeled by the elements in the cell. The array consists of four corners of a $CODLS(n + a, a)$ if the cell is $(i, i), 0 \leq i < m$ (or $0 \leq i < m - 1$). For the other cases, the cell will be filled with a modified $COILS(n)$, that is, by permuting the columns, the main diagonal of the $COILS(n)$ becomes its back diagonal (or leave the cell $(m - 1, m - 1)$ empty). We then obtain the upper left part of an $ICOILS(mn + a, a)$ (or $ICOILS(mn + a, n + a)$) using the standard Wilson construction on Latin squares (see, for example, [14]).

The design $CODLS(mn + a, a)$ comes from the resulting $ICOILS(mn + a, a)$ by permuting rows and columns and entries with the permutation σ_1 given by

$$\sigma_1(i) = \begin{cases} i & \text{if} \quad 0 \leq i < mn/2 \\ i + a & \text{if} \quad mn/2 \leq i < mn \\ i - mn/2 & \text{if} \quad mn \leq i < mn + a \end{cases}$$

For the design $CODLS(mn + 2, m)$, we first fill the size $n + 2$ hole in the lower right corner of the resulting $ICOILS(mn + 2, n + 2)$ with a $COTLS^*(n + 2, 2)$ to form a $COTLS(mn + 2, m)$ with a sub-$CODLS(m)$ which consists of the cells in $\{(in + (n - 1)/2, jn + (n - 1)/2) : 0 \leq i, j < m\}$. We then obtain the desired design by permuting rows and columns and entries with the permutation σ_2 given by

$$\sigma_2(i) = \begin{cases} i - k + 1 & \text{if } kn \le i < kn + (n-1)/2, \\ & \quad 0 \le k < m/2 \\ i - k & \text{if } kn + (n-1)/2 < i < (k+1)n, \\ & \quad 0 \le k < m/2 \\ i - k + m & \text{if } kn \le i < kn + (n-1)/2, \\ & \quad m/2 \le k < m \\ i - k + m + 1 & \text{if } kn + (n-1)/2 < i < (k+1)n, \\ & \quad m/2 \le k < m \\ m(n-1)/2 + k + 1 & \text{if } i = kn + (n-1)/2, 0 \le k < m \\ 0 & \text{if } i = mn \\ mn + 1 & \text{if } i = mn + 1 \end{cases} \qquad \square$$

Construction 4.4 If there exist a COILS(m), a COILS(n) and a CODLS($n + a, a$), then there exist a CODLS($mn + a, n + a$) and a CODLS($mn + a, a$).

Construction 4.5 Suppose there exist a CODLS(m, n) and a CODLS(n, a), then there exists a CODLS(m, a).

We now state the final set of constructions. The first construction is well known, and the second is essentially a variation of the block design analogue of [12, Theorem 9] and its verification is also straightforward.

Construction 4.6 Suppose there exists a PBD $B(K \cup \{n*\}, 1; v)$ with a block N of size n and for each $k \in K$ there exists a COILS(k). Furthermore suppose the v-set X of objects of the design can be partitioned into disjoint subsets:

$$X = S_1 \cup S_2 \cup \cdots \cup S_m \cup N$$

such that each S_i has even order and is a subset of some block B_i, and for each i there exists a CODLS($|B_i|, |B_i| - |S_i|$). Then there exists a CODLS(v, n).

Proof Let us denote $|S_i|$ by s_i and $|B_i|$ by b_i. We first relabel the elements of X so that

$$\begin{aligned} G_1 &= \{0, 1, \cdots, s_1/2 - 1, v - s_1/2, \cdots, v - 1\}, \\ G_2 &= \{s_1/2, s_1/2 + 1, \cdots, s_1/2 + s_2/2 - 1, \\ & \qquad v - s_1/2 - s_2/2, \cdots, v - s_1/2 - 1\}, \end{aligned}$$

and so on; N contains the central n elements. Then the standard PBD construction on Latin squares (see, for example, [8]) produces the required square, provided that the Latin squares used to correspond to each of the block B_i is a CODLS($b_i, b_i - s_i$) and each of the other blocks of size k is a COILS(k).   □

Construction 4.7 Suppose $(X, \mathcal{B})$ is a PBD $B(K, 1; v)$ such that X admits t partitions $B_{i_1}, B_{i_2}, \cdots B_{i_{k_i}}, 1 \leq i \leq t$, into disjoint blocks and all the blocks in these partitions are distinct. For every $i, 1 \leq i \leq t$, suppose that there is an integer n_i such that an ICOILS($|B_{ij}| + n_i, n_i$) exists for every $j, 1 \leq j \leq k_i$. Let $\mathcal{B}'$ be the collection of blocks not belonging to any partition. Suppose the v-set X of objects of the design can be partitioned into disjoint subsets:

$$X = S_1 \cup S_2 \cup \cdots \cup S_u$$

such that each S_i has even order and is a subset of some block B_i in $\mathcal{B}'$, and for each i there exists a CODLS($|B_i|, |B_i| - |S_i|$). Then there exists a CODLS($v + n, n$) if there is a COILS(m) for every block but $B_i, 1 \leq i \leq u$ in $\mathcal{B}'$ of size $m \in K$, where $n = n_1 + n_2 + \cdots + n_t$.

Constructions 4.6 and 4.7 will prove quite applicable in the use of GDDs in most of our constructions. For example, from Construction 4.6 we have

Construction 4.8 Let K be a set of positive integers. Suppose there exists a K-GDD of group-type $m_1 m_2 \cdots m_n$ and

1 for every $k \in K$ there exists a COILS(k),

2 for every $i < n$ there exists a CODLS(m_i) and m_i is even.

Then there exists a CODLS(v, m_n), where $v = \Sigma_{1 \leq i \leq n} m_i$.

Combining the constructions in Constructions 4.6 and 4.7, we have the following construction whose proof is similar to the proof of Lemma 2.7 in [4].

Construction 4.9 Suppose there exists a RGD[$k, 1, m; km$] and $0 \leq t \leq m$, where m is odd. For $1 \leq i \leq t$, suppose there exists ICOILS($k + n_i, n_i$), where $n_1 + n_2 + \cdots + n_t = n - 1$. Further suppose the following exist: COILS(k), CODLS(n) and CODLS($m + 1, 2$). Then there exists a CODLS($km + n, k + 1$).

5. A General Bound

We shall assume that the reader is familiar with the standard terminal of group-divisible designs (GDDs) and Wilson's "Fundamental Construction" (see, for example, [15]). Of course, a GD$[k, 1, n; kn]$ is equivalent to $k - 2$ pairwise orthogonal Latin squares of order n, briefly $k - 2$ POLS(n).

Lemma 5.1 If prime power $q \geq 7$, then there exists an FCOLS(1^q) with a transversal which consists of the positions in $\{(q-1, q-1), (i, q-i-2) : 0 \leq i < q - 1\}$.

Proof For the case q is an even prime power, see [1]. For the case q is an odd prime power, see the proof of Lemma 3.4. □

We need the following recursive construction for FCOLS. Before stating it, we define a *weighting* of a GDD $(X, \mathcal{G}, \mathcal{A})$ to be any mapping $w : X \to Z^+ \cup \{0\}$.

Lemma 5.2([3]) Suppose that $(X, \mathcal{G}, \mathcal{A})$ is a GDD and let $w : X \to Z^+ \cup \{0\}$ be a weighting of the GDD. For every $x \in X$, let S_x be the multiset of $w(x)$ copies of x. For each block $A \in \mathcal{A}$, assume an FCOLS of type $\{S_x : x \in A\}$ is given. Then there are FCOLS of type $\{\sum_{x \in G} w(x) : G \in \mathcal{G}\}$.

We now state the main constructions.

Lemma 5.3 Suppose n and k are positive integers, m odd, $2 \leq n \leq 4m - 4$, $1 \leq k < 2m$, k odd and $k \neq 3$ and $k \neq 21$ if $m = 11$, such that there exists a GD$[12, 1, m; 12m]$. Then there exists a CODLS($9m + n + k, n$). Further, for $9m + n + 5 \leq v \leq 11m + n$ and $v - n$ even, there exists a CODLS(v, n).

Proof In all groups but three of the GD$[12, 1, m; 12m]$, we give the points weight 1. In the third-last group, we give s (s odd) points weight 1 and the remaining points weight 0. In the second-last group, we give t points weight 1 and the remaining points weight 0. We observe that if $s + t = k \geq 1$ and $k \neq 3$ and $k \neq 21$ if $m = 11$, then we can choose s and t such that both CODLS(s) and CODLS(t) exist. In the last group, we give weight 0, 1, 2, 3 or 4 such that the total weight is n. We can apply Lemma 5.2 with the necessary input designs from Lemma 2.3, in which one size eleven input block is an FCOLS(1^{11}) which comes from Lemma 5.1. This construction produces an FCOLS($m^9 s^1 t^1 n^1$) with a sub-FCOLS(1^{11}) which consists of the cells in $\{(i, j) : i, j = 9m + (s -$

$1)/2, 9m+s, km+(m-1)/2 \ (0 \le k < 9)\}$. We then fill the size m holes with $\mathrm{CODLS}(m,1)$, the size s holes with $\mathrm{CODLS}(s,1)$ and the size t holes with $\mathrm{CODLS}(t)$, and obtain the required design by permuting rows and columns and entries with the permutation σ given in the following table.

$\square$

$\sigma(i)$	condition
$i - k(m+1)/2 + 5$	$km \le i < km + (m-1)/2,$ $0 \le k < 9$
k	$i = km + (m-1)/2, 0 \le k < 5$
$9m + s + t + n + k - 10$	$i = km + (m-1)/2, 5 \le k < 9$
$i + 8(m-1) - k(3m-1)/2 +$ $s + t + n + 3$	$km + (m-1)/2 < i < (k+1)m,$ $0 \le k < 9$
$i - 9(m+1)/2 + 5$	$9m \le i < 9m + (s-1)/2$
$9m + s + t + n - 1$	$i = 9m + (s-1)/2$
$i - 9(m+1)/2 + t + n + 5$	$9m + (s-1)/2 < i < 9m + s$
$i - 9(m+1)/2 - (s+1)/2 + 5$	$9m + s \le i < 9m + s + t/2$
$i - 9(m+1)/2 - (s+1)/2 + n + 5$	$9m + s + t/2 \le i < 9m + s + t$
$i - 9(m+1)/2 - (s+t+1)/2 + 5$	$9m + s + t \le i < 9m + s + t +$ $n - 1$

Lemma 5.4 Suppose n, m and k are positive integers, m even, $2 \le n \le 4m$, $0 \le k \le 2m$, k even and $k \ne 2$, such that there exists a $\mathrm{GD}[12, 1, m; 12m]$. Then there exists a $\mathrm{CODLS}(9m+n+k, n)$. Further, for $9m+n+8 \le v \le 11m+n$ and $v-n$ even, there exists a $\mathrm{CODLS}(v, n)$.

Proof In all groups but three of the $\mathrm{GD}[12, 1, m; 12m]$, we give the points weight 1. In the third-last group, we give s (s even) points weight 1 and the remaining points weight 0. In the second-last group, we give t points weight 1 and the remaining points weight 0. We observe that if $s + t = k \ge 1$ and $k \ne 2, 6$ or 10, then we can choose s and t such that both $\mathrm{CODLS}(s)$ and $\mathrm{CODLS}(t)$ exist. In the last group, we give weight 0, 2, 3 or 4 such that the total weight is n. We can apply Lemma 5.2 with the necessary input designs from Lemma 2.3 to obtain an $\mathrm{FCOLS}(m^9 s^1 t^1 n^1)$. We then fill the size m holes with $\mathrm{CODLS}(m)$, the size s holes with $\mathrm{CODLS}(s)$ and the size t holes with $\mathrm{CODLS}(t)$, and obtain the required design by permuting rows and columns.

For the case $k = 6$ and 10, we can choose s and t such that both CODLS$(s + 1)$ and CODLS$(t + 1)$ exist. In the last group, we give weight 0, 1, 2, 3 or 4 such that the total weight is $n - 1$. We can apply Lemma 5.2 with the necessary input designs from Lemma 2.3 to obtain an FCOLS$(m^9 s^1 t^1 (n - 1)^1)$. We then fill the size m holes with CODLS$(m + 1, 1)$, the size s holes with CODLS$(s + 1, 1)$ and the size t holes with CODLS$(t+1, 1)$, and obtain the required design by permuting rows and columns. □

Observing that the condition requiring a GD$[12, 1, m; 12m]$ in Lemma 5.3 can be replaced by a GD$[10, 1, m; 10m]$, we have the following construction.

Lemma 5.5 Suppose n and k are positive integers, m odd, $2 \leq n \leq 2m - 2$, $1 \leq k < 2m$, k odd and $k \neq 3$ and $k \neq 21$ if $m = 11$, such that there exists a GD$[10, 1, m; 10m]$. Then there exists a CODLS$(7m + n + k, n)$. Further, for $7m + n + 5 \leq v \leq 9m + n$ and $v - n$ even, there exists a CODLS(v, n).

Lemma 5.6 Suppose n, m and k are positive integers, m even, $2 \leq n \leq 2m - 2$, $0 \leq k \leq m$, k even and $k \neq 2$, such that there exists a GD$[9, 1, m; 9m]$. Then there exists a CODLS$(7m + n + k, n)$.

Proof In all groups but two of the GD$[9, 1, m; 9m]$, we give the points weight 1. In the second-last group, we give k points weight 1 and the remaining points weight 0. In the last group, we give weight 0, 1 or 2 such that the total weight is $n - 1$. We can apply Lemma 5.2 with the necessary input designs from Lemma 2.3 to obtain an FCOLS$(m^7 k^1 (n - 1)^1)$. We then fill the size m holes with CODLS$(m + 1, 1)$, the size k holes with CODLS$(k+1, 1)$ and obtain the required design by permuting rows and columns. □

Lemma 5.7 Suppose n, m and k are positive integers, m odd, $2 \leq n \leq m$, $0 \leq k \leq m$ and k odd and $k \neq 21$ if $m = 11$, $k \neq 21$ if $m = 11$, such that there exists a GD$[8, 1, m; 8m]$. Then there exists a CODLS$(7m + n + k, n)$. Further, for $7m + n \leq v \leq 8m + n$ and $v - n$ even, there exists a CODLS(v, n).

Proof In all groups but two of the GD$[8, 1, m; 8m]$, we give the points weight 1. In the second-last group, we give k points weight 2 and the remaining points weight 1. In the last group, we give n points weight 1 and the remaining points weight 0. We can apply Lemma 5.2 with the necessary input designs from Lemma 2.3, in which one size seven input block is a CODLS$(7, 1)$ which comes from FCOLS(1^7) in Lemma 5.1. This construction produces an FCOLS$(m^6 (m + k)^1 n^1)$. We then fill the size m holes with CODLS($m, 1$), the size $m+k$ holes with CODLS(m+k, 1) and obtain the required design by permuting rows and columns. □

Lemma 5.8([10]) There is a series of positive integers

$$
\begin{aligned}
M &= \{m_i : i = 1, 2, 3, \cdots\} \\
&= \{11, 13, 16, 17, 19, 23, 25, 27, 29, 31, 32, 37, 41, 43, 47, \cdots\}
\end{aligned}
$$

such that $m_{i+1} - m_i \le 10$ and there exists a GD[12, 1, m_i; 12m_i] for all $i \ge 1$.

We are now in a position to prove Theorem 1.4.

The proof of Theorem 1.4: First of all, apply Lemma 5.5 with $m = 17, 19$ and 29, we know that there exists a CODLS($120 + n, n$) for $2 \le n \le 32$, CODLS($146 + n, n$) for $2 \le n \le 36$ and CODLS($210 + n, n$) for $2 \le n \le 56$.

Our proof relies heavily on Lemmas 5.3 and 5.4. For any fixed $n \ge 41$, there exists an $i \ge 1$ such that $4m_i < n \le 4m_{i+1}$. Thus we have $4m_{i+1} - n < 4m_{i+1} - 4m_i \le 40$ and $m_{i+1} \le (39 + n)/4$. Apply Lemmas 5.3 and 5.4 recursively, we know that there exists a CODLS(v, n) whenever $v \ge 9m_{i+1} + n + 5$. Therefore there exists a CODLS(v, n) whenever $v \ge 9(39 + n)/4 + n + 5$, that is, whenever $v \ge 13n/4 + 93$.

For $7 \le n \le 40$, we have from Construction 5.3 that there exist CODLS(v, n) for $104 + n \le v \le 118 + n$, therefore for $v \ge 104 + n$. Since $13n/4 + 93 \ge 104 + n$ for $7 \le n \le 44$, there exists a CODLS(v, n) whenever $v \ge 13n/4 + 93$.

For $2 \le n \le 6$, the result comes from the following lemma. □

Lemma 5.9 For any positive integer $2 \le n \le 6$, if $v \ge 48 + n$ and $v \ne 58 + n$, then there exists a CODLS(v, n) for $v - n$ even.

Proof Apply Lemmas 5.3 and 5.4 with $m \in M$, from the proof of Theorem 1.4 we know that the result is true for $v \ge 104 + n$. Apply Lemmas 5.5 with $m \in \{9, 11, 13\}$, the result is true for $68 + n \le v \le 104 + n$. Apply Lemma 5.6 with $m = 8$, the result is true for $60 + n \le v \le 68 + n$. Apply Lemma 5.7 with $m = 7$, the result is true for $50 + n \le v < 56 + n$.

Finally, from GD[5, 1, 12; 60], we delete $12 - n$ points in one group to obtain a GD[{4, 5}, 1, {12, n}; 48+n], so the result is true for $v = 48 + n$. □

6. The Case $2 \leq n \leq 6$

Lemma 6.1 There exists a CODLS$(v, 2)$ for all $8 \leq v \leq 48$ and v even and $v = 60$.

Proof For the case $v \in \{8, 10, 12, 14, 16, 20, 24\}$, see Lemma 3.3. For the case $v \in \{22, 30, 34, 38, 42, 46\}$, write $v = 4n + 2, n = 5, 7 - 11$, and apply Construction 4.3. For the case $v = 26$, write $v = 4 \times 6 + 2$ and apply Construction 4.2.

For the case $v = 18$, we start with a GD[5, 1, 4; 20] and delete two points in a group to obtain a GD[{4, 5}, 1, {4, 2*}; 18], and apply Construction 4.8.

For the case $v \in \{28, 32, 36, 40, 44, 48, 60\}$, we start with an ICOILS $(v/4, 2)$ and filling CODLS(4) in every cell, to obtain a CODLS$(v, 8)$ by permuting rows and columns and entries same as in Construction 4.1. The result follows from Construction 4.5. □

Lemma 6.2 There exists a CODLS$(v, 3)$ for all $11 \leq v \leq 49$ and v odd and $v = 61$, except possibly $v = 11$.

Proof For the case $v \in \{13, 15, 17, 21, 25, 29, 37\}$, see Lemma 3.3. For the case $v \in \{49, 61\}$, write $v = 12m + 1, m = 4, 5$, and apply Constructions 4.4 and 4.5. For the case $v = 43$, write $43 = 4 \times 10 + 3$ and apply Construction 4.3. For the case $v = 27$, we start with a GD[{4, 5}, 1, {4, 3*}; 27], which exists according to [3], and apply Construction 4.8. For the case $v = 45$, we start with a GD[{5, 7}, 1, {7, 3*}; 45], which exists according to [11], and apply Construction 4.6. The disjoint subsets come from the groups and a block of size 7.

For the case $v \in \{19, 23, 31, 35, 39, 43, 47\}$, we start with a GD[5, 1, n; 5n], $n = 4 - 5, 7 - 9, 11$, and delete $n - 3$ points in a group to obtain a GD[{4, 5}, 1, {n, 3*}; 4n + 3], and apply Construction 4.6. The disjoint subsets consist of the groups, if n is even. If n is odd, the disjoint subsets come from the groups and a block of size five, where we take the first four groups minus the intersection points with the five block, the intersection points of the first four groups with the five block, and the unique group of size three.

For the case $v = 33$, we start with an RGD[5, 1, 7; 35] and delete a block to obtain a GD[{4, 5}, 1, 6; 30] with a parallel class of blocks of size five. We then adjoin three infinite points to this GDD and apply Construction 4.7. For the parallel class consisting of the blocks of size five, we fill in ICOILS(7, 2), and for the parallel class consisting of the groups, we fill in CODLS(7, 1). In the resulting construction of the CODLS(33, 3), the hole of size three consists of the three infinite points.

For the case $v = 41$, we start with an FCOLS($1^7 3^1$) and fill COILS(4) in every cell to obtain an FCOLS($4^7 12^1$). We then adjoin an infinite point to this design to obtain a CODLS(41, 13), by filling in the holes of size four with CODLS(5, 1) and permuting rows, columns, and entries as in Construction 4.1. The result follows from Construction 4.5. □

Lemma 6.3 There exists a CODLS(v, 4) for all $14 \leq v \leq 50$ and v even and $v = 62$.

Proof For the case $v \in \{14, 18, 26\}$, see Lemma 3.3. For the case $v \in \{16, 20, 24, 28, 32, 36, 40, 44, 48\}$, write $v = 4n$ and apply Construction 4.1. The necessary input designs FCOLS(4^n) come from Lemma 2.2. For the case $v \in \{22, 30, 38, 46, 50, 62\}$, write $v = 4n + 2$ and apply Construction 4.3. The input designs COTLS*$(n + 2, 2)$ come from Lemma 3.4.

For the case $v = 34$, we start with an RGD[5, 1, 7; 35] and delete a block to obtain a GD[$\{4, 5\}$, 1, 6; 30] with a parallel class of blocks of size five. We then adjoin four infinite points to this GDD and apply Construction 4.7. For the parallel class consisting of the blocks of size five, we fill in ICOILS(7, 2), and for the parallel class consisting of the groups, we fill in CODLS(8, 2) In the resulting construction of the CODLS(34, 4), the hole of size four consists of the four infinite points.

For the case $v = 42$, we start with a PBD B($\{4, 5, 7, 9\}$, 1; 42), which exists according to [11]. Observing that it is also a GD[$\{4, 5, 7, 9\}$, 1, $\{4, 5\}$; 42] with group-type $4^3 5^6$ and a block of size seven which intersects each of the groups of size five, we then apply Construction 4.6 to obtain the desired design. The disjoint subsets come from the groups and the block of size seven. □

Lemma 6.4 There exists a CODLS(v, 5) for all $17 \leq v \leq 51$ and v odd and $v = 63$.

Proof For the case $v = 19, 23, 27$, see Lemma 3.3. For the case $v \in \{17, 21, 25, 29, 33, 37, 41, 45, 49\}$, write $v = 4n + 1$, and apply Construction 4.2. The necessary input designs FCOLS(4^n) come from Lemma 2.2. For the case $v = 35$, we start with a GD[5, 1, 7; 35] and apply Construction 4.6. The disjoint subsets come from the groups and a block of size five.

For the case $v \in \{31, 39, 47\}$, we start with an RGD$[6, 1, n; 6n], n = 7, 9, 11$, and delete all but one point from each of two groups to obtain a GD$[\{4, 5, n\}, 1, \{4, 6*\}; 4n + 2]$. We adjoin one infinite point to the GDD and apply Construction 4.6. The disjoint subsets come from the blocks which come from the groups of the GDD.

For the case $v = 43, 63$, we start with an RGD$[5, 1, n; 5n], n = 9, 13$ and delete two points in a group to obtain a GD$[\{4, 5\}, 1, \{n, (n - 2)*\}; 5n - 2]$, and apply Construction 4.6. The disjoint subsets come from the groups and a block of size five.

For the case $v = 51$, we start with an GD$[5, 1, 11; 55]$ and delete four points in a group to obtain a GD$[\{4, 5\}, 1, \{11, 7*\}; 51]$, and apply Construction 4.6. The disjoint subsets come from the groups and a block of size five. □

Lemma 6.5 There exists a CODLS$(v, 6)$ for all $22 \leq v \leq 52$, and v even and $v = 64$.

Proof For the case $v \in \{20, 24, 28\}$, see Lemma 3.3. For the case $v \in \{36, 40, 44, 48, 52, 64\}$, write $v = 5m + n$, $m = 7, 9, 11$, and apply Construction 4.9.

For the case $v \in \{34, 38, 42, 50\}$, we start with a GD$[5, 1, n; 5n]$, $n = 7 - 9, 11$, and delete $n - 6$ points in a group to obtain a GD$[\{4, 5\}, 1, \{n, 6*\}; 4n + 6]$, and apply Construction 4.6. The disjoint subsets consist of the groups, if n is even, or come from the groups and one block of size five, if n is odd.

For the case $v \in \{22, 30, 46\}$, we start with a GD$[6, 1, n; 6n], n = 5, 7, 11$, and delete all but one point from each of two groups to obtain a GD$[\{4, 5, 6*\}, 1, \{n, 1\}; 4n + 2]$, and apply Construction 4.6. The disjoint subsets come from the groups and the unique block of size six.

For the case $v = 26$, we start with an RGD$[5, 1, 5; 25]$ and adjoin an infinite point, say x, to the groups, and then delete a point different from x from the resulting design. This produces a GD$[\{5, 6\}, 1, \{4, 5*\}; 25]$, where x belongs to each block of size six and the unique group of size 5. In this GDD we give x weight 2 and give all other points of the GDD weight 1. Using FCOLS(1^5) and FCOLS$(1^5 2^1)$ as input, this produces an FCOLS$(4^5 6^1)$. By filling in the holes of size four with CODLS(4) and permuting rows and columns, we get the desired CODLS$(26, 6)$.

For the case v $= 32$, we start with an RGD$[5, 1, 7; 35]$ and delete four points in a block of some parallel class $\mathcal{P}$ of blocks of size five to obtain a

GD[$\{4,5\}, 1, \{6,7\}; 31$]. In this GDD, we select another block of size five, say B, from the parallel class $\mathcal{P}$. We then adjoin an additional point, say x, to this GDD and construct CODLS(8, 2) or FCOLS($1^5 2^1$) on each of the groups with the additional point x, where the hole of size two is formed from the point x and the point of intersection of each group with B. We construct CODLS(5, 1) on each of the remaining 5 blocks of the parallel class $\mathcal{P}$ and FCOLS(1^4) or FCOLS(1^5) on the other blocks of the GDD. We finally obtain the desired CODLS(32, 6) by permuting rows and columns. The hole of size six consists of the points in the block B and the additional point x. □

We are now in a position to prove Theorem 1.5.

The proof of Theorem 1.5: Theorem 1.5 comes from Lemmas 6.1 - 6.5 and Lemma 5.9. □

Acknowledgments

The research of the first author was partially supported by NSERC Grant OGP 0005320. The research of the third author was partially supported by NSF Grant CCR-0098093. The authors wish to thank an anonymous referee for his comments on improving the presentation of the paper.

References

[1] F.E. Bennett, B. Du, and H. Zhang, *Existence of conjugate orthogonal diagonal Latin squares*, J. Combin. Designs, **5** (1997), 449-461.

[2] F.E. Bennett, B. Du, and H. Zhang, *Existence of self-orthogonal diagonal Latin squares with a missing subsquare*, Discrete Math., to appear.

[3] F.E. Bennett, L. Wu, and L. Zhu, *Conjugate orthogonal Latin squares with equal-sized holes*, Ann. Discrete Math., **34** (1987), 65-80.

[4] F.E. Bennett, L. Wu and L. Zhu, *Further results on incomplete* $(3,2,1)$-*conjugate orthogonal idempotent Latin squares*, Discrete Math., **84** (1990), 1-14.

[5] F.E. Bennett and L. Zhu, *Incomplete conjugate orthogonal idempotent Latin squares*, Discrete Math., **65** (1987), 5-21.

[6] F.E. Bennett and L. Zhu, *Conjugate-orthogonal Latin squares and related structures*, J. Dinitz & D. Stinson (Editors), Contemporary design theory: A collection of surveys, Wiley, New York, 1992, pp.41-96.

[7] C.J. Colbourn and J.H. Dinitz, *The CRC handbook of combinatorial designs*, CRC Press, Inc., Boca Raton, 1996.

[8] J. Dénes and A.D. Keedwell, *Latin squares and their applications*, Academic Press, New York and London, 1974.

[9] J. Doyen and R.M. Wilson, *Embeddings of Steiner triple systems*, Discrete Math., **5** (1973), 229-239.

[10] B. Du, *Self-orthogonal diagonal Latin square with missing subsquare*, JCMCC, **37**(2001), 193-203.

[11] M. Greig, *Designs from projective planes and PBD bases*, J. Combin. Designs 7(1999), 341-374.

[12] M.J. Pelling and D.G. Rogers, *Stein quasigroups I: combinatorial aspects*, Bull. Austral. Math. Soc., **18** (1978), 221-236.

[13] S.K. Stein, *On the foundations of quasigroups*, Trans. Amer. Math. Soc., **85** (1957), 228-256.

[14] R.M. Wilson, *Concerning the number of mutually orthogonal Latin squares*, Discrete Math., **9** (1974), 181-198.

[15] R.M. Wilson, *Constructions and uses of pairwise balanced designs*, Math. Centre Tracts, **55**(1974), 18-41.

[16] H. Zhang and F.E. Bennett, *Existence of some $(3,2,1)$-HCOLS and $(3,2,1)$-ICOILS*, JCMCC, **27**(1998), 53-64.

Chapter 3

COMBINATORIAL TRADES: A SURVEY OF RECENT RESULTS

Elizabeth J. Billington

Department of Mathematics
The University of Queensland
Brisbane, Qld 4072
Australia

Abstract   The concept of a trade in a combinatorial structure has existed for some years now. However, in the last five years or so there has been a great deal of activity in the area. This survey paper builds upon the one by Khosrovshahi, Maimani and Torabi which appeared in Discrete Applied Mathematics (Volume 95, pp. 361–176) in 1999. In the short time since that survey appeared, the number of papers in the area has almost doubled. Trades are used in designs and latin squares; they also crop up in graph theory. In this paper the most recent work on trades is surveyed, with applications given.

1.   Introduction and Preliminaries

Work on trades in design theory originated in the 1960s (see Hedayat [30], Section 3), although the idea behind a trade was used much earlier than this, in other forms. Back in 1916, White, Cole and Cummings [50] used "quadrangular and hexagonal transformations" while finding the Steiner triple systems of order 15. These were essentially what we now call trades of volumes 4 and 6, with block size 3, or $(3, 2)$ trades of volumes 4 and 6.

Formally, a (k, t) trade of volume m and foundation size v (sometimes referred to as a (v, k, t) trade) is a pair $\{T_1, T_2\}$ of sets of blocks (subsets) of size k, based on a v-set, such that T_1 and T_2 each contain m blocks of size k, with $T_1 \cap T_2 = \emptyset$, and so that each t-set chosen from the v-set occurs exactly the same number of times in blocks of T_1 as it does in blocks of T_2. All v elements occur (necessarily each at least twice) in T_1 in order for the foundation size to be v. The trade is called *Steiner* if each t-set occurs at most once in each T_i, $i = 1, 2$. The trade is *simple* if there are no repeated blocks in T_1 or in T_2. Clearly a Steiner trade must be simple, but the converse is not true.

A trade is called *fundamental* ([23]) if it contains no proper trade, that is, if a trade $\{T_1, T_2\}$ is such that there is no proper subset $\emptyset \neq S_1 \subset T_1$ with mate S_2 forming a trade $\{S_1, S_2\}$. The term *minimal* has also been used here (see [26]), but we reserve the term minimal for use with trade volumes.

Example 1.1
A Steiner $(9, 3, 2)$ trade of volume 8:

T_1		T_2	
1 3 6	2 3 9	1 3 9	2 3 6
1 4 8	2 4 6	1 4 6	2 4 7
1 5 9	2 5 7	1 5 8	2 5 9
3 4 7	3 5 8	3 4 8	3 5 7

A comprehensive survey of trades appears in [38]. Here we build upon that survey by concentrating on the considerable amount of work done in this area in the last few years. We include work done since that survey, or (in a few cases) earlier work not mentioned in [38]. It is interesting to note that most of this work appears in papers with at least one author from either Queensland or Iran!

In subsequent sections, four kinds of trades will be considered. The type just described is perhaps the most common: trades in designs. We shall also look at trades in latin squares, at trades in graphs, and at trades derived from "latin representations" used in tripartite graph decompositions.

2.   Trades in designs

Almost all the recent work on trades in the block design context has been with $(k, 2)$-trades. So for this section we take $t = 2$.

2.1 Possible Steiner trade volumes, regardless of foundation

A series of three papers, [28], [27] and [35], with publication dates 2000, 1999 and 1998 (in reverse order of submission dates!) have solved the question of determining all possible trade volumes for a Steiner $(k, 2)$ trade for any block size k, when the foundation size is of no importance.

Possible trade volumes are related to possible block intersection sizes. The intersection problem for combinatorial designs asks how many possible common blocks two designs D_1 and D_2 can have, where D_1 and D_2 have the same parameters and are based on the same set of points. Clearly if the blocks common to D_1 and D_2 are deleted, the remaining blocks T_1 in D_1 and T_2 in D_2 will form a trade $\{T_1, T_2\}$. So the possible volume $|T_1|$ of the trade is intimately connected with the intersection problem for block designs; see [7] for an old survey. It was known from such intersection work that Steiner trade volumes which cannot be achieved for block size three are 1, 2, 3 and 5, while trade volumes which cannot be achieved for block size four are 1, 2, 3, 4, 5 and 7.

Let $X(k)$ denote the set of unachievable or forbidden trade volumes for a Steiner $(k, 2)$-trade. Thus early work on design intersection problems showed that $X(3) = \{1, 2, 3, 5\}$ and $X(4) = \{1, 2, 3, 4, 5, 7\}$. Also let $\text{TS}(k)$ denote the set of achievable $(k, 2)$ trade volumes, or the *trade spectrum*. Thus $X(k) \cup \text{TS}(k) = \mathbf{N} \cup \{0\}$, and $X(k) \cap \text{TS}(k) = \emptyset$.

The culmination of [28], [27] and [35] resulted in the following.

Theorem 1 *The forbidden $(k, 2)$ Steiner trade volumes are given by*

$$X(k) = \{1, 2, \ldots, 2k - 3\} \cup \{2k - 1, 2k + 1, \ldots, 3k - \frac{5}{4}\}$$

(whichever of $3k - 5$, $3k - 4$ is odd), except $\text{TS}(2) = \{0\}$ and $15 \notin X(7)$.

Thus the number of forbidden trade volumes increases linearly with the block size.

2.2 Possible trade volumes: non-Steiner case

A (v, k, t) trade is *simple* if it has no repeated blocks, and it has *index* λ if some t-subset occurs in λ blocks of T_1, but no t-subset occurs in more than λ blocks.

Ramsey [46] investigates the trade spectrum for simple $(k, 2)$ trades of index λ. The problem divides into three obvious parts; he obtains many results for block size $k = 3$ and for $k \geq 5$, and complete results for $k = 4$. Let $X_\lambda(k)$ denote the set of unachievable or forbidden trade volumes for a simple $(k, 2)$ trade of index λ. Then for block size 4, the following result gives all the forbidden trade volumes.

Theorem 2 [46]

$$
\begin{aligned}
X_1(4) &= \{1, 2, 3, 4, 5, 7\} \quad \text{(the Steiner case)}; \\
X_2(4) &= \{1, 2, 3, 5\}; \\
X_3(4) = X_4(4) &= \{1, 2, 3, 4, 5\}; \\
X_5(4) = X_6(4) &= \{1, 2, 3, 4, 5, 6, 7\}; \\
X_\lambda(4) &= \{1, 2, \ldots, s - 1\} \quad \text{where } s = \lceil 7\lambda/6 \rceil,
\end{aligned}
$$

except $(m, \lambda) \in X_\lambda(4)$ for the pairs: $(9, 7)$, $(11, 9)$, $(12, 10)$, $(13, 11)$, $(14, 12)$, $(19, 16)$, $(20, 17)$, $(21, 18)$, $(27, 23)$.

For $k \geq 5$, when the index is 2, odd volumes between 7 and $2k - 1$ are unknown; are they in the trade spectrum or in $X_2(k)$? (The difficulty is that there are no Steiner trades of volume less than $2k - 2$.)

For $k = 3$, spectrum results are incomplete when the index λ is at least 5.

2.3   Possible Steiner $(v, 3, 2)$ trade volumes with fixed foundation

Although [13] by Bryant appeared in 1997, we include its results here, for completeness in connection with the next subsection.

The possible volumes of trades with foundation size v, block size $k = 3$ and $t = 2$, were found, as tabulated below. Note that the entry o denotes a missing trade volume; the expected minimum volume cannot be achieved when the foundation size is 1, 3 or 4 modulo 6, and the smallest but one expected volume cannot be achieved when the foundation size is a multiple of 6. Below, $M(v)$ denotes the number of triples in a maximum packing of K_v; the value of this varies slightly, according to the congruence class of v modulo 6.

STEINER $(v, 3, 2)$ TRADE VOLUMES

foundation v	volume m
$v \leq 5$	$\emptyset$
$v \equiv 0 \pmod 6$	$\frac{2v}{3}$, $\circ$ $\frac{2v}{3} + 2, \ldots, M(v)$
$v \equiv 1 \pmod 6$	$\circ$ $\frac{2v+1}{3} + 1, \ldots, M(v)$
$v \equiv 2 \pmod 6$, $v > 8$	$\frac{2v+2}{3}, \ldots, M(v)$
$v = 8$	6, 8
$v \equiv 3 \pmod 6$	$\circ$ $\frac{2v}{3} + 1, \ldots, M(v)$
$v \equiv 4 \pmod 6$	$\circ$ $\frac{2v+1}{3} + 1, \ldots, M(v)$
$v \equiv 5 \pmod 6$	$\frac{2v+2}{3}, \ldots, M(v)$

2.4 Steiner and non-Steiner $(v, 3, 2)$ trades of minimum volume

In 1999, Eslami and Tayfeh-Rezaie [23] investigated the possible *minimum* $(v, 3, 2)$ trade volumes and also their structure, in both the Steiner and the non-Steiner cases. The existence of such trades of minimum volume in the Steiner case follows from work two years earlier in [13]; the classifications and the non-Steiner minimum volume results are new. The non-Steiner case with foundation size 5 modulo 6 is incomplete.

We tabulate the results on the minimum volume and also the number of non-isomorphic trades of such volumes.

	STEINER		NON-STEINER	
Found'n v (mod 6)	Min vol m	Number of trades	Min vol m	Number of trades
0	$\frac{2v}{3}$	1	$\frac{2v}{3} + 2$ $(v > 12)$	7
1	$\frac{2v+1}{3} + 1$ $(v > 13)$	4	$\frac{2v+1}{3} + 1$	1
2	$\frac{2v+2}{3}$	1	$\frac{2v+2}{3} + 1$	1
3	$\frac{2v}{3} + 1$	1	$\frac{2v}{3} + 2$ $(v > 9)$	10
4	$\frac{2v+1}{3} + 1$ $(v > 16)$	9	$\frac{2v+1}{3} + 1$	1
5	$\frac{2v+2}{3}$	1		??

Exceptions to the number of non-isomorphic trades of minimum volume m with foundation v are given by pairs (v, m) as follows:

Steiner case: $(7, 1), (13, 3), (10, 4), (16, 8)$.
Non-Steiner case: $(6, 1), (12, 5), (9, 6)$.

2.5   Simple, non-Steiner $(v, 3, 2)$ trades of maximum volume

Recall that a simple trade contains no repeated blocks. Since a trade $\{T_1, T_2\}$ also satisfies $T_1 \cap T_2 = \emptyset$, it follows that $T_1 \cup T_2$ must be a subset of the set of all $\binom{v}{3}$ triples from a v-set. So, roughly speaking, to obtain a simple $(v, 3, 2)$ trade of maximum possible volume, the set of all $\binom{v}{3}$ triples from a v-set needs to be partitioned into sets T_1 and T_2, together with a "discard set" (as small as possible).

Khosrovshahi and Torabi [40] find such maximum possible trade volumes for all foundations v except $v \equiv 5 \pmod{6}$; in this latter case an upper bound on the maximum trade volume is given; equality remains to be shown for values of v greater than 17.

2.6   Strong Steiner trades

Recall that a Steiner (k, t) trade $\{T_1, T_2\}$ has each t-set occurring no more than once in any block of T_1. Such a Steiner trade is said to be *strong* if any block of T_1 meets any block of T_2 in at most t points. Hamilton and Khodkar [29] investigate minimum possible volumes of strong Steiner $(k, 2)$ trades; in this case with $t = 2$, for the strong property, blocks in T_1 meet blocks in T_2 in at most *two* points. Of course when the block size k is 3, since $T_1 \cap T_2 = \emptyset$, a Steiner trade is automatically strong. So [29] concentrates on block size $k \geq 4$.

In particular, it is shown in [29] that a strong Steiner $(q + 1, 2)$ trade has at least q^2 blocks if q is even, and $q^2 + q$ blocks if q is odd. Hamilton and Khodkar construct such a strong Steiner trade of minimum volume q^2 for any $q = 2^\alpha$, with foundation size $q^2 + q$, and also one on $(q + 1)^2$ points of volume $q^2 + q$ for each q with $q + 1 = 2^\alpha$. Indeed, if $\{T_1, T_2\}$ is a $(v, q + 1, 2)$ strong Steiner trade of volume q^2, then T_1 and T_2 are the duals of affine planes.

Other results include:
(a) There exists a $(q^2 + q + 1, q + 1, 2)$ strong Steiner trade of volume $q^2 + q + 1$ for every prime power q.
(b) If q is a power of 2, then there exists a $(v, q + 1, 2)$ strong Steiner trade of volume m for every $m \geq q^2(q^2 + q + 1)$.

(c) If q is a power of 2 and $q-1$ is a prime power then there exists a $(v, q, 2)$ strong Steiner trade of volume m for every $m \geq (q^2-q)(q^2-q+1)$.

2.7 Directed trades

In a directed t-design, the ordered block $x_1 x_2 \ldots x_k$ is said to contain the ordered t-tuple $x_{i(1)} x_{i(2)} \ldots x_{i(t)}$ for all $i(1) < i(2) < \ldots < i(t)$.

A (v, k, t) directed trade of volume m is a pair $\{T_1, T_2\}$ of disjoint collections of m directed blocks, with the number of occurrences of any directed t-tuple in blocks of T_1 exactly the same as the number its occurrences in blocks of T_2.

In [47], Soltankhah investigates directed trades. In particular she shows that in any (v, k, t) directed trade of volume $m > 0$, the minimum foundation v is k, and the minimum volume is $2^{\lfloor t/2 \rfloor}$, and such minimum foundation and minimum volume directed trades exist. Furthermore, in the cases $t = 2$ and 3, she shows existence of directed (v, k, t) trades of all volumes $m \geq 2$, except there is no directed $(v, 4, 3)$ trade of volume 3.

3. Trades in latin squares

A *latin trade* is a pair $\{L_1, L_2\}$ of partial latin rectangles with precisely the same m filled cells, such that

(i) L_1 and L_2 contain different elements in each filled cell (i, j);

(ii) in each occupied row i, L_1 and L_2 contain set-wise the same symbols;

(iii) in each occupied column j, L_1 and L_2 contain set-wise the same symbols.

Example 3.1
A latin trade of volume 8:

1			2		2			1
2	3	1			1	2	3	
	2	3	1			3	1	2

Note that by labelling the three occupied rows of the above latin rectangles with 3, 4 and 5, and the four occupied columns with 6, 7, 8 and 9, and replacing each occupied cell by the triple corresponding to row, column, entry, we obtain the Steiner $(9, 3, 2)$ trade of volume 8 given in Example 1.1 above. The symbol 3 is used both as a row label and an entry symbol, since the row labelled 3 does not contain the entry 3. Note that considered as a block element in Example 1.1, the element 3 occurs in *four* blocks of T_1. (See Lemma 3 of [5].)

Other terminology has been used in the literature. Fu and Fu [25] called a latin trade $\{L_1, L_2\}$ "disjoint and mutually balanced" partial latin squares; Keedwell [33] used the term "critical partial latin square" and Donovan et al. [20] used "latin interchange". The volume of a latin trade, which is the number m of filled cells, has also been called the *size* of the trade. A latin trade of volume 4 is also called an *intercalate*. In [20] all latin trades of volume at most 11 are found.

The paper [5] raises an interesting question relating latin trades of volume $6m$ and Steiner $(3, 2)$ (triple) trades of volume m. The Steiner $(3, 2)$ (triple) trades of volumes at most 9 have been classified by Khosrovshahi and Maimani [37]. Take such a Steiner $(3, 2)$ triple trade $\{T_1, T_2\}$. For each triple $x\,y\,z$ in T_1, place six entries in a partial latin rectangle L_1 as follows: put x in row y and column z, and put x in row z and column y; similarly place y in cells (x, z) and (z, x); and place z in cells (x, y) and (y, x). Do the same with the triples in T_2, forming a second partial latin rectangle L_2. Then $\{L_1, L_2\}$ is a latin trade of volume $6m$.

The paper [5] asks which such latin trades of volume $6m$ may be partitioned (or decomposed) into six latin trades of volume m. This question is answered for all the Steiner $(3, 2)$ trades of volume at most 9, and a sufficient condition for decomposability of the latin trade of volume $6m$ is given. Example 1.1 above is a Steiner $(3, 2)$ trade of volume 9 which gives rise to a decomposable latin trade, but which does not satisfy the sufficient condition given in Lemma 2 of [5]; it was also given in [5] as an example. A necessary and sufficient condition for decomposability of the latin trade of volume $6m$ arising from a Steiner $(3, 2)$ trade of volume m remains to be found!

4.   Trade uses

The original use of a trade as defined by Hedayat [30] was to avoid some undesirable blocks in an experimental design, while retaining the same parameter set and variety set. However, as mentioned above, the earliest use was probably in the construction of the Steiner triple systems of order 15 by White, Cole and Cummings [50] back in 1916. Here we just briefly mention some of the various uses of trades in more recent times.

The *intersection problem* for two combinatorial structures asks how many common blocks, entries or objects the two structures, based on the same underlying set and with the same parameters, may have. In the case of two block designs, once the blocks common to the two designs have been removed, the remaining two sets of blocks necessarily form a trade.

Furthermore, by starting with one design, and "trading" sets of its blocks, a second design is obtained, having the "untraded" blocks in common with the first design. So trades have been extensively used to solve various intersection problems.

In the block design context, a *defining set* for a design is a set of blocks which has unique completion, in one way only, to the design. It has long been known that for a set of blocks to be a defining set of a design, the set must intersect every trade in the design (or otherwise there could not be unique completion; a non-intersecting trade T_1 could be traded with its mate T_2); see [49] and references therein.

Trades in latin squares are equally important in connection with *critical sets*. A critical set is a partial latin square C of order n which is contained in a *unique* latin square L of order n such that any proper subset of C is contained in at least two latin squares of order n. As in the design theory case, any critical set C must meet *all* latin trades in L, or otherwise there would not be *unique* completion from C to L.

These facts, and more, are used in many papers on defining sets and critical sets, such as [4, 15, 19, 34, 45] to name but a few.

The *support* of a block design is the set of all *distinct* blocks in the design, and the support size is the size of this set. A balanced incomplete block design with parameters (v, k, λ) must necessarily have $\lambda > 1$ for the support size to possibly be less than the total number of blocks.

Hedayat [30] pointed out that different support sizes are important in the statistical applications of designs. Algorithms such as that given in [36] use trades in order to obtain designs with minimal support. See also Section 4.2.2 of the survey paper [38].

Trades have also been extensively used in construction of t-designs; see for example [21, 22, 31, 39].

A $(v, k, t\lambda)$ balanced incomplete block design (BIBD) is called *irreducible* or *indecomposable* if it contains no $(v, k, s\lambda)$ BIBD among its blocks, with $s < t$; otherwise the design is reducible or decomposable.

Trades have been used in the construction of irreducible (or indecomposable) designs; for instance see the constructions in [6]; see also [48] for a survey of irreducible designs.

A further use of a kind of trade is sufficiently different to warrant its own section; we consider this next.

5.   Latin representations and trades used in tripartite decompositions

It is well-known that a complete tripartite graph has a decomposition into triangles if and only if the three vertex parts of the graph all contain the same number of vertices. There is then an exact 1–1 correspondence between each triangle in a decomposition of $K_{n,n,n}$ and each entry in a latin square of order n: the three vertex parts can be labelled $\{1_w, 2_w, \ldots, n_w\}$, $\{1_c, 2_c, \ldots, n_c\}$ and $\{1_e, 2_e, \ldots, n_e\}$ where w, c and e refer to the row, column and entry for the latin square of order n. (We reserve r for later use!) Then an entry k in row i and column j will correspond to the triangle $\{i_w, j_c, k_e\}$.

Now consider a complete tripartite graph $K_{r,s,t}$ where the three parts may have different sizes; suppose $r \leq s \leq t$. There is still a corresponding "latin representation" which we may take, corresponding to all the edges in this graph $K_{r,s,t}$. (See Fig. 5.1.)

With vertices $V_1 = \{1_w, 2_w, \ldots, r_w\}$, $V_2 = \{1_c, 2_c, \ldots, s_c\}$ and $V_3 = \{1_e, 2_e, \ldots t_e\}$ (where w, c, e denote row, column, entry) in the above latin representation, A is a latin rectangle, $[AC]$ is row latin in the t symbols, and $[AB]^T$ is column latin in the t symbols. Each entry in A corresponds to three edges in the tripartite graph (these edges form a triangle); each entry in B corresponds to an edge between V_2 and V_3; each entry in C corresponds to an edge between V_1 and V_3.

In the paper [18], various "trades" are used to obtain 5-cycles. For example, the two triangles and one 4-cycle:

$$(i_w, k_c, 1_e), \ (j_w, \ell_c, 1_e), \ (i_w, 2_e, j_w, 3_e)$$

correspond in the latin representation to cells (i, k) and (j, ℓ) in A containing symbol 1, together with the four entries from C, 2 and 3 in rows

i and j (see Fig. 5.1). These trade with the following two 5-cycles, in that exactly the same 10 edges are used:

$$(i_w, k_c, 1_e, j_w, 2_e), \quad (i_w, 1_e, \ell_c, j_w, 3_e).$$

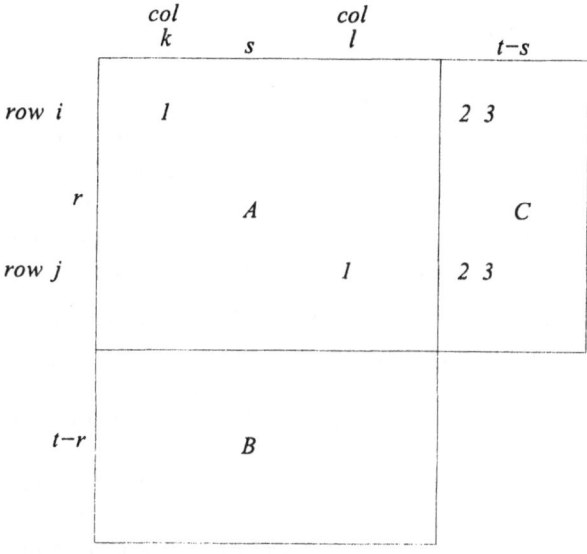

Figure 5.1   Latin representation for $K_{r,s,t}$

At least four other similar trades are used in [18], enabling 5-cycles to be traded from various 3-cycles, 4-cycles and even 6-cycles.

The idea of *trading* various entries in such a latin representation in order to obtain decompositions of complete tripartite graphs was first used in [8], where complete tripartite graphs were decomposed into specified numbers of 3-cycles and 4-cycles. The same ideas were extended in [18] to yield more results on decompositions of complete tripartite graphs into 5-cycles. The question asking whether the necessary conditions for decomposing a complete tripartite graph $K_{r,s,t}$ into 5-cycles are also sufficient conditions was posed in [43]; the problem at this stage remains open, although Cavenagh has further work in progress using trades in latin representations. Cavenagh [17] has also used latin representations to decompose complete tripartite graphs into triangles with a pendant edge; other related work appears in his MSc Thesis [16]. Furthermore, detailed trades from latin representations are used in [12] to obtain *gregarious* decompositions of tripartite graphs into 4-cycles. A gregarious 4-cycle decomposition of a complete tripartite graph is a decomposition into 4-cycles where each 4-cycle has vertices in all three parts. Similarly, a gregarious decomposition of a complete n-partite graph into 4-cycles,

into 4-cycles where each 4-cycle has vertices in all three parts. Similarly, a gregarious decomposition of a complete n-partite graph into 4-cycles, $n \geq 4$, is a decomposition into 4-cycles where each cycle has vertices in four distinct parts of the graph.

6. μ-way trades

Until very recently, all work on trades has involved what are basically 2-way trades: one set of blocks (or entries in a latin square), T_1, may be traded with a disjoint set of blocks (or entries in a latin square), T_2, preserving the structure of the (possibly partial) design or latin square. In [3], in the case of partial latin squares, μ squares rather than 2 are considered; the idea of a μ-*way latin trade* is conceived.

Formally, a μ-way latin trade with $\mu \geq 2$, of volume m, is a set of μ partial latin rectangles containing precisely the same m filled cells, such that:
(i) if cell (i, j) is filled, it contains a different entry in each of the μ partial latin rectangles;
(ii) in each occupied row i, each of the m partial latin rectangles contains, set-wise, the same symbols;
(iii) in each occupied column j, each of the m partial latin rectangles contains, set-wise, the same symbols.
The paper [3] concentrates on possible volumes m for any given μ. In particular, let S_μ denote the set of all possible volumes of a set of μ-way latin trades. Then [24] shows that $S_2 = N \setminus \{1, 2, 3, 5\}$, while [2] shows that $S_3 = N \setminus \{1, 2, \ldots, 8, 10, 11, 13, 14\}$. Then the following results are obtained in [3]:

Theorem 3 [3] *(a) $S_4 = N \setminus \{1, 2, \ldots, 15, 17, 18, 19, 21, 22, 26\}$.*
(b) $S_5 = N \setminus \{1, 2, \ldots, 24, 26, 29, 31, 32, 33, 37, 38\}$.
(c) S_μ contains mn for any $m, n \geq \mu$, and also contains all integers greater than or equal to $3\mu^2 + \mu - 1$.
(d) The only integers less than $(\mu + 1)^2$ which S_μ contains are:
$\mu^2, \mu(\mu + 1), (\mu + 1)^2 - 2$ and $(\mu + 1)^2 - 1$.
(e) If $\mu \geq 4$, then S_μ does not contain any integers in the set
$\{(\mu + 1)^2 + 1, \ldots, \mu(\mu + 3) - 2\}$.
(f) $(\mu + 1)^2 - 2 \in S_\mu$ for $\mu = 2, 4, 5, 6$ and not for $\mu = 3$.
(g) $\mu(\mu + 3) - 1 \in S_\mu$ for $\mu = 2, 3, 4, 5$.

Note that the question of whether S_μ contains $(\mu+1)^2 - 2$ is undecided for $\mu \geq 7$, and whether it contains $\mu(\mu + 3) - 1$ is undecided for $\mu \geq 6$.

When $\mu = 4$, $(\mu + 1)^2 - 2$ is 23; a 4-way latin trade of volume 23 is given next.

Example 6.1
A 4-way trade of volume 23.

	3	1	2	4
5		2	3	1
3	4	5	1	2
2	1	4	5	3
1	2	3	4	5

	1	3	4	2
2		1	5	3
1	2	4	3	5
5	3	2	1	4
3	4	5	2	1

	2	4	3	1
3		5	1	2
5	1	2	4	3
1	4	3	2	5
2	3	1	5	4

	4	2	1	3
1		3	2	5
2	3	1	5	4
3	2	5	4	1
5	1	4	3	2

The concept of μ-way trades in the design context has not as yet really been investigated thoroughly. For cycle systems, the idea of μ-way trades was used in [1]. Also Milici and Quattrocchi [44] looked at the intersection problem for sets of *three* Steiner triple systems, and found the set of integers k for which there exist three Steiner triple systems of the same order and based on the same underlying set, having precisely k common triples. This is closely related to finding the possible trade volumes of 3-way triple trades. An example of a 3-way triple trade of volume 6 is given next.

Example 6.2
A 3-way triple system trade of volume 6.

1 4 5	1 4 7	1 4 6
1 6 7	1 5 6	1 5 7
2 4 6	2 4 5	2 4 7
2 5 7	2 6 7	2 5 6
3 4 7	3 4 6	3 4 5
3 5 6	3 5 7	3 6 7

The problem of finding possible μ-way trade volumes for triple systems and for other block sizes remains open. It is easy to see that 6 is the smallest 3-way trade volume for block size 3.

7.   Graphical trades

In 2001, the concept of a trade was generalised from that of a $(k, 2)$ trade with blocks of size k to the case where a "block" is a simple graph G. Of course a block of size k, in the 2-design case, can be regarded as a complete graph K_k with its $\binom{k}{2}$ edges corresponding to all the pairs.

Formally, a G-trade of volume m is a pair $\{T_1, T_2\}$ where each T_i consists of m graphs, pairwise edge-disjoint, each isomorphic to G, with the m copies in T_1 distinct from the m copies in T_2, and the edge-set of the graphs in T_1 being identical to the edge-set of the graphs in T_2, and forming a simple graph say H. In other words,

$$T_1 = \{G_1, G_2, \ldots, G_m\}, \quad T_2 = \{G'_1, G'_2, \ldots, G'_m\},$$

with $G_i \simeq G'_i \simeq G$, $1 \leq i \leq m$, $\bigcup_{i=1}^m G_i = \bigcup_{i=1}^m G'_i = H$, where H is a simple graph, and with $G_i \neq G'_j$, $1 \leq i, j \leq m$, and $\bigcup_{i=1}^m E(G_i) = \bigcup_{i=1}^m E(G'_i)$.

In the case that G is K_k, this definition amounts to that of a Steiner $(k, 2)$ trade of volume m; the Steiner property follows from the fact that H is a simple graph.

The first paper [10] to consider trades in a graphical context concentrated chiefly on finding possible trade volumes, especially of graphs G that were "almost" complete graphs in some sense. The size of the foundation (the number of vertices in the simple graph H) was unrestricted. In Section 2.1 above, Theorem 1 on possible trade volumes for Steiner $(k, 2)$ trades shows that the number of *forbidden* trade volumes is approximately two and a half times the block size k:

$$X(K_k) = \{1, 2, \ldots, 2k - 3\} \cup \{2k - 1, 2k + 1, \ldots, \lceil \frac{3k - 4}{2} \rceil + \lfloor \frac{3k - 5}{2} \rfloor\}$$

except $\mathrm{TS}(K_2) = \{0\}$, and $15 \notin X(K_7)$.

Motivation for the investigation of trade volumes of "almost" complete graphs arose partly from [9] where (apart from two exceptions for order $v = 11$) it was shown that two $K_4 - e$ designs (with b blocks) of order v exist, having s common blocks, for all s except $s = b - 1, b - 2$. In [10] it is shown that $X(K_k - e) = \{1, 2\}$, for all $k \geq 4$. Thus the removal of *one* edge from K_k (from a "complete" block of size k), results in a drop of the forbidden trade volumes from around two and a half k values to just two values!

Other results in [10] show that removal of a path of length two from a complete graph results in a graph whose forbidden trade volume is just

1, while removal of two disjoint edges yields a graph with forbidden trade volumes 1 and 2 for foundation at least 5. Even removal of a complete subgraph from K_k, or removal of a path, results in a graph having trades of volumes 2 upwards, or else 3 upwards. Removal of a cycle from K_k also yields similar results, although the result for $K_k - C_6$ where $k \geq 7$ is incomplete at this stage; it is known that

$$\{1\} \subseteq X(K_k - C_6) \subseteq \{1, 3, 5\}, \quad \text{for } k \geq 7.$$

A complete classification of those graphs G having $X(G) = \{1\}$ or $X(G) = \{1, 2\}$ remains to be found. Partial results are given in [10]: in particular, any graph G possessing an independent set of vertices whose removal disconnects the graph has trades of all volumes greater than or equal to 2. This, for example, shows that all bipartite graphs and all cycles of length at least 4 have only the single forbidden trade volume 1: they have trades of all volumes from 2 upwards. It also shows that the Petersen graph P has $X(P) = \{1\}$! However, the prism graph, K_6 minus a hamilton cycle, has no such independent set of vertices, yet nevertheless it possesses trades of volumes 2 and 3, and hence of all volumes except 1.

A subsequent paper, [11], generalises results for trade volumes for a complete graph to trade volumes for a complete partite graph. Of course a complete graph may in a sense be regarded as a complete partite graph where all the parts have size 1. Once again the forbidden trade volumes do *not* increase with the order of the complete partite graph, when the graph is not a complete graph K_k. The largest set of forbidden trade volumes to arise is $\{1, 2, 3, 4, 5\}$.

The following is a slight update on the results in [11], thanks to new work by an Honours' student [32].

Theorem 4 [11], [32]. *Let G be a complete partite graph.*

(i) *If G has n parts of size 2, then*
$X(G) = \{1\}$ *if $n = 2$;* $X(G) = \{1, 2, 3\}$ *if $n = 3$;*
$\{1, 2, 3, 4\} \subseteq X(G) \subseteq \{1, 2, 3, 4, 5\}$ *if $n = 4$;*
for $n \geq 5$, $X(G) = \{1, 2, 3, 4, 5\}$.

(ii) *If G has n parts of size $p \geq 3$, then*
$X(G) = \{1\}$ *if $n = 2$;* $\{1, 2\} \subseteq X(G) \subseteq \{1, 2, 3, 5\}$ *if $n = 3$;*
$\{1, 2, 3\} \subseteq X(G) \subseteq \{1, 2, 3, 4, 5\}$ *if $n \geq 4$.*

(iii) *If G has parts of different sizes, then $X(G) = \{1, 2\}$, except $2 \in$ $TS(G)$ when G has just two parts or when G has at least 3 parts of sizes a_i, $1 \leq i \leq n$, with, for some i, j, $2a_j = a_i + \sum_{k=1}^{n} a_k$.*

To complete the work on possible trade volumes of complete partite graphs, it thus remains to determine whether: (a) $5 \in X(G)$ when G is $K_{2,2,2,2}$; (b) $3, 5 \in X(G)$ when G is $K_{p,p,p}$, $p \geq 3$; and (c) $4, 5 \in X(G)$ when G has at least four parts all of size $p \geq 3$.

Further work, besides [10, 11], on trades in graphs, has investigated the range of possible Steiner trade volumes, for fixed foundation v, when G is C_3, C_4 or C_5; see [13], [14], [42] respectively. (The case of C_3 is of course the same as K_3.)

Since the trades are Steiner, in each case an upper bound on the possible trade volume with foundation v is given by the number of cycles C_3, C_4 or C_5 in a maximum packing of K_v with these cycles. The lower bound is, roughly speaking, about $2v/3$ for C_3, $v/3$ for C_4 and $v/4$ for C_5. In the case of cycles of length 3, since C_3 is the same as K_3, we have less freedom for "small" trade volumes. Each point in the foundation has to belong to at least two triples and so has degree at least four. So in a Steiner C_3-trade $\{T_1, T_2\}$, the simple graph H consisting of all the 3-cycles in T_1 has minimum degree 4. Thus $4v$ is a lower bound on the degree sum of H, which must equal twice the number of edges in H, or $6m$ if $|T_1| = m$. Hence $m \geq \lceil 2v/3 \rceil$.

In the case of cycles of length k where $k \geq 4$, it is possible to have a trade in which, if the volume m is even, only m points have degree 4 while $(k-2)$ points in each cycle have degree 2; in this case there are km edges, and the degree sum $4kv/(k-1)$ is equal to twice this number of edges. Hence a lower bound for the possible volume m of a k-cycle trade on v points, with $k \geq 4$, is given by $m \geq \lceil v/(k-1) \rceil$. This is $\lceil v/3 \rceil$ for 4-cycles and $\lceil v/4 \rceil$ for 5-cycles.

The papers [13, 14, 42] each exhibit trades of "small", "intermediate" and "large" size, thus showing that all Steiner C_k trade volumes with given foundation v and volume m can be achieved, when $k = 3, 4, 5$, except for four values which cannot be attained in the case of 3-cycles.

8.   Future directions

In the field of *graphical trades* there are many fascinating unanswered questions. At present the only graph known with its number of forbidden trade volumes dependent upon its order (that is, the number of vertices in the graph) is a complete graph. However, Wormald [51] believes that random graphs will behave like complete graphs, in that the forbidden trade volumes of a random graph will most likely increase in number with the order of the random graph.

A complete classification of those graphs G having $X(G) = \{1\}$ or $\{1,2\}$ also awaits discovery, although this is probably a very hard problem.

Graphical trades which are non-Steiner have not yet been investigated. In this case, the trade could still be a simple trade, (so no exact duplicates in T_1 or in T_2) with the graph G being a simple graph too, but with the graph $H = \bigcup G_i$ (where $G_i \in T_1$) *not* simple, so the trade will not be Steiner.

The possible μ-*way trades* in the block design situation, with $\mu \geq 3$, is another area almost untouched as yet. So far the main work on μ-way trades has been with μ-way latin trades. There remain several open questions there too, with regard to possible μ-way latin trade volumes.

For trades consisting of blocks, many questions remain. In particular, for trades consisting of triples, completion of work on $(v, 3, 2)$ trades when $v \equiv 5 \pmod 6$ remains; see Sections 2.4 and 2.5.

The further use of *latin representations* and trades of that type, exchanging entries in latin representations for various small graphs, will doubtless yield further results relating to tripartite graph decompositions. Perhaps 3-dimensional latin cubes or latin cube representations could be used for 4-partite graph decompositions!

There is no doubt that trades in their various guises form an area of research which is rich in the diversity of the uses and applications which arise; thus work on trades generally is set to continue.

References

[1] P. Adams, E.J. Billington, D.E. Bryant and A. Khodkar, *The μ-way intersection problem for m-cycle systems*, Discrete Mathematics **231** (2001), 27–56.

[2] P. Adams, E.J. Billington, D.E. Bryant and E.S. Mahmoodian, *The three-way intersection problem for latin squares*, Discrete Math. (to appear).

[3] P. Adams, E.J. Billington, D.E. Bryant and E.S. Mahmoodian, *On the possible volumes of μ-way latin trades*, Aequationes Mathematicae (to appear).

[4] P. Adams, A. Khodkar and C. Ramsay, *Smallest defining sets of some STS(19)*, J. Combin. Math. Combin. Comput. **38** (2001), 225–230.

[5] R. Bean, D. Donovan, A. Khodkar and A.P. Street, *Steiner trades that give rise to completely decomposable latin interchanges*, Proc. Eleventh Australasian Workshop on Combinatorial Algroithms, University of Newcastle, NSW, 2000, pp. 17–30.

[6] E.J. Billington, *Further constructions of irreducible designs*, Congressus Numerantium **35** (1982), 77–89.

[7] E.J. Billington, *The intersection problem for combinatorial designs*, Congressus Numerantium **92** (1993), 33–54.

[8] E.J. Billington, *Decomposing complete tripartite graphs into cycles of lengths 3 and 4*, Discrete Mathematics **197/198** (1999), 123–135.

[9] E.J. Billington, M. Gionfriddo and C.C. Lindner, *The intersection problem for $K_4 - e$ designs*, J. Statistical Planning and Inference **58** (1997), 5–27.

[10] E.J. Billington and D.G. Hoffman, *Trades and graphs*, Graphs and Combinatorics **17** (2001), 39–54.

[11] E.J. Billington and D.G. Hoffman, *Trade spectra of complete partite graphs*, Discrete Mathematics (to appear).

[12] E.J. Billington and D.G. Hoffman, *Decomposition of complete tripartite graphs into gregarious 4-cycles*, Discrete Mathematics (to appear)

[13] D.E. Bryant, *On the volume of trades in triple systems*, Australasian Journal of Combinatorics **15** (1997), 161–176.

[14] D.E. Bryant, M. Grannell, T. Griggs and B. Maenhaut, *On the volume of 4-cycle trades*, submitted.

[15] D.E. Bryant and B.M. Maenhaut, *Defining sets of G-designs*, Australasian Journal of Combinatorics **17** (1998), 257–266.

[16] N.J. Cavenagh, *Graph decompositions of complete tripartite graphs using trades*, M.Sc. Thesis, University of Queensland, 1998.

[17] N.J. Cavenagh, *Decompositions of complete tripartite graphs into triangles with an edge attached*, Utilitas Mathematica (to appear).

[18] N.J. Cavenagh and E.J. Billington, *On decomposing complete tripartite graphs into 5-cycles*, Australasian Journal of Combinatorics **22** (2000), 41–62.

[19] C. Delaney, B.D. Gray, K. Gray, B.M. Maenhaut, M.J. Sharry and A.P. Street, *Pointwise defining sets and trade cores*, Australasian Journal of Combinatorics **16** (1997), 51–76.

[20] D. Donovan, A. Howse and P. Adams, *A discussion of latin interchanges*, J. Combin. Math. Combin. Comput. **23** (1997), 161–182.

[21] Z. Eslami and G.B. Khosrovshahi, *A complete classification of 3-(11, 4, 4) designs with nontrivial automorphism group*, Journal of Combinatorial Designs **8** (2000), 419–425.

[22] Z. Eslami and G.B. Khosrovshahi, *Some new 6-(14, 7, 4) designs*, J. Combinatorial Theory series A **93** (2001), 141–152.

[23] Z. Eslami and B. Tayfeh-Rezaie, *On 2-(v, 3) trades of minimum volume*, Australasian Journal of Combinatorics **19** (1999), 239–251.

[24] H.-L. Fu, *On the construction of certain type of latin squares with prescribed intersections*, Ph.D. Thesis, Auburn University, 1980.

[25] C.-M. Fu and H.-L. Fu, *The intersection problem of Latin squares*, Graphs, designs and combinatorial geometries (Catania, 1989), J. Combin. Inform. System Sci. **15** (1990), 89–95.

[26] B.D. Gray, *The maximum number of trades of volume four in a symmetric design*, Utilitas Mathematica **52** (1997), 193–203.

[27] B.D. Gray and C. Ramsay, *On the spectrum of Steiner (v, k, t) trades II*, Graphs and Combinatorics **15** (1999), 405–415.

[28] B.D. Gray and C. Ramsay, *On the spectrum of Steiner (v, k, t) trades I*, J. Combin. Math. Combin. Comput. **34** (2000), 133–158.

[29] N. Hamilton and A. Khodkar, *On minimum possible volumes of strong Steiner trades*, Australasian Journal of Combinatorics **20** (1999), 197–203.

[30] A.S. Hedayat, *The theory of trade-off for t-designs*, Coding Theory and Design Theory, Part II, IMA Vol. Math. Appl., **21**, 101–126, Springer, New York, 1990.

[31] A. Hedayat and H.L. Hwang, *Construction of BIB designs with various support sizes — with special emphasis for v = 8 and k = 4*, J. Combinatorial Theory, series A **36** (1984), 163–173.

[32] P.D. Jenkins, *Combinatorial trades*, University of Queensland Honours Thesis, 2001.

[33] A.D. Keedwell, *Critical sets for latin squares, graphs and block designs: a survey*, Congessus Numerantium **113** (1996), 231–245.

[34] A. Khodkar, *Smallest defining sets for the* 36 *non-isomorphic twofold triple systems of order nine*, J. Combin. Math. Combin. Comput. **17** (1995), 209–215.

[35] A. Khodkar and D.G. Hoffman, *On the non-existence of Steiner* $(v, k, 2)$ *trades with certain volumes*, Australasian Journal of Combinatorics **18** (1998), 303–311.

[36] G.B. Khosrovshahi and E.S. Mahmoodian, *A linear algebraic algorithm for reducing the support size of t-designs and to generate a basis for trades*, Communications in Statistics, Simulation and Computation **16** (1987), 1015–1038.

[37] G.B. Khosrovshahi and H.R. Maimani, *On* 2-$(v, 3)$ *Steiner trades of small volumes*, Ars Combinatoria **52** (1999), 199–220.

[38] G.B. Khosrovshahi, H.R. Maimani and R. Torabi, *On trades: an update*, Discrete Applied Mathematics **95** (1999), 361–376.

[39] G.B. Khosrovshahi, A. Nowzari-Dalini and R. Torabi, *Trading signed designs and some new* 4-$(12, 5, 4)$ *designs*, Designs, Codes and Cryptography **11** (1997), 279–288.

[40] G.B. Khosrovshahi and R. Torabi, *Maximal trades*, Ars Combinatoria **51** (1999), 211–223.

[41] G.B. Khosrovshahi and H. Yousefi-Azari, *Octahedrals in Steiner quadruple systems*, Utilitas Mathematica **51** (1997), 33–39.

[42] B.M. Maenhaut, *On the volume of* 5-*cycle trades*, Graphs and Combinatorics **17** (2001), 315–328.

[43] E.S. Mahmoodian and M. Mirzakhani, *Decomposition of complete tripartite graphs into* 5-*cycles*, in Combinatorics Advances (eds. C.J. Colbourn and E.S. Mahmoodian), Kluwer Academic Publishers (1995), 235–241.

[44] S. Milici and G. Quattrocchi, *On the intersection problem for three Steiner triple systems*, Ars Combinatoria **24A** (1987), 174–194.

[45] C. Ramsay, *An algorithm for enumerating trades in designs, with an application to defining sets*, J. Combin. Math. Combin. Comput. **24** (1997), 3–31.

[46] C. Ramsay, *On the index of simple trades*, Australasian Journal of Combinatorics **20** (1999), 207–221.

[47] N. Soltankhah, *On directed trades*, Australas. J. Combin. **11** (1995), 59–66.

[48] A.P. Street, *A survey of rreducible balanced incomplete block designs*, Ars Combinatoria **19A** (1985), 43–60.

[49] A.P. Street, *Trades and defining sets*, in C.J. Colbourn and J.H. Dinitz, CRC Handbook of Combinatorial Designs, CRC Publishing Co., **IV.46** (1996), 474–478.

[50] H.S. White, F.N. Cole and L.D. Cummings, *Complete classification of the triad systems on fifteen elements*, Memoirs of the National Academy of Sciences 14 (1916), 1–189.

[51] N. Wormald, private communication, 2001.

Chapter 4

TWO-STAGE GENERALIZED SIMULATED ANNEALING FOR THE CONSTRUCTION OF CHANGE-OVER DESIGNS

Yuk W. Cheng

School of Mathematics,
University of New South Wales.

Deborah J. Street

Department of Mathematical Sciences,
University of Technology, Sydney.

William H. Wilson

School of Computer Science and Engineering,
University of New South Wales.

Abstract   Kunert (1983) has found sufficient conditions for the optimal estimation of direct treatment effects from change-over designs. In this paper, a two-stage generalized simulated annealing algorithm uses these conditions to construct large change-over designs where $t|p$ and $t\backslash n$ and where $t\backslash p$ and $t|n$.

1.   Introduction

In some experimental situations the experimental units are expensive or difficult to obtain, for example fruit trees, livestock or patients with a particular disease. In such circumstances, if treatments are not expected to have any permanent effect on the units, it may be appropriate to apply, sequentially, more than one treatment to each experimental unit. Such designs are called change-over designs.

More formally, a change-over design $COD(t, n, p)$ is a design on n experimental units to compare t treatments. Each unit receives a sequence of p, not necessarily distinct, treatments during the course of the experiment. Each treatment in the sequence is applied for one period and the response noted. We will represent the design by a $p \times n$ array containing entries from $\{1, 2, ..., t\}$. We will use $\Omega_{t,n,p}$ to denote the set of all $COD(t, n, p)$s on n units with p periods and t treatments. Table 1.1 shows two $COD(3, 6, 7)$s.

1	2	3	2	3	1		1	1	2	2	3	3
3	1	2	3	1	2		1	1	3	2	2	3
2	3	1	1	2	3		2	3	3	1	2	1
2	3	1	1	2	3		2	2	1	3	3	1
3	1	2	3	1	2		3	3	2	1	1	2
1	2	3	2	3	1		3	2	1	3	1	2
1	2	3	2	3	1		2	1	3	3	2	1

Table 1.1. Two $COD(3, 6, 7)$s

A *preperiod* is sometimes applied prior to the commencement of the experiment so that all observations have a residual treatment effect. Let $\tilde{\Omega}_{t,n,p}$ be the set of all change-over designs with preperiod.

One common linear model assumes that the observation Y_{ku}, made on unit u during period k, is the sum of a period effect (α_k), a unit effect (β_u), a direct treatment effect ($\tau_{k,u}$), a first order residual treatment effect ($\rho_{k-1,u}$) (for any period other than the first) and an error term E_{ku}. The observations are assumed to be independent of each other, so $\mathrm{corr}(E_{ku}, E_{gw}) = 0$ for all pairs $(k, u) \neq (g, w)$. The variance of the error terms is assumed to be constant, so $\mathrm{var}(E_{ku}) = \sigma^2$. Thus

$$Y_{ku} = \alpha_k + \beta_u + \tau_{k,u} + \rho_{k-1,u} + E_{ku},$$

$$1 \leq k \leq p, 1 \leq u \leq n, \rho_{0,u} = 0.$$

In the rest of this paper, we will introduce a two-stage generalized simulated annealing algorithm which uses the conditions in Kunert [12] to construct optimal change-over designs with $t|p$ and $t\backslash n$ and with $t\backslash p$ and $t|n$. Many of the designs constructed have small t but large n and p as these are used frequently in marketing. Here t is the number of products, n is the number of shops and p is the number of weeks that the products are on display in various different arrangements (Federer [8]). Each product will be displayed at a particular location (say at the end of an aisle, near the check-out, in the aisle) each week and the volume of sales recorded. This allows the benefits of particular store display positions to be quantified.

2.   Notation and Definitions

We adopt the following notation from C.S. Cheng and Wu [2]. For a design $d \in \Omega_{t,n,p} \cup \tilde{\Omega}_{t,n,p}$, let h_{ik} = the number of appearances of treatment i in period k, $\tilde{h}_{ik}$ = the number of appearances of treatment i in period $k-1$, $\tilde{h}_{i1} = 0$ if the design is in $\Omega_{t,n,p}$, n_{iu} = the number of appearances of treatment i on unit u in periods 1 to p, $\tilde{n}_{iu}$ = the number of appearances of treatment i on unit u in periods 1 to $(p-1)$, m_{ij} = the number of appearances of treatment i preceded by treatment j, r_i = the number of appearances of treatment i in periods 1 to p, and $\tilde{r}_i$ = the number of appearances of treatment i in periods 1 to $(p-1)$, where $1 \leq u \leq n$, $1 \leq k \leq p$, $1 \leq i, j \leq t$.

So $r_i = \sum_{u=1}^{n} n_{iu}$ and $\tilde{r}_i = \sum_{u=1}^{n} \tilde{n}_{iu}$. We now collect these constants into matrices and let $D = diag(r_1, r_2, ..., r_t)$, $\tilde{D} = diag(\tilde{r}_1, \tilde{r}_2, ..., \tilde{r}_t)$, $M = (m_{ij})$, $N_p = (h_{ik})$, $\tilde{N}_p = (\tilde{h}_{ik})$, $N_u = (n_{iu})$, $\tilde{N}_u = (\tilde{n}_{iu})$, $N_p \tilde{N}_p^T = (f_{ij})$ and $N_u \tilde{N}_u^T = (g_{ij})$.

Recall that a matrix, A, say, is *completely symmetric* if $A = aI + bJ$, for some constants a and b, where I is the identity matrix and J is the matrix with all entries equal to 1.

Definition 1 A *generalized Youden design (GYD)* has $D = rI$ for some r, $N_u N_u^T$ completely symmetric with all entries in N_u being either $[p/t]$ or $[p/t]+1$ and $N_p N_p^T$ completely symmetric with all entries in N_p being either $[n/t]$ or $[n/t]+1$, where $[x]$ is the largest integer less than or equal to x.

The two designs in Table 1.1 are *GYD*s.

The optimality of any design is usually given as some function of the variance-covariance matrix of parameter estimates. In the case of a change-over design there are two sets of parameter estimates that are of interest, the direct treatment effects and the residual treatment effects.

The variance-covariance matrix for the estimation of direct treatment effects is given by C_D^-, the generalised inverse of C_D, where

$$C_D = C_{11} - C_{12}C_{22}^- C_{12}^T$$

and

$$
\begin{aligned}
C_{11} &= D - \frac{1}{n}N_p N_p^T - \frac{1}{p}N_u N_u^T + \frac{1}{np}N_u J N_u^T \\
C_{22} &= \tilde{D} - \frac{1}{n}\tilde{N}_p \tilde{N}_p^T - \frac{1}{p}\tilde{N}_u \tilde{N}_u^T + \frac{1}{np}\tilde{N}_u J \tilde{N}_u^T \\
C_{12} &= M - \frac{1}{n}N_p \tilde{N}_p^T - \frac{1}{p}N_u \tilde{N}_u^T + \frac{1}{np}N_u J \tilde{N}_u^T.
\end{aligned}
$$

The variance-covariance matrix for the estimation of residual treatment effects is given by C_R^-, where

$$C_R = C_{22} - C_{12}^T C_{11}^- C_{12}.$$

A design is said to be A-optimal if it has the minimum sum for the eigenvalues of the relevant variance-covariance matrix. This means that the corresponding design has the minimum average variance of the elementary contrasts of treatment effects. A design is said to be D-optimal if it has the minimum product for the eigenvalues of the relevant variance-covariance matrix. This means that the corresponding design has the minimum generalised variance of treatment estimates.

Kunert [12] has found conditions for a *COD* to be optimal for the estimation of direct treatment effects in terms of the matrices defined above. A design $d \in \Omega_{t,n,p} \cup \tilde{\Omega}_{t,n,p}$ is D- or A- optimal for the estimation of direct effects over $\Omega_{t,n,p} \cup \tilde{\Omega}_{t,n,p}$ if it is a *GYD* and has

$$
m_{ij} = \begin{cases}
r_j/t, & (\ t|p \text{ and } t|n), \\
f_{ij}/n, & (\ t|p \text{ and } t\nmid n), \\
g_{ij}/p, & (\ t\nmid p \text{ and } t|n).
\end{cases}
$$

For this reason we find *GYDs* with these properties in this paper. These designs turn out to have very good optimality properties for residual treatment effects as well.

3. A Two Stage Generalized Simulated Annealing Algorithm

Many researchers have used generalized simulated annealing as a tool in the development of optimal experimental designs and in searching for some classes of Latin squares. Recent examples include Meyer and Nachitsheim [13], Elliott and Gibbons [7], John and Whitaker [10], Venables and Eccleston [14], Jansan, Douven and Van Berkum [9] and Eccleston and Whitaker [6]. A general discussion of the construction of efficient simulated annealing algorithms may be found in Duque-Anton [4].

We will use a two-stage generalized simulated annealing to find optimal $CODs$. The first stage is to find a feasible design where $n|f_{ij}$ for $t|p$ and $t\backslash n$ or with $p|g_{ij}$ for $t\backslash p$ and $t|n$. The second stage will make use of the feasible design to search for optimal $CODs$ with $f_{ij}/n = m_{ij}$ $(g_{ij}/p = m_{ij})$. The advantage of a two-stage approach is that it can reduce the number of possibilities in searching for the optimal design if the change in each step in the second stage will not affect the objective function value of the final design from the first stage.

When searching for $CODs$, the interchange steps within each iteration must ensure the new design is still a GYD. Random interchange of entries in the design is thus impossible and we use some chain properties of $GYDs$, described below, to determine entries which can be interchanged while ensuring that the new designs are still $GYDs$.

For the GYD, d, we let $d(k, u)$ denote the treatment given to unit u during period k.

Definition 2 A *closed unit chain of length l* in a GYD is a set of distinct periods $a_1, a_2, ..., a_l$ chosen from units x and y such that $d(a_{i+1}, x) = d(a_i, y)$, $1 \leq i \leq l-1$, $d(a_1, x) = d(a_l, y)$ and $d(a_i, x) \neq d(a_i, y)$, $1 \leq i \leq l$.

Example 1 In the COD of Table 3.2, units $x = 1$, $y = 2$ and periods $a_1 = 10$, $a_2 = 15$, $a_3 = 4$, $a_4 = 1$, $a_5 = 2$, $a_6 = 5$ in the $COD(3, 4, 15)$ form a closed unit chain of length 6. When $i = 1$, for example, $d(15, 1) = d(10, 2) = 2$.

Definition 3 A *closed period chain of length l* in a GYD is a set of distinct units $b_1, b_2, ..., b_l$ chosen from periods q and r such that $d(q, b_{i+1}) = d(r, b_i)$, $1 \leq i \leq l - 1$, $d(q, b_1) = d(r, b_l)$ and $d(q, b_i) \neq d(r, b_i)$, $1 \leq i \leq l$.

$$
\begin{bmatrix}
2 & 1 & 1 & 3 & 3 & 2 & 2 & 3 & 2 & 2 & 3 & 3 & 1 & 1 & 1 \\
1 & 3 & 1 & 2 & 2 & 3 & 2 & 2 & 3 & 1 & 2 & 1 & 1 & 3 & 3 \\
3 & 2 & 2 & 1 & 3 & 1 & 1 & 3^* & 1 & 2 & 1^* & 2 & 3 & 3 & 2 \\
2 & 3 & 3 & 2 & 1 & 2 & 3 & 1^* & 1 & 3 & 3^* & 1 & 2 & 2 & 1
\end{bmatrix}^T
\qquad
M = \begin{bmatrix}
6 & 6 & 7 \\
6 & 6 & 6 \\
6 & 7 & 6
\end{bmatrix}
$$

Table 3.2. A $COD(3,4,15)$ with examples of a closed unit chain (in italics), a dual unit closed chain (underlined) and a binary closed chain (indicated by *), and the corresponding M matrix of the design.

Definition 4 A *dual closed unit (period) chain* in a GYD is a closed unit (period) chain in which all the entries on the same units (periods) are different. Thus $l \le t$.

Example 2 The underlined entries in the $COD(3,4,15)$ of Table 3.2 form a dual closed unit chain with units $x = 1$, $y = 2$, $l = t = 3$ and periods $a_1 = 2$, $a_2 = 11$, $a_3 = 1$.

Definition 5 A *binary closed chain* in a GYD is a collection of treatments in a closed unit chain or a closed period chain with $d(a,q) = d(b,r)$ and $d(a,r) = d(b,q)$, $1 \le a,b \le p$ and $1 \le q,r \le n$.

Example 3 The *-ed entries in the $COD(3,4,15)$ in Table 3.2 form a binary closed chain with units $q = 3$, $r = 4$ and periods $a = 8$, $b = 11$.

We construct the initial GYD by the patchwork method of Kiefer [11] or using the tables and computational constructions given in Ash [1]. Both of these authors used Latin squares and balanced incomplete block designs in their constructions. Our computational experience suggests that it is better to use different Latin squares in the patchwork construction of the initial GYD because it then takes fewer hill climbing moves to converge to the optimal COD.

When finding optimal $CODs$ with $t|p$ and $t\nmid n$, we use $\Psi((f_{ij})) = \sum_{i=1}^{t} \sum_{j=1}^{t} S_{ij}$ as the objective function in the first stage, where

$$
S_{ij} = \begin{cases} mod(f_{ij}, n) & \text{if } 0 \le mod(f_{ij}, n) \le \frac{n}{2} \\ n - mod(f_{ij}, n) & \text{otherwise} \end{cases}
$$

and $\Psi((m_{ij})) = \sum_{i=1}^{t} \sum_{j=1}^{t} |f_{ij} - m_{ij} \times n|$ as the objective function in the second stage.

When finding optimal $CODs$ with $t\backslash p$ and $t|n$, we use the same objective functions at each stage but we replace f_{ij} by g_{ij} and n by p.

Analytical considerations suggest that a reasonable initial value of T should be such that $0.5 \leq P \leq 0.9$ holds, where $P = exp(-\Delta E/T)$, $T = \frac{1}{\theta} = \frac{log(step)+1}{k}$ and k is a constant, $\Delta E = \frac{\Delta\Psi(.)}{\Psi(.)}$. This is because probabilities too close to unity lead to inefficient searches (apparently every step is accepted). It requires too many function evaluations to escape from a local minimum if P is less than 0.5.

We allow relatively few climbing moves in stage 1 as there are many designs that satisfy the optimality criterion of the first stage. However in the second stage there are relatively few designs that satisfy the criterion and so we allow more climbing steps. We found that the algorithm usually moves away from a local minimum after two or three steps.

4. Performance of the Two-Stage GSA

Table 4.3 shows the A-optimal values for direct and residual treatment effects, as well as the A-efficiency (the ratio of the realised A-optimal value and the theoretically best possible A-optimal value) for residual treatment effects for a number of designs constructed by our algorithm. We see that all of these designs perform extremely well for residual effects in addition to being optimal for the estimation of direct effects.

To obtain the lower bound for the A-optimal value for residual treatment effects we need a bound on $tr(C_R)$. As $C_{12} = 0$ for all our designs, $C_R = C_{22}$. Hence the optimal design will have the largest $tr(C_{22})$ and have all the eigenvalues equal.

For $t|p$ and $t\backslash n$, we can evaluate $tr(C_R)$ explicitly and we get that

$$
\begin{aligned}
tr(C_R) = \; & n(p-1) + ((q+1)t - n)(\frac{np}{t} - q)^2 + (n - qt)(\frac{np}{t} - (q+1))^2 \\
& - \frac{p-1}{n}[((q+1)t - n)q^2 + (n - qt)(q+1)^2] \\
& - \frac{n}{p}[(t-1)(\frac{p^2}{t} + (\frac{p}{2} - 1)^2],
\end{aligned}
$$

where $qt < n \leq (q+1)t$.

For $t\backslash p$ and $t|n$, we can get upper and lower bounds on $\mathrm{tr}(C_R)$. Suppose that $qt < p < (q+1)t$. Then the bounds are given by

$$n(p-1) + \frac{n(p-1)^2}{tp} - \frac{n(p-1)}{t} - \frac{t}{p}(2pq + p - tq^2 - tq - 2q - 1)$$

and

$$n(p-1) + \frac{n(p-1)^2}{tp} - \frac{n(p-1)}{t} - \frac{t}{p}(2pq + p - tq^2 - tq - 2q + 1).$$

t	n	p	direct A-opt value	residual A-opt value	residual A-efficiency
3	4	15	$\frac{8}{15}$	0.1149	1.00
3	4	18	$\frac{4}{45}$	0.0944	1.00
3	4	21	$\frac{8}{105}$	0.0802	1.00
3	4	24	$\frac{1}{15}$	0.0697	1.00
3	5	12	$\frac{10}{96}$	0.1146	0.999
3	8	18	$\frac{8}{189}$	0.0450	1.00
4	2	12	$\frac{3}{4}$	0.8429	0.978
4	3	12	$\frac{3}{8}$	0.4138	0.996
4	5	24	$\frac{15}{144}$	0.1089	1.00
5	4	20	$\frac{4}{15}$	0.2817	0.999
3	12	5	$\frac{5}{48}$	$\frac{1}{7}$	0.921
3	12	7	$\frac{7}{96}$	$\frac{1}{12}$	1.00
3	12	8	$\frac{4}{64}$	$\frac{1}{13}$	0.945
3	15	5	$\frac{1}{12}$	$\frac{1}{19}$	1.00
3	21	5	$\frac{5}{84}$	$\frac{1}{13}$	0.978
3	21	8	$\frac{16}{441}$	$\frac{2}{47}$	0.977
3	24	5	$\frac{5}{96}$	$\frac{1}{14}$	0.922

Table 4.3. A-optimal values for direct and residual treatment effects for designs constructed by the 2-stage GSA algorithm

We are aware of three other algorithms explicitly written for the construction of optimal change-over designs. One is due to Eccleston and Street [5], one due to Donev [3] and one due to Eccleston and Whitaker [6].

The algorithm in Eccleston and Street [5] begins by finding optimal row-column designs then within this class of designs searches for the design or designs with the minimum value of trace of MM^T, a measure related to the optimality of the designs for estimating residual effects. If there is more than one design selected at this stage, then they will choose the one or ones with the best A-optimal value for direct effects. Thus they need a lower bound on the A-optimal value and they give such a bound in the paper. Because $tr(MM^T)$ is minimised it is unlikely that the Eccleston-Street algorithm will find designs with entries in the M matrix that are very unequal.

The algorithm in Donev [3] finds A-optimal *COD*s for a variety of possible models, including the one that we discuss in this paper. He describes his algorithm as a 'block exchange algorithm' where a block is a sequence of treatments applied to a subject. The algorithm begins with a randomly chosen design and at each iteration the exchange of a sequence in the design for one not in the design is carried out if the exchange results in the maximum reduction in the A-optimal value. The iterations stop when no further improvement is possible. Donev recommends that a number of different starting designs be used to increase the likelihood of obtaining a design near to the A-optimality bound for the design size in question.

Donev's algorithm can construct designs with $p \leq 6$ and $n \leq 20$ although it also appears to require that $np \leq 70$. Hence we have compared the algorithms for the case where $t = 3, n = 12$ and $p = 5$. A sample design from each program, together with the A-optimal value for direct effects, is given in Table 4.4. The algorithm given here ran in about 1/1000 of the time and gave an optimal, rather than near-optimal, design.

The algorithm given by Eccleston and Whitaker [6] is based on nested simulated annealing using two objective functions. The designs that they obtain by this approach are often very good, although sometimes the algorithm performs poorly. The efficiencies that they quote range from 0.8943 to 0.4813 for direct effects and 0.9307 to 0.5238 for residual effects.

Thus our algorithm is preferable if a *GYD* exists for the situation for which a change-over design is required.

We have coded the GSA algorithm in ANSI FORTRAN 77. Copies of the source code are available from the first author `hcheng@fish.wa` `.gov.au`. Modifications that produce optimal *COD*s with additional

	Design												A-opt value
Donev	2	2	1	3	3	2	1	3	1	1	3	3	0.10503
	3	2	3	1	3	2	1	3	1	1	3	2	
	1	1	3	2	1	1	3	2	3	2	1	3	
	3	3	1	2	1	1	3	2	2	3	1	1	
	1	3	2	1	2	3	2	1	3	3	2	1	
GSA	2	2	3	1	1	1	1	3	3	2	3	2	0.10417
	1	2	3	1	2	1	3	3	1	2	2	3	
	1	3	1	2	2	3	2	1	2	1	3	3	
	3	3	1	2	3	3	2	2	1	1	2	1	
	2	1	2	3	3	2	1	2	3	3	1	1	

Table 4.4. Comparison of two $COD(3, 12, 5)$'s.

properties, such as circular CODs in which the last period is equal to the first period, are also available.

Acknowledgments

This work was partially supported by the Australian Research Council.

References

[1] Ash, A. (1981). Generalized Youden designs : Constructions and tables. *J. Statistical Planning and Inference* 5, 1–25.

[2] Cheng, C. S. and Wu, C. F. (1980). Balanced repeated measurements designs. *Annals of Statistics* 8, 1272–1283.

[3] Donev, A. N. (1997). An algorithm for the construction of crossover trials. *Applied Statistics* 46, 288–298.

[4] Duque-Anton, M. (1997). Constructing efficient simulated annealing algorithms. *Discrete Applied Mathematics* 77, 139–159.

[5] Eccleston, J. A. and Street, D. J.(1994). An algorithm for the construction of optimal or near optimal change-over designs. *Australian Journal of Statistics* 36, 371– 378.

[6] Eccleston, J. A. and Whitaker, D. (1999). On the design of optimal change-over experiments through multi-objective simulated annealing. *Statistics and Computing* 9, 37–42.

[7] Elliott, J. R. and Gibbons, P. B. (1992). The construction of sub-square free Latin squares by simulated annealing. *Australasian Journal of Combinatorics* **5**, 209–228.

[8] Federer, W.T. (1995). Personal communication.

[9] Jansen, J., Douven, R. C. M. H. and Van Berkum, E. E. M. (1992). An annealing algorithm for searching optimal block designs. *Biometrics* **34**, 529–538.

[10] John, J. A. and Whitaker D. (1993). Construction of resolvable row-column design using simulated annealing. *Australian Journal of Statistics* **35**, 237–245.

[11] Kiefer, J. (1975). Construction and optimality of generalized Youden designs. In *A Survey of Statistical Designs and Linear Models,* Edited by J. N. Srivastava. Amsterdam: North-Holland, 333–353.

[12] Kunert, J. (1983). Optimal designs and refinement of the linear model with applications to repeated measurement designs. *Annals of Statistics* **11**, 247–57.

[13] Meyer, R. K. and Nachitsheim, C. J. (1988). Constructing exact D-optimal experimental designs by simulated annealing. *American Journal of Mathematical and Management Sciences* **8**, 329–359.

[14] Venables, W. N. and Eccleston, J. A. (1994). Randomized search strategies for finding optimal or near optimal block and row-column designs. *Australian Journal of Statistics* **35**, 371–382.

Chapter 5

NEW LOWER BOUNDS ON THE MAXIMUM NUMBER OF MUTUALLY ORTHOGONAL STEINER TRIPLE SYSTEMS

J. H. Dinitz

Department of Mathematics, University of Vermont,
Burlington, VT USA 05401

P. Dukes

Department of Mathematics, California Institute of Technology,
Pasadena, CA USA 91125

Abstract   Two Steiner triple systems (STS) are orthogonal if their sets of triples are disjoint, and two disjoint pairs of points defining intersecting triples in one system fail to do so in the other. We define the quantity $\sigma(n)$ as the size of a maximum collection of pairwise orthogonal STS of order n. Special starters in the finite fields are used to improve the best known lower bounds on $\sigma(n)$ for prime-powers $n \equiv 1 \pmod 6$, $n < 500$. Additionally, hill-climbing and isomorphisms are used together to show $\sigma(n) \geq 3$ or 4 for certain other small n, including some orders $n \equiv 3 \pmod 6$. Asymptotic existence for three mutually orthogonal STS is a consequence.

1.   Introduction

A *Steiner triple system* (STS) of order n is a pair $(V, \mathcal{B})$, where V is an n-set of elements and $\mathcal{B}$ is a collection of 3-subsets (triples) of V such that every pair of elements in V is contained in a unique triple of $\mathcal{B}$.

The necessary numerical condition $n \equiv 1$ or 3 (mod 6) is well-known to be sufficient [13].

Two STS, $(V, \mathcal{B}_1)$ and $(V, \mathcal{B}_2)$, are *orthogonal* if
   (i) $\mathcal{B}_1 \cap \mathcal{B}_2 = \emptyset$, and
   (ii) for u, v, x, y distinct, $\{u, v, a\}, \{x, y, a\} \in \mathcal{B}_1$ and $\{u, v, w\}, \{x, y, z\}$
        $\in \mathcal{B}_2$ implies $w \neq z$.

Every STS defines a *third element function* $\Theta : \binom{V}{2} \rightarrow V$ given by $\Theta(\{u, v\}) = w$ if and only if $\{u, v, w\}$ is a triple. Two STS $(V, \mathcal{B}_1)$ and $(V, \mathcal{B}_2)$ with third element functions Θ_1 and Θ_2, respectively, are orthogonal if and only if for each $c \in V$, the list $(\Theta_2(\{u, v\}) | \Theta_1(\{u, v\}) = c)$ consists of distinct elements none of which equal c. This verification is called the *orthogonality certificate*.

It can be checked by hand that there exists a pair of orthogonal STS of order 7 but that there does not exist such a pair for orders 3 or 9. The reader is referred to [1] for a discussion of the history of orthogonal Steiner triple systems. We note here that they were first introduced by O'Shaughnessey [11] in 1968 as a means to finding Room squares. After much work the spectrum problem for orthogonal STS was completely solved [1] in 1994. The paper [7] gives a description of the algorithms used to find many small orders. We build upon these algorithms in this present work. Let $\sigma(n)$ denote the size of a maximum collection of mutually orthogonal Steiner triple systems of order n. We record the result in [1] for later reference.

Theorem 1 [1] $\sigma(n) \geq 2$ for all $n \equiv 1, 3$ (mod 6), $n \neq 3, 9$.

Several other results concerning $\sigma(n)$ are known. These are summarized in the following three theorems.

Theorem 2 [14] $\sigma(n) \geq 3$ for $n = 2^{6k\pm1} - 1$ and $\sigma(127) = 6$.

Theorem 3 [8] $\sigma(31) \geq 6$, $\sigma(43) \geq 4$, $\sigma(61) \geq 3$, $\sigma(67) \geq 6$, $\sigma(103) \geq 5$, $\sigma(109) \geq 3$, $\sigma(139) \geq 8$, $\sigma(151) \geq 12$, $\sigma(157) \geq 8$, and $\sigma(163) \geq 6$.

The results of this paper will improve upon many of the bounds in Theorem 3.

Theorem 4 [8] $\lim_{m \to \infty} \sigma(6m + 1) = \infty$.

An interesting question is to consider upper bounds on $\sigma(n)$. From the condition of disjoint block sets alone, there can be no more than $\binom{n}{3}/|\mathcal{B}_i| = n - 2$ pairwise orthogonal STS $(V, \mathcal{B}_i)$ of order n. But also the existence of k mutually orthogonal STS of order n implies k mutually orthogonal symmetric Latin squares of order n. Hence $\sigma(n) \leq (n-1)/2$ may be a better conjectured bound. More discussion on this upper bound can be found in [4].

In the next section, we merge the algebraic methods of Dinitz [2] and Gross [8] to improve the best known bounds on $\sigma(n)$ for $n \equiv 1 \pmod 6$ a prime-power less than 500. Later, hill-climbing is used to handle some small values (in particular, we will show that $\sigma(19) \geq 3$) and we will also exhibit the first known collections of three pairwise orthogonal STS of 3 (mod 6) order. We are quite pleased with these results as it was originally conjectured in [11] that there would never be orthogonal Steiner triple systems for any order congruent to 3 modulo 6.

2. Starters over Finite Fields

2.1 One parameter systems

In what follows, the sum (product) of two sets is defined to be the set of all sums (products) of elements by taking one from each set. Similarly, the sum (product) of an element with a set is the set of all sums (products) of that element with every member of the set. The following construction is essentially that of Gross [8].

Let K be a finite field of order $q = 6t + 1$, t odd (so $q \equiv 7 \pmod{12}$). Let $U \subset Q$ be subgroups of K^* of index 6 and 3, respectively. Choose $a \in K^* \backslash Q$ such that $a(a+1) \in Q$. Now $U\{\{0, a, a+1\}\}$ is a set of triples whose $6t$ total differences exhaust K^*. So, $(K, K + U\{\{0, a, a+1\}\})$ is a Steiner triple system of order q developed under K. This STS will be denoted as S_a.

Example: Let K be the field of order 19 (with generator 2). Then $U = \{1, 7, 11\}$ and $Q = U \cup \{8, 18, 12\}$. Then $a = 3$ satisfies the criterion above. The triples forming the base blocks of S_3 (an STS(19)) are thus $U\{\{0, 3, 4\}\}$ $\{\{0, 3, 4\}, 7\{0, 3, 4\}, 11\{0, 3, 4\}\} = \{\{0, 3, 4\}, \{0, 2, 9\}, \{0, 14, 6\}\}$.

Now consider two STS, say from S_a and S_b. A result in [8] states that these STS are orthogonal if ab^{-1} and $(a+1)(b+1)^{-1}$ are both in U. At this time, we note a generalization of this which, though not as elegant, offers a check of orthogonality easily implementable on a computer.

Lemma 5 S_a and S_b are orthogonal if and only if x, y, z are in distinct cosets of U in K^*, where $x = a - b$,

$$y = \begin{cases} ab^{-1} - 1, & \text{if } ab^{-1} \in U, \\ 1 - a - ab^{-1}, & \text{if } ab^{-1} \in -U, \\ (a+1)(b+1)b^{-1} - a, & \text{if } (a+1)b^{-1} \in U, \\ 1 - (a+1)(b+1)b^{-1}, & \text{if } (a+1)b^{-1} \in -U, \end{cases}$$

$$\text{and} \quad z = \begin{cases} 1 - (a+1)(b+1)^{-1}, & \text{if } (a+1)(b+1)^{-1} \in U, \\ 1 - (a+1)b(b+1)^{-1}, & \text{if } (a+1)(b+1)^{-1} \in -U, \\ ab(b+1)^{-1} - a - 1, & \text{if } a(b+1)^{-1} \in U, \\ -ab(b+1)^{-1} - 1, & \text{if } a(b+1)^{-1} \in -U. \end{cases}$$

Proof. First note that it is straightforward to show that y and z given above are indeed well defined. Let Θ_a and Θ_b be the third element functions for the two STS. Clearly, since $\{0, b, b+1\} \in S_b$,

$$\Theta_b(\{a, a+1\}) = a - b = x.$$

Suppose $ab^{-1}, (a+1)(b+1)^{-1} \in U$. A straightforward computation gives

$$\Theta_b(\{-1, -1-a\}) = \Theta_b(\{0, -a\}) - 1 = ab^{-1}\Theta_b(\{0, -b\}) - 1 = ab^{-1} - 1 = y,$$

$$\Theta_b(\{1, -a\}) = 1 + (a+1)(b+1)^{-1}\Theta_b(\{0, -(b+1)\}) = 1 - (a+1)(b+1)^{-1} = z.$$

The other 15 cases are similar. So the list $(\Theta_b(\{u, v\}) | \Theta_a(\{u, v\}) = 0)$ consists of distinct elements if and only if x, y, z are in distinct cosets of U in K^*. By construction, these STS are invariant under additive shifts, so it is sufficient to merely check one such list.   □

Remarks: If $ab^{-1}, (a+1)(b+1)^{-1} \in U$, it is clear that x, y, z as in the lemma are always in distinct cosets of U in K^*, hence the result in [8]. However, this lemma is more general. For example, in the field of order 79, S_3 and S_{49} are orthogonal by Lemma 5, but not by the test in [8].

When $(q - 1)/6$ is even, there are similar conditions for orthogonality (but with more cases). This will be treated in Section 2.2.

The following approach is now used to build sets of mutually orthogonal STS of order q.
  (1) Determine the set of all $a \in K^* \setminus Q$ such that $a(a + 1) \in Q$.
  (2) Use Lemma 5 to check orthogonality of each pair of resulting STS.

(3) Create a graph G of orthogonality between these systems and find a large clique.

Our results are summarized in Table 6 below. The column "primitive" lists the primitive polynomials used, (in the prime case $x - g$, where g is a generator of K^*). The next four columns list the number of systems considered, the total number of orthogonal pairs, the size of a largest clique in the orthogonality graph G (giving a lower bound for $\sigma(q)$), and the values of a forming this clique. We have listed only those orders $q \equiv 7 \pmod{12}$, $q < 500$, for which an improvement on the best known bound for $\sigma(q)$ is obtained. A table with bounds on $\sigma(n)$ for other n will be given in the conclusion.

2.2   Higher parameter systems

Here a richer collection of Steiner triple systems is introduced analogous to the 2^s quotient starters in [2]. Let K be a finite field of order $q = 3 \cdot 2^{s+1}t + 1$ with t odd and $s \geq 1$. As before, let $U \subset Q$ be subgroups of K^* of index $3 \cdot 2^{s+1}$ and 3, respectively. Suppose g is a generator for K^* and let $h = g^3$. Let $T = \{1, h, h^2, \ldots, h^{2^s-1}\}U$. Then T is a transversal of ± 1 (or half-set) in Q; in other words, T and $-T$ partition Q. The same conditions on an element a as before, namely that $a \in K^* \setminus Q$ and $a(a + 1) \in Q$, guarantee $T\{\{0, a, a + 1\}\}$ is a difference family over K generating an STS.

Now consider 2^s (not neccessarily distinct) elements $a_i \in K^* \setminus Q$, $0 \leq i < 2^s$, and suppose the set $\cup_i \pm h^i\{1, a_i, a_i + 1\}$ exhausts a set of coset representatives for U in K^*. Then $U\{h^i\{0, a_i, a_i + 1\}|0 \leq i < 2^s\}$ is also a difference family over K^* generating an STS. For $\mathbf{a} = (a_0, a_1, \ldots)$ we will write $S_{\mathbf{a}} = (K, K + U\{h^i\{0, a_i, a_i + 1\}|0 \leq i < 2^s\}$. The third element function for this STS will be denoted by $\Theta_{\mathbf{a}}$.

Example: Let K be the field of order 61 (with generator $g = 2$). Note $s = 1$ since $61 = 6 \cdot 2^1 \cdot 5 + 1$, and $h = 8$. Then $U = \langle h^4 \rangle = \{1, 9, 20, 58, 34\}$, $T = \{1, 8\}U$, and $Q = \pm T$. The field elements 5 and 6 belong to the coset $2Q$, while 35 and 36 belong to $4Q$. Furthermore, $\pm U\{5, 6\}$ and $\pm 8U\{35, 36\}$ exhaust the respective cosets of Q in K^* (with no repetition). So $\mathbf{a} = (a_0, a_1) = (5, 35)$ generates an STS. The 10 base blocks for $S_{\mathbf{a}}$ are

$$\{1, 9, 20, 58, 34\} \cdot \{\{0, 5, 6\}, 8\{0, 35, 36\}\}.$$

Table 6 *New bounds on $\sigma(q)$ for some $q \equiv 7 \pmod{12}$, with q a prime power and $q < 500$.*

q	primitive	systems	edges	$\sigma(G) \geq$	clique
67	$x - 2$	14	70	8	$38, 36, 37, 28, 12, 29, 54, 30$
79	$x - 3$	14	34	4	$3, 49, 59, 30$
103	$x - 5$	26	199	8	$57, 86, 58, 56, 15, 46, 6, 47$
151	$x - 6$	38	484	14	$88, 22, 47, 128, 32, 95, 55, 11, 62,$ $118, 105, 45, 103, 139$
163	$x - 2$	42	594	9	$2, 32, 151, 72, 87, 11, 113, 75, 107$
199	$x - 3$	42	495	9	$3, 46, 71, 14, 162, 72, 169, 53, 70$
211	$x - 2$	50	811	14	$180, 201, 204, 173, 189, 49, 196, 9,$ $154, 47, 14, 163, 176, 21$
223	$x - 3$	56	979	12	$85, 96, 185, 211, 152, 204, 92, 160,$ $78, 80, 19, 21$
271	$x - 6$	54	774	9	$36, 142, 204, 77, 108, 133, 179, 182, 21$
283	$x - 3$	56	1033	11	$185, 179, 22, 159, 248, 238, 165, 236,$ $11, 44, 193$
307	$x - 5$	72	1578	10	$25, 284, 286, 222, 293, 126, 276, 169,$ $65, 133$
331	$x - 3$	74	1774	11	$3, 81, 195, 109, 295, 255, 112, 222,$ $288, 169, 197$
343	$x^3 + 3x + 2$	72	1431	9	$2x^2 + 6x + 1, 2x^2 + 2x + 6,$ $x^2 + 3x + 1, 5x^2 + 3x + 3,$ $5x^2 + 6x + 5, x^2 + 4x + 3,$ $6x^2 + 5x + 5, 4x^2 + 5x, 6x^2$
367	$x - 6$	74	1666	11	$282, 42, 139, 223, 284, 153, 283, 165,$ $70, 83, 65$
379	$x - 2$	78	1881	11	$2, 46, 291, 278, 358, 17, 230, 31, 89,$ $152, 74$
439	$x - 15$	104	3514	14	$180, 169, 106, 160, 338, 212, 267, 38,$ $204, 325, 171, 409, 45, 334$
463	$x - 3$	98	3100	15	$3, 335, 217, 310, 263, 16, 258, 21, 195,$ $142, 369, 166, 441, 354, 372$
487	$x - 3$	114	4215	16	$3, 9, 239, 372, 350, 252, 109, 180, 155,$ $52, 68, 153, 466, 88, 192, 326$
499	$x - 7$	104	3616	17	$57, 397, 170, 43, 317, 455, 139, 407,$ $328, 178, 84, 320, 359, 34, 441, 91, 464$

In practice, it is sometimes the case that s is too large for a full consideration of 2^s parameters. Computations are simplified here by taking $|\{a_i : 0 \leq i < 2^s\}| = 1, 2$, or 4, with the a_i periodic in i. The resulting STS will be called a one, two, or four parameter system, respectively. For a fixed order, many more systems may result from considering two or four parameters than from just considering one parameter systems. However, the orthogonality test generalizing Lemma 5 is more stringent when $s > 0$.

Theorem 7 *Let K be a field of order $q = 3 \cdot 2^{s+1}t+1$ with t odd. For K^*, suppose g is a generator, $h = g^3$, and U is a subgroup of order t. Assume* **a** *and* **b** *generate the STS $S_\mathbf{a}$ and $S_\mathbf{b}$ in K^*. Define $\beta_j, \beta'_j \in \pm\{b_j, b_j+1\}$, $f(i), f'(i) \in \{0, 1, \dots, 2^s - 1\}$, and $u_i, u'_i \in U$ according to*

$$-a_i h^i = u_i h^{f(i)}\beta_{f(i)} \text{ and } -(a_i + 1)h^i = u'_i h^{f'(i)}\beta'_{f'(i)}.$$

For each $i \in \{0, \dots, 2^s - 1\}$, let

$$x_i = h^i(a_i - b_i), \ y_i = -h^i + u_i h^{f(i)}\theta(\beta_{f(i)}), \text{ and } z_i = h^i + u'_i h^{f'(i)}\theta(\beta'_{f'(i)}),$$

where $\theta(b_j) = b_j + 1$, $\theta(b_j + 1) = b_j$, $\theta(-b_j) = 1$, and $\theta(-b_j - 1) = -1$. Then $S_\mathbf{a}$ and $S_\mathbf{b}$ are orthogonal if and only if the $3 \cdot 2^s$ elements x_i, y_i, z_i are in distinct cosets of U in K^*.

Proof: Observe first that the notation introduced is well-defined because the base blocks of $S_\mathbf{b}$ form a difference family. Since the systems $S_\mathbf{a}$ and $S_\mathbf{b}$ are invariant under additive translation, they are orthogonal if and only if $(\Theta_\mathbf{b}(\{u, v\})|\Theta_\mathbf{a}(\{u, v\}) = 0)$ consists of distinct elements of K^*; that is, if

$$\Theta_\mathbf{b}(\{a_i h^i, (a_i + 1)h^i\}), \quad \Theta_\mathbf{b}(\{-(a_i + 1)h^i, -h^i\}), \quad \text{and}$$
$$\Theta_\mathbf{b}(\{-a_i h^i, h^i\}), \quad \quad \text{with } 0 \leq i < 2^s$$

are all in distinct cosets of U in K^*. Using the notation in the statement of the theorem, these may be computed:

$$\begin{aligned}
\Theta_\mathbf{b}(\{a_i h^i, (a_i + 1)h^i\}) &= a_i h^i + \Theta_\mathbf{b}(\{0, h^i\}) = a_i h^i - b_i h^i = x_i, \\
\Theta_\mathbf{b}(\{-(a_i + 1)h^i, -h^i\}) &= -h^i + \Theta_\mathbf{b}(\{0, -a_i h^i\}) \\
&= -h^i + u_i \Theta_\mathbf{b}(\{0, h^{f(i)}\beta_{f(i)}\}) \quad\quad (1) \\
&= -h^i + u_i h^{f(i)}\theta(\beta_{f(i)}) = y_i,
\end{aligned}$$

and similarly $\Theta_\mathbf{b}(\{-a_i h^i, h^i\}) = z_i$. □

Luckily, there is not total chaos among the elements x_i, y_i, z_i. Suppose that the parameter period is p, so that $a_i = a_{i+p}$ and similarly for the b_i, where the indices are mod 2^s. Then clearly $x_{i+p} = h^p x_i$ for all i. Now if for some i it turns out that $f(i) < 2^s - p$, then

$$
\begin{aligned}
u_{i+p} h^{f(i+p)} \beta_{f(i+p)} &= -a_{i+p} h^{i+p} = h^p(-a_i h^i) \\
&= h^p(u_i h^{f(i)} \beta_{f(i)}) = u_i h^{f(i)+p} \beta_{f(i)}.
\end{aligned}
$$

Thus $h^{f(i+p)}\beta_{f(i+p)}$ and $h^{f(i)+p}\beta_{f(i)}$ are in the same cosets of U in K^*. By periodicity of the b_j and the difference family condition, it must be that $f(i+p) = f(i) + p$ and $\beta_{f(i+p)} = \beta_{f(i)}$, which in turn implies that $u_{i+p} = u_i$. So $y_{i+p} = h^p y_i$, and similarly for the z_i. Note that when $f(i) \geq 2^s - p$, there is a small issue with β_i and β_i' switching sign. To summarize, the distinct coset test of the theorem is less stringent for larger periods p.

Example: For $q = 97 = 6 \cdot 2^4 + 1$, we generated all one, two, and four parameter STS S_a and automated Theorem 7 to build an orthogonality graph. The number of vertices, edges, and edge density of these graphs are presented in below. In each case, the maximum clique size found was 3. (In Lemma 16 we show $\sigma(97) \geq 4$.)

period	vertices	edges	density
1	26	47	1.45×10^{-1}
2	111	106	1.74×10^{-2}
4	1786	359	2.25×10^{-4}

Under the rough estimation that both the number of B parameter systems and edge density of the B parameter orthogonality graph are proportional to the Bth power of those quantities for one parameter systems, it seems reasonable to expect little improvement when considering higher parameter systems. After all, the expected clique size of a random graph with N vertices and edge probability p is $\log_{1/p} N$. However, it is certainly the case that individual orders may turn out to admit larger collections of mutually orthogonal STS when more parameters are considered.

Two or four parameters were attempted for all 1 (mod 12) prime power orders less than 500. For some of these orders, however, a largest clique was found simply among one parameter systems. Some small orders belong to this class but still larger mutually orthogonal collections are presented later. The results for the remaining orders in this class are presented in the following table. Field element a in the column "clique" represents that $S_{\mathbf{a}}$, where $a_i = a$ for all $i = 0, 1, \ldots, 2^s - 1$, is among the systems present in the maximum clique.

Table 8 *New bounds on $\sigma(q)$ using one parameter systems for some $q \equiv 1 \pmod{12}$, with q a prime power and $q < 500$.*

q	s	primitive	systems	edges	$\sigma(G) \geq$	clique
49	3	$x^2 + x + 3$	14	23	4	$2, 4, 5x + 5, 2x + 1$
61	1	$x - 2$	14	29	5	$4, 12, 47, 13, 46$
121	2	$x^2 + x + 7$	32	127	5	$x, 8x + 7, 4x + 7, 7x + 3, 3x + 10$
169	2	$x^2 + x + 2$	38	159	4	$12x + 11, 4x, 5x + 12, 6x + 5$
193	5	$x - 5$	38	69	3	$46, 47, 84$
241	3	$x - 7$	50	214	4	$12, 19, 221, 228$
289	4	$x^2 + x + 3$	72	299	3	$x, 13x + 4, 11x + 4$
313	2	$x - 10$	62	445	5	$119, 76, 101, 214, 62$
337	3	$x - 10$	74	403	4	$227, 128, 109, 208$
361	2	$x^2 + x + 2$	78	711	5	$7x + 10, 3x + 5, 2x + 16,$ $7x + 12, 3x + 17$
409	2	$x - 21$	98	1044	6	$206, 164, 96, 355, 17, 141$
433	3	$x - 5$	96	675	4	$25, 22, 49, 95$
457	2	$x - 13$	104	1220	6	$361, 323, 98, 133, 358, 149$

In the next table, we give improved bounds on $\sigma(q)$ when two parameters are considered. In all these cases $s = 1$. The column "clique" gives vectors **a** generating pairwise orthogonal STS. A singleton entry a in this column means $a_i = a$ for $i = 0, 1$.

Table 9 *New bounds on $\sigma(q)$ using two parameter systems for some $q \equiv 1 \pmod{12}$, with q a prime power and $q < 500$.*

q	primitive	systems	edges	$\sigma(G) \geq$	clique
109	$x - 6$	120	1433	6	79, (30, 6), (20, 88), (58, 39), (14, 94), (84, 24)
181	$x - 2$	359	13968	7	57, (148, 119), (147, 43), (165, 15), (20, 164), (80, 37), (17, 163)
229	$x - 6$	616	42981	7	6, 182, 134, 69, 166, (94, 99), (133, 110)
277	$x - 5$	616	44659	8	116, 160, (78, 71), (100, 55), (10, 219), (248, 85), (209, 115), (136, 35)
349	$x - 2$	1233	187283	9	122, 226, (141, 12), (189, 188), (93, 191), (23, 243), (164, 2), (50, 219), (55, 325)
397	$x - 5$	1776	359971	10	307, 202, 339, 57, 194, 74, 89, 322, (338, 58), (308, 88)
421	$x - 2$	1827	391517	10	400, 277, 20, 192, (300, 134), (191, 239), (382, 38), (326, 280), (398, 380), (237, 57)

One surprisingly nice construction using four parameter systems is now given.

Proposition 10 *There exist four mutually orthogonal STS of order 25.*

Proof: Let $K = \mathbf{Z}_5[x]/(x^2 + x + 2)$, with generator $g = x$ and $h = g^3 = 4x + 2$. Note $s = 2$ for $q = 25$. The four systems with parameter lists $(4x + 4, x, x, x), (2x + 3, 2x + 3, 3x + 1, 3x + 1), (3x, 3x, 3x, 3x)$ and $(2x + 2, 2x + 2, 2x + 2, 2x + 2)$ are claimed to be pairwise orthogonal. The four sets of base blocks are given below.

I : $\{\{0, 4x + 4, 4x\}\} \cup \{h, h^2, h^3\}\{\{0, x, x + 1\}\}$,

II : $\{1, h\}\{\{0, 2x + 3, 2x + 4\}\} \cup \{h^2, h^3\}\{\{0, 3x + 1, 3x + 2\}\}$,

III : $\{1, h, h^2, h^3\}\{\{0, 3x, 3x + 1\}\}$, and

IV : $\{1, h, h^2, h^3\}\{\{0, 2x + 2, 2x + 3\}\}$.

We will check that the first two sets of starter blocks S_a and S_b (I and II above) form a difference family over K and that the resulting STS are orthogonal. The pairs occurring with 0 in S_a, their third elements in S_b, and the corresponding pairs occurring with 0 in S_b are given in the table below. Note the left and right columns exhaust K^* (the difference family condition) and the middle column consists of distinct elements of K^* (the orthogonality condition). Analysis of the other pairs of systems is similar. □

$\{u, v\} \in \Theta_{\mathbf{a}}^{-1}(0)$	$\Theta_{\mathbf{b}}(\{u, v\})$	$w + \{u, v\} \in \Theta_{\mathbf{b}}^{-1}(0)$
$4x + 4, 4x$	$2x + 1$	$2x + 3, 2x + 4$
$1, x + 1$	$2x + 2$	$3x + 4, 4x + 4$
$4x + 1, 3x + 4$	$4x + 3$	$3, 4x + 1$
$3x + 2, 2x + 4$	2	$3x, 2x + 2$
$3x + 1, x + 3$	$3x + 2$	$4, 3x + 1$
$x, 4$	$4x + 4$	$2x + 1, x$
$2x, 2x + 2$	$x + 3$	$x + 2, x + 4$
$3x + 3, 3$	$2x$	$x + 3, 3x + 3$
$4x + 2, 2x + 3$	$4x + 1$	$1, 3x + 2$
$x + 4, 4x + 3$	3	$x + 1, 4x$
$x + 2, 2x + 1$	$3x + 4$	$4x + 3, 2$
$2, 3x$	x	$4x + 2, 2x$

Remark: The first two STS can be described more compactly as

$$\{1, h^5, h^6, h^7\}\{\{0, 4x + 4, 4x\}\}, \text{ and } \{1, h, h^5, h^6\}\{\{0, 2x + 3, 2x + 4\}\}.$$

In this alternative description, the a_i and b_i are constant (period one), but the transversals used deviate from the usual $\{1, h, h^2, h^3\}$.

Skew-orthogonal STS

It was shown in [11] that pairs of orthogonal STS correspond to Room squares. For this Room square to be skew, [5], the following additional condition must hold.

(iii) $\{u, v, a\}, \{x, y, w\} \in \mathcal{B}_1$ and $\{u, v, z\}, \{x, y, a\} \in \mathcal{B}_2$ implies $w \neq z$. (Note $u = x$ is possible).

Two orthogonal STS $(V, \mathcal{B}_1)$ and $(V, \mathcal{B}_2)$ satisfying (iii) are *skew-orthog-onal*. The algebraic methods introduced so far can be easily modi-

fied to handle this stronger relationship. It is an easy consequence of
the definitions that two abelian group generated STS with third element functions Θ_1 and Θ_2 are skew-orthogonal if and only if the set
$(\pm\Theta_2(\{u,v\})|\Theta_1(\{u,v\}) = 0)$ exhausts the nonzero group elements. In
other words, the orthogonality certificate must form a *half-set* in the
group. Theorem 7 (and similarly Lemma 5) may be modified for skew-
orthogonality by stipulating that all the elements $\pm x_i, \pm y_i, \pm z_i$ lie in
distinct cosets of U. Of course, it is equally easy to automate this test
on computer and find cliques in the resulting skew-orthogonal subgraphs
of the original graphs. It is not our intention to conduct a thorough analysis here. However, it should be reported that we often found cliques of
size at least three. It is known that skew-orthogonal pairs of STS arise
from Mullin-Nemeth starters in the finite fields. In the table below, a
few lower bounds for the maximum number of pairwise skew-orthogonal
STS of order q, denoted $\sigma'(q)$, which improve upon this are noted.

q	31	43	61	67	79	103	109	127	139	151	157	163
$\sigma'(q) \geq$	3	3	3	4	3	5	3	4	4	5	3	6

3.   Automorphisms and Hill-climbing

In this section we will describe the use of hill-climbing techniques to
find sets of orthogonal Steiner triple systems.

3.1   Small 3 (mod 6) orders

An *automorphism* of an STS $(V, \mathcal{B})$ is a function $f : V \to V$ such
that $f\mathcal{B} = \mathcal{B}$. If such an f (viewed as a permutation) consists of three
cycles of equal length $n/3$, the STS is called *3-cyclic*. Such a system is
determined completely by $(n-1)/2$ *base triples*, or orbit representatives
for f. See [1] or [7] for more on automorphisms of triple systems. We
now outline a construction for 3 mutually orthogonal 3-cyclic STS of
order $n \equiv 3 \pmod 6$. Take the pointset $V = \mathbf{Z}_{n/3} \times \mathbf{Z}_3$, with generating
automorphism $x_i \mapsto (x + 1)_i$ and define the map $\alpha : x_i \mapsto x_{i+1}$. As
usual, subscripts represent the second coordinate in the product. A
standard hill-climb (like the algorithm in [7]) is used to construct a
set of $(n-1)/2$ base triples, which when developed form a set $\mathcal{B}$ of
blocks so that $(V, \mathcal{B})$ is orthogonal to $(V, \alpha\mathcal{B})$. A randomly chosen base
block $B = \{x, y, z\}$ either augments a partial STS or replaces a block

already covering one of the pairs in B, provided the new system causes no conflict with either orthogonality condition (i) or (ii). Call the third element relation in the partial design Θ. Condition (i) simply amounts to checking $\alpha^{\pm 1}x \neq \Theta(\{\alpha^{\pm 1}y, \alpha^{\pm 1}z\})$. Verifying (ii) is where the majority of the computing time lies, since it involves a loop over all points c and the tests (when Θ is defined)

$$\Theta(\alpha x, \alpha y) \neq \Theta(\alpha c, \alpha \Theta(c, z)),$$

as well as for α^{-1} and the other rearrangements of x, y, z.

Orthogonality of two STS is clearly invariant under a common map. So these base triples actually produce three pairwise orthogonal STS: (V, B), $(V, \alpha B)$, and $(V, \alpha^2 B)$. We have had success with this method for $27 \leq n \leq 81$, with computation times ranging between seconds and many days. The time before finding a solution depends partly on a threshold parameter which restarts the hill-climb after repeated failure to augment the partial design. This parameter was set roughly proportional to the total number of base blocks possible for the given order. Computing time also varies according to the optimization in compiling and the platform used. An exhaustive search of all 3-cyclic STS of order 21 revealed no design with the desired condition.

Lemma 11 $\sigma(n) \geq 3$ *for* $n \in \{27, 33, 39, 45, 51, 57, 63, 69, 75, 81\}$.

Proof: Three pairwise orthogonal STS of order n are presented below by specifying the $(n-1)/2$ base triples for one system. The other systems are obtained by shifting indices. $\square$

$n = 27$: $\{0_0, 2_1, 8_2\}$, $\{0_1, 1_0, 2_0\}$, $\{0_0, 2_2, 7_2\}$, $\{0_0, 3_1, 4_2\}$, $\{0_2, 6_1, 7_1\}$, $\{0_1, 0_0, 4_0\}$, $\{0_1, 3_1, 5_1\}$, $\{0_2, 6_0, 8_0\}$, $\{0_0, 0_2, 6_2\}$, $\{0_0, 6_1, 5_2\}$, $\{0_1, 4_2, 5_2\}$, $\{0_1, 5_0, 8_0\}$, $\{0_1, 0_2, 7_2\}$

$n = 33$: $\{0_0, 10_1, 7_2\}$, $\{0_1, 4_0, 8_0\}$, $\{0_2, 3_0, 9_0\}$, $\{0_1, 9_2, 10_2\}$, $\{0_1, 0_0, 2_0\}$, $\{5_1, 8_1, 9_1\}$, $\{0_0, 8_1, 10_2\}$, $\{0_0, 2_1, 4_1\}$, $\{0_2, 0_0, 8_0\}$, $\{0_1, 5_0, 6_0\}$, $\{0_0, 1_1, 6_2\}$, $\{0_2, 0_1, 5_1\}$, $\{0_1, 4_2, 7_2\}$, $\{0_0, 4_2, 9_2\}$, $\{0_1, 1_2, 3_2\}$, $\{0_0, 1_2, 5_2\}$

$n = 39$: $\{0_0, 0_2, 4_2\}$, $\{0_0, 3_1, 11_1\}$, $\{0_2, 3_0, 10_0\}$, $\{0_2, 5_0, 8_0\}$, $\{0_1, 7_0, 11_0\}$, $\{0_2, 0_1, 2_1\}$, $\{0_2, 1_1, 4_1\}$, $\{0_0, 10_1, 12_2\}$, $\{5_2, 10_2, 11_2\}$, $\{0_1, 3_2, 6_2\}$, $\{0_0, 0_1, 9_1\}$, $\{0_0, 12_1, 7_2\}$, $\{0_2, 7_0, 12_0\}$, $\{0_1, 8_0, 9_0\}$, $\{0_0, 7_1, 8_1\}$, $\{0_0, 1_1, 2_2\}$, $\{0_1, 5_2, 7_2\}$, $\{0_2, 2_0, 4_0\}$, $\{0_2, 3_1, 9_1\}$

$n = 45$: $\{0_2, 0_1, 4_1\}$, $\{0_0, 4_2, 14_2\}$, $\{0_0, 2_2, 3_2\}$, $\{0_0, 2_1, 9_1\}$, $\{0_0, 6_1, 12_1\}$, $\{0_2, 11_1, 13_1\}$, $\{0_0, 7_1, 8_2\}$, $\{0_0, 3_1, 4_1\}$, $\{0_2, 5_1, 8_1\}$, $\{0_2, 2_0, 3_0\}$, $\{0_2, 0_0, 10_0\}$,

$\{0_0, 13_1, 11_2\}$, $\{0_1, 1_0, 4_0\}$, $\{0_0, 6_2, 10_2\}$, $\{0_0, 1_1, 7_2\}$, $\{0_0, 4_0, 13_0\}$,
$\{0_1, 9_2, 12_2\}$, $\{0_0, 5_1, 10_1\}$, $\{0_1, 0_0, 7_0\}$, $\{0_1, 8_2, 14_2\}$, $\{0_0, 1_2, 9_2\}$, $\{0_1, 3_2, 5_2\}$

$n = 51$: $\{0_0, 14_1, 3_2\}$, $\{0_0, 6_1, 12_1\}$, $\{0_0, 2_1, 1_2\}$, $\{0_2, 2_1, 10_1\}$, $\{0_1, 1_0, 8_0\}$,
$\{1_0, 4_0, 12_0\}$, $\{0_1, 11_2, 13_2\}$, $\{0_2, 3_1, 8_1\}$, $\{0_0, 10_1, 12_2\}$, $\{0_0, 7_2, 15_2\}$,
$\{0_0, 0_1, 15_1\}$, $\{0_1, 4_0, 16_0\}$, $\{0_0, 0_2, 4_2\}$, $\{0_0, 5_1, 8_1\}$, $\{0_1, 13_0, 14_0\}$, $\{0_2, 4_0, 8_0\}$,
$\{0_2, 9_1, 13_1\}$, $\{0_0, 11_1, 16_2\}$, $\{0_0, 5_2, 10_2\}$, $\{0_0, 7_1, 2_2\}$, $\{0_2, 0_1, 16_1\}$,
$\{3_2, 4_2, 10_2\}$, $\{0_2, 9_0, 11_0\}$, $\{0_0, 11_2, 14_2\}$, $\{0_2, 7_1, 14_1\}$

$n = 57$: $\{2_2, 5_2, 12_2\}$, $\{0_1, 5_0, 10_0\}$, $\{0_0, 1_1, 2_2\}$, $\{0_1, 0_0, 16_0\}$, $\{0_1, 4_0, 11_0\}$,
$\{0_2, 13_0, 15_0\}$, $\{0_0, 4_1, 16_2\}$, $\{0_2, 15_1, 17_1\}$, $\{0_0, 11_2, 15_2\}$, $\{10_1, 11_1, 17_1\}$,
$\{0_1, 5_2, 18_2\}$, $\{2_1, 6_1, 11_1\}$, $\{0_0, 2_1, 18_1\}$, $\{0_0, 10_1, 7_2\}$, $\{0_1, 6_2, 14_2\}$,
$\{0_1, 15_2, 17_2\}$, $\{0_1, 2_0, 13_0\}$, $\{0_0, 7_1, 17_2\}$, $\{0_0, 16_1, 10_2\}$, $\{0_2, 10_0, 11_0\}$,
$\{8_0, 12_0, 18_0\}$, $\{0_0, 13_1, 3_2\}$, $\{0_0, 12_1, 1_2\}$, $\{0_0, 0_2, 14_2\}$, $\{0_0, 12_2, 13_2\}$,
$\{0_0, 11_1, 18_2\}$, $\{0_0, 5_1, 5_2\}$, $\{0_2, 8_1, 16_1\}$

$n = 63$: $\{0_1, 4_1, 9_1\}$, $\{0_0, 1_2, 3_2\}$, $\{0_0, 10_1, 13_2\}$, $\{0_0, 1_1, 12_1\}$, $\{0_0, 17_1, 10_2\}$,
$\{0_0, 19_1, 20_1\}$, $\{0_2, 1_1, 14_1\}$, $\{0_0, 7_2, 8_2\}$, $\{0_1, 8_0, 14_0\}$, $\{0_0, 11_2, 20_2\}$,
$\{0_2, 0_0, 17_0\}$, $\{0_0, 15_1, 17_2\}$, $\{0_0, 14_1, 15_2\}$, $\{0_2, 0_1, 2_1\}$, $\{0_0, 2_0, 14_0\}$,
$\{0_0, 11_1, 5_2\}$, $\{0_0, 6_2, 16_2\}$, $\{0_2, 9_0, 12_0\}$, $\{0_0, 9_1, 14_2\}$, $\{0_1, 11_2, 16_2\}$,
$\{0_0, 2_1, 19_2\}$, $\{0_1, 10_2, 13_2\}$, $\{0_1, 5_0, 18_0\}$, $\{0_1, 16_0, 17_0\}$, $\{0_2, 3_0, 19_0\}$,
$\{0_2, 12_1, 15_1\}$, $\{0_1, 8_2, 12_2\}$, $\{0_2, 3_1, 17_1\}$, $\{0_0, 0_1, 6_1\}$, $\{0_2, 7_2, 13_2\}$,
$\{0_1, 3_0, 13_0\}$

$n = 69$: $\{0_1, 11_2, 18_2\}$, $\{3_0, 9_0, 11_0\}$, $\{0_0, 2_1, 16_1\}$, $\{0_0, 19_1, 20_1\}$, $\{0_0, 4_2, 10_2\}$,
$\{0_0, 1_1, 22_2\}$, $\{0_0, 12_2, 21_2\}$, $\{0_0, 18_1, 14_2\}$, $\{0_1, 2_0, 15_0\}$, $\{0_0, 13_1, 20_2\}$,
$\{0_1, 6_0, 17_0\}$, $\{0_2, 6_0, 7_0\}$, $\{0_2, 0_1, 21_1\}$, $\{0_0, 5_1, 6_2\}$, $\{0_2, 5_0, 10_0\}$, $\{0_0, 9_1, 22_1\}$,
$\{0_0, 4_1, 10_1\}$, $\{0_2, 3_1, 7_1\}$, $\{0_2, 10_1, 17_1\}$, $\{0_1, 8_2, 9_2\}$, $\{0_0, 11_1, 14_1\}$,
$\{1_2, 9_2, 13_2\}$, $\{0_0, 2_2, 7_2\}$, $\{0_2, 0_0, 20_0\}$, $\{0_2, 13_1, 18_1\}$, $\{0_2, 8_1, 19_1\}$,
$\{0_2, 4_0, 18_0\}$, $\{0_1, 16_0, 20_0\}$, $\{0_0, 9_2, 11_2\}$, $\{0_1, 14_2, 17_2\}$, $\{0_1, 12_2, 22_2\}$,
$\{0_0, 12_1, 15_2\}$, $\{0_0, 0_1, 15_1\}$, $\{0_2, 15_0, 22_0\}$

$n = 75$: $\{0_0, 20_1, 24_2\}$, $\{2_0, 9_0, 14_0\}$, $\{0_2, 19_1, 23_1\}$, $\{0_2, 9_1, 20_1\}$, $\{0_0, 23_1, 9_2\}$,
$\{0_2, 11_0, 14_0\}$, $\{0_0, 8_2, 18_2\}$, $\{0_1, 1_0, 20_0\}$, $\{0_1, 4_0, 19_0\}$, $\{0_0, 9_1, 15_1\}$,
$\{0_0, 7_1, 2_2\}$, $\{0_0, 6_2, 20_2\}$, $\{0_1, 12_0, 14_0\}$, $\{0_1, 13_0, 17_0\}$, $\{0_1, 6_0, 7_0\}$,
$\{0_1, 14_2, 19_2\}$, $\{0_1, 0_0, 9_0\}$, $\{0_1, 10_2, 22_2\}$, $\{6_1, 21_1, 23_1\}$, $\{0_0, 1_2, 23_2\}$,
$\{0_0, 4_1, 22_1\}$, $\{0_2, 0_0, 8_0\}$, $\{0_0, 17_1, 15_2\}$, $\{0_0, 7_2, 13_2\}$, $\{0_0, 3_2, 5_2\}$,
$\{0_2, 4_0, 15_0\}$, $\{0_0, 2_1, 3_1\}$, $\{0_0, 14_1, 22_2\}$, $\{0_2, 1_1, 10_1\}$, $\{0_0, 12_2, 16_2\}$,
$\{0_0, 1_1, 4_2\}$, $\{0_1, 1_2, 17_2\}$, $\{0_2, 4_1, 7_1\}$, $\{0_2, 0_1, 12_1\}$, $\{0_0, 10_1, 19_2\}$,
$\{0_2, 13_1, 18_1\}$, $\{5_2, 22_2, 23_2\}$

$n = 81$: $\{0_1, 4_2, 20_2\}$, $\{0_0, 4_1, 23_1\}$, $\{0_0, 22_1, 24_2\}$, $\{0_2, 12_0, 26_0\}$, $\{0_0, 19_2, 20_2\}$,
$\{0_2, 2_1, 22_1\}$, $\{0_0, 5_1, 10_1\}$, $\{0_2, 14_1, 24_1\}$, $\{0_0, 18_1, 19_1\}$, $\{0_0, 4_2, 6_2\}$,
$\{0_0, 3_1, 10_2\}$, $\{0_2, 1_0, 9_0\}$, $\{0_0, 2_2, 25_2\}$, $\{0_2, 0_0, 18_0\}$, $\{0_0, 1_1, 17_2\}$,
$\{0_1, 8_2, 11_2\}$, $\{0_2, 8_2, 15_2\}$, $\{0_2, 4_0, 15_0\}$, $\{0_0, 13_1, 15_1\}$, $\{0_0, 12_1, 24_1\}$,

$\{0_0, 8_1, 21_1\}, \{0_1, 0_0, 20_0\}, \{0_0, 1_0, 3_0\}, \{0_0, 9_1, 25_1\}, \{0_0, 12_0, 17_0\},$
$\{0_1, 9_2, 19_2\}, \{0_2, 9_1, 13_1\}, \{0_0, 16_1, 13_2\}, \{0_2, 16_0, 20_0\}, \{0_0, 6_1, 5_2\},$
$\{0_2, 17_1, 26_1\}, \{0_0, 3_2, 21_2\}, \{0_1, 10_0, 16_0\}, \{0_0, 14_1, 14_2\}, \{0_1, 15_2, 21_2\},$
$\{0_0, 8_2, 22_2\}, \{0_0, 2_1, 26_1\}, \{0_0, 20_1, 16_2\}, \{0_1, 17_2, 22_2\}, \{0_2, 15_1, 21_1\}$

Note that the examples above are the first for values of $n \equiv 3 \pmod 6$ where $\sigma(n) \geq 3$. One easy consequence of Theorem 4, Lemma 11, and the PBD-closure of the existence of k mutually orthogonal STS is now given.

Theorem 12 *There exists N such that $\sigma(n) \geq 3$ for all $n \equiv 1, 3 \pmod 6$, $n \geq N$.*

A construction of arbitrarily large sets of mutually orthogonal STS of order 3 (mod 6) is not known at this time. However, the following seems very reasonable.

Conjecture 13 *For $n \equiv 3 \pmod 6$, $\lim_{n \to \infty} \sigma(n) = \infty$.*

3.2   Multiplicative images of cyclic STS

An STS of order n with an automorphism consisting of a single cycle of length n is called *cyclic*. When $n \equiv 1 \pmod 6$, the design is determined by $(n-1)/6$ base triples. The cyclic group $V = \mathbf{Z}_n$ with generating automorphism $x \mapsto x + 1$ will be used for the points of a cyclic STS here. The base triples can then be chosen as *zero-sum* triples, by an appropriate additive shift. Let $\mu : V \to V$ be multiplication by $m \in \mathbf{Z}_n^*$, and define μ to act on each block of $\mathcal{B}$ elementwise. Suppose the multiplicative order of m in $\mathbf{Z}_n^*$ is t. It is of interest when $\mu^i \mathcal{B}, 0 \leq i < t$, form block sets for t pairwise orthogonal STS of order n. The case when $m = -1$ (and thus $t = 2$) was studied by Schrieber in [12]. The resulting objects are called *opposite orthogonal* STS. There is a simple characterization of when a cyclic STS $(V, \mathcal{B})$ is orthogonal to $(V, -\mathcal{B})$.

Proposition 14 [12] *A cyclic STS $(V, \mathcal{B})$ is opposite orthogonal if and only if the zero-sum triples of $\mathcal{B}$ have among them no repeated elements of $V \setminus \{0\}$.*

This fact can be exploited in the search for several pairwise orthogonal STS in many ways. The most fruitful method we found was to first hill-climb to an opposite orthogonal STS $(V, \mathcal{B})$, and then backtrack or hill-climb to another cyclic STS subject to it being orthogonal to both $(V, \mathcal{B})$ and $(V, -\mathcal{B})$. This produces a set of three pairwise orthogonal STS. We have had success at this method for $31 \leq n \leq 205$. The small examples were found in seconds, but the largest ones took a few days on a parallel computer. In general, when the number of base triples is similar, this algorithm appears to be comparable in efficiency to the algorithm for 3 (mod 6) orders presented earlier.

Lemma 15 $\sigma(n) \geq 3$ *for all* $n \in \{37, 55, 91, 115, 133, 145, 175\}$.

Proof: The three systems are given for each order by specifying the $(n-1)/6$ zero-sum base triples for an opposite orthogonal STS, followed by $(n-1)/6$ base triples for a cyclic STS orthogonal to both the first system and its negative. Note in each case the first collection of triples are disjoint. □

$n = 37$: $\{13, 26, 35\}, \{10, 12, 15\}, \{2, 14, 21\}, \{18, 24, 32\}, \{4, 8, 25\}, \{33, 34, 7\}$;
    $\{8, 12, 17\}, \{2, 10, 25\}, \{23, 25, 26\}, \{2, 8, 27\}, \{14, 25, 35\}, \{11, 28, 35\}$

$n = 55$: $\{29, 11, 15\}, \{39, 9, 7\}, \{12, 22, 21\}, \{33, 36, 41\}, \{49, 34, 27\}, \{44, 16, 50\}$,
    $\{52, 23, 35\}, \{54, 38, 18\}, \{30, 19, 6\}$;
    $\{20, 18, 17\}, \{38, 16, 1\}, \{40, 13, 2\}, \{43, 36, 31\}, \{46, 37, 27\}, \{46, 42, 22\}$,
    $\{48, 35, 27\}, \{52, 36, 22\}, \{54, 31, 25\}$

$n = 91$: $\{11, 15, 65\}, \{8, 22, 61\}, \{64, 67, 51\}, \{72, 7, 12\}, \{66, 5, 20\}, \{3, 50, 38\}$,
    $\{41, 1, 49\}, \{87, 81, 14\}, \{73, 63, 46\}, \{19, 78, 85\}, \{10, 52, 29\}, \{23, 44, 24\}$,
    $\{30, 2, 59\}, \{35, 90, 57\}, \{28, 37, 26\}$;
    $\{1, 18, 30\}, \{8, 24, 79\}, \{83, 85, 41\}, \{6, 82, 17\}, \{21, 3, 49\}, \{90, 30, 67\}$,
    $\{67, 58, 62\}, \{27, 17, 41\}, \{25, 26, 18\}, \{50, 69, 11\}, \{48, 18, 21\}, \{55, 89, 23\}$,
    $\{7, 45, 85\}, \{5, 26, 74\}, \{68, 18, 12\}$

$n = 115$: $\{2, 64, 49\}, \{100, 21, 109\}, \{93, 86, 51\}, \{16, 40, 59\}, \{72, 75, 83\}$,
    $\{10, 99, 6\}, \{31, 52, 32\}, \{50, 107, 73\}, \{42, 17, 56\}, \{45, 62, 8\}, \{112, 81, 37\}$,
    $\{102, 92, 36\}, \{103, 29, 98\}, \{96, 68, 66\}, \{87, 5, 23\}, \{63, 34, 18\}, \{79, 12, 24\}$,
    $\{13, 76, 26\}, \{53, 15, 47\}$;
    $\{41, 29, 18\}, \{2, 57, 89\}, \{108, 34, 113\}, \{13, 28, 19\}, \{4, 54, 58\}, \{75, 99, 18\}$,
    $\{60, 39, 105\}, \{109, 110, 2\}, \{91, 14, 62\}, \{30, 93, 50\}, \{93, 63, 90\}, \{16, 2, 58\}$,
    $\{50, 89, 67\}, \{66, 13, 26\}, \{9, 78, 60\}, \{55, 39, 29\}, \{13, 32, 57\}, \{21, 101, 103\}$,
    $\{64, 101, 33\}$

$n = 133$ $\{116, 109, 41\}, \{92, 79, 95\}, \{114, 89, 63\}, \{74, 112, 80\}, \{24, 19, 90\}$,
    $\{77, 3, 53\}, \{98, 10, 25\}, \{33, 111, 122\}, \{17, 103, 13\}, \{47, 75, 11\}, \{14, 35, 84\}$,
    $\{91, 45, 130\}, \{76, 105, 85\}, \{121, 81, 64\}, \{88, 78, 100\}, \{55, 22, 56\}$,
    $\{38, 129, 99\}, \{118, 65, 83\}, \{120, 18, 128\}, \{127, 31, 108\}, \{72, 70, 124\}$,
    $\{26, 67, 40\}$;

$\{20, 108, 114\}$, $\{9, 62, 110\}$, $\{50, 26, 91\}$, $\{24, 120, 15\}$, $\{103, 69, 83\}$,
$\{87, 95, 43\}$, $\{108, 103, 85\}$, $\{88, 109, 2\}$, $\{19, 34, 30\}$, $\{129, 80, 55\}$,
$\{46, 122, 92\}$, $\{95, 31, 66\}$, $\{54, 116, 16\}$, $\{24, 87, 8\}$, $\{120, 59, 69\}$, $\{76, 18, 54\}$,
$\{84, 126, 6\}$, $\{50, 69, 38\}$, $\{55, 98, 105\}$, $\{72, 75, 73\}$, $\{17, 94, 34\}$, $\{107, 80, 14\}$

$n = 145$: $\{115, 44, 131\}$, $\{59, 94, 137\}$, $\{23, 18, 104\}$, $\{144, 129, 17\}$, $\{11, 61, 73\}$,
$\{47, 60, 38\}$, $\{134, 15, 141\}$, $\{2, 16, 127\}$, $\{53, 50, 42\}$, $\{32, 126, 132\}$,
$\{138, 31, 121\}$, $\{13, 43, 89\}$, $\{83, 4, 58\}$, $\{1, 54, 90\}$, $\{66, 110, 114\}$,
$\{88, 67, 135\}$, $\{68, 8, 69\}$, $\{20, 77, 48\}$, $\{102, 37, 6\}$, $\{140, 116, 34\}$,
$\{124, 84, 82\}$, $\{139, 64, 87\}$, $\{81, 91, 118\}$, $\{107, 3, 35\}$;
$\{137, 38, 45\}$, $\{91, 127, 17\}$, $\{86, 143, 58\}$, $\{126, 44, 65\}$, $\{62, 76, 50\}$,
$\{115, 53, 40\}$, $\{124, 106, 87\}$, $\{5, 46, 84\}$, $\{67, 125, 36\}$, $\{48, 3, 19\}$, $\{97, 38, 93\}$,
$\{8, 105, 14\}$, $\{122, 26, 137\}$, $\{46, 111, 122\}$, $\{126, 3, 53\}$, $\{57, 24, 125\}$,
$\{74, 113, 49\}$, $\{41, 81, 14\}$, $\{30, 25, 28\}$, $\{81, 72, 82\}$, $\{4, 12, 36\}$, $\{66, 117, 14\}$,
$\{132, 34, 4\}$, $\{140, 120, 97\}$

$n = 175$: $\{116, 150, 84\}$, $\{90, 159, 101\}$, $\{169, 88, 93\}$, $\{72, 113, 165\}$,
$\{148, 58, 144\}$, $\{89, 132, 129\}$, $\{85, 55, 35\}$, $\{76, 59, 40\}$, $\{57, 173, 120\}$,
$\{168, 161, 21\}$, $\{125, 17, 33\}$, $\{138, 171, 41\}$, $\{3, 64, 108\}$, $\{94, 100, 156\}$,
$\{152, 4, 19\}$, $\{149, 139, 62\}$, $\{63, 75, 37\}$, $\{114, 7, 54\}$, $\{31, 122, 22\}$,
$\{141, 117, 92\}$, $\{124, 45, 6\}$, $\{109, 95, 146\}$, $\{155, 11, 9\}$, $\{56, 136, 158\}$,
$\{47, 68, 60\}$, $\{172, 52, 126\}$, $\{153, 43, 154\}$, $\{50, 98, 27\}$, $\{163, 42, 145\}$;
$\{76, 97, 14\}$, $\{147, 74, 161\}$, $\{90, 132, 5\}$, $\{70, 111, 104\}$, $\{23, 62, 173\}$,
$\{88, 78, 66\}$, $\{35, 130, 48\}$, $\{110, 148, 156\}$, $\{125, 44, 101\}$, $\{149, 34, 79\}$,
$\{79, 56, 85\}$, $\{137, 163, 107\}$, $\{132, 25, 121\}$, $\{168, 2, 101\}$, $\{43, 149, 39\}$,
$\{87, 115, 26\}$, $\{124, 73, 174\}$, $\{117, 74, 42\}$, $\{107, 70, 125\}$, $\{87, 35, 34\}$,
$\{171, 80, 113\}$, $\{30, 77, 108\}$, $\{76, 56, 92\}$, $\{126, 107, 161\}$, $\{38, 104, 87\}$,
$\{4, 139, 67\}$, $\{24, 68, 83\}$, $\{114, 111, 109\}$, $\{85, 58, 162\}$

$\{97, 153, 124\}$, $\{9, 16, 162\}$, $\{40, 41, 106\}$, $\{63, 57, 67\}$,
$\{81, 121, 172\}$, $\{163, 22, 2\}$, $\{69, 155, 150\}$, $\{3, 91, 93\}$,
$\{29, 47, 111\}$, $\{5, 103, 79\}$, $\{101, 8, 78\}$, $\{7, 21, 159\}$, $\{161, 186, 27\}$,
$\{167, 84, 123\}$, $\{0, 39, 132\}$, $\{90, 148, 21\}$, $\{121, 75, 5\}$, $\{10, 48, 90\}$,
$\{139, 74, 88\}$, $\{119, 43, 36\}$, $\{118, 166, 54\}$, $\{159, 44, 175\}$,
$\{105, 146, 113\}$, $\{125, 165, 103\}$, $\{27, 109, 183\}$, $\{100, 81, 135\}$,
$\{93, 68, 178\}$ $\{12, 146, 47\}$ $\{200, 136, 74\}$, $\{129, 62, 14\}$
$\{134, 56, 15\}$, $\{198, 77, 135\}$ $\{29, 116, 60\}$, $\{125, 36, 44\}$,
$\{104, 133, 173\}$ $\{159, 153, 98\}$, $\{10, 143, 52\}$, $\{117, 149, 144\}$
$\{53, 83, 69\}$, $\{180, 103, 127\}$ $\{99, 177, 109\}$, $\{167, 27, 173\}$
$\{102, 137, 153\}$, $\{76, 129, 6\}$, $\{105, 31, 38\}$ $\{143, 134, 37\}$,
$\{193, 97, 21\}$, $\{29, 57, 109\}$ $\{47, 10, 81\}$, $\{79, 102, 41\}$, $\{130, 98, 90\}$
$\{170, 201, 11\}$, $\{42, 98, 38\}$ $\{134, 189, 61\}$, $\{194, 78, 147\}$,

Another method for finding three pairwise orthogonal STS is available when $3 \mid \phi(n)$. In this case, there is a primitive third root of unity $m \in \mathbf{Z}_n$.

Hill-climbing on a cyclic STS subject to orthogonality to its m and m^2 multiplicative shifts has worked for small n as well.

Using a square root of -1 and hill-climbing to find four orthogonal STS in two opposite orthogonal pairs has also worked, but produced only minor improvements to our earlier work. The examples below each took on the order of a few hours due to the large number of restarts needed.

Lemma 16 $\sigma(n) \geq 4$ for $n \in \{73, 85, 97\}$.

Proof: For each of these orders, we give a subgroup H of order 4 in $\mathbf{Z}_n^*$, followed by a list of $(n-1)/6$ base triples for which the cyclic systems generated under multiplication by H are pairwise orthogonal. Note in each case the collection of triples is disjoint. □

$n = 73$: $H = \{\pm 1, \pm 27\}$
  $\{9, 68, 69\}$, $\{33, 41, 72\}$, $\{17, 63, 66\}$, $\{14, 20, 39\}$, $\{18, 25, 30\}$, $\{21, 54, 71\}$,
  $\{36, 45, 65\}$, $\{12, 23, 38\}$, $\{22, 57, 67\}$, $\{5, 26, 42\}$, $\{40, 44, 62\}$, $\{28, 58, 60\}$

$n = 85$: $H = \{\pm 1, \pm 13\}$
  $\{14, 15, 56\}$, $\{43, 49, 78\}$, $\{11, 76, 83\}$, $\{32, 68, 70\}$, $\{18, 22, 45\}$, $\{33, 58, 79\}$,
  $\{36, 50, 84\}$, $\{7, 31, 47\}$, $\{19, 29, 37\}$, $\{41, 60, 69\}$, $\{3, 25, 57\}$, $\{16, 27, 42\}$,
  $\{9, 12, 64\}$, $\{21, 26, 38\}$

$n = 97$: $H = \{\pm 1, \pm 22\}$
  $\{3, 31, 63\}$, $\{13, 87, 94\}$, $\{41, 76, 77\}$, $\{9, 29, 59\}$, $\{1, 25, 71\}$, $\{12, 20, 65\}$,
  $\{47, 69, 78\}$, $\{7, 24, 66\}$, $\{21, 36, 40\}$, $\{57, 67, 70\}$, $\{10, 89, 95\}$, $\{5, 39, 53\}$,
  $\{22, 27, 48\}$, $\{2, 35, 60\}$, $\{23, 80, 91\}$, $\{28, 82, 84\}$

3.3 Rotational automorphisms for $n = 19$ and 21

An STS of order $n = kt + 1$ with an automorphism consisting of a fixed point and t cycles of equal length on the remaining points is called *t-rotational*. The pointset $(\mathbf{Z}_k \times \mathbf{Z}_t) \cup \{\infty\}$ is typically used, with the second coordinate in the product represented with subscripts. The automorphism is then $(\infty) \prod_{0 \leq i < t} (0_i \ 1_i \ \ldots \ (k-1)_i)$. As with cyclic STS, a set of base blocks is enough to specify the design, and it is often relatively fast to hill-climb to a t-rotational STS that is orthogonal to some orbit shift of itself. For $n = 19$ and 21, these orders fail to admit sets of more than two pairwise orthogonal STS by the earlier methods. So a range of techniques were attempted using 6-rotational and 4-rotational STS, respectively. We were successful in showing $\sigma(19) \geq 3$

by first hill-climbing to a 6-rotational STS that is orthogonal to a shift of its orbits, and then hill-climbing again to an orthogonal mate of both. This worked surprisingly fast but several different initial pairs had to be tried before a mate to both was found.

Proposition 17 *There exist three pairwise orthogonal STS of order* 19.

Proof: Below are the 19 base blocks for a 6-rotational STS orthogonal to its orbit shift defined by $x_i \mapsto x_{i+3}$, followed by base blocks for another 6-rotational STS orthogonal to each of these. Label the resulting designs IA, IB, and II. The orthogonality certificates follow in a table according to this labeling. □

$$\{0_5, 1_2, 2_1\}, \{\infty, 0_1, 2_0\}, \{0_2, 1_2, 1_0\}, \{\infty, 0_4, 2_2\}, \{0_0, 1_0, 0_5\}, \{0_4, 0_2, 2_1\}, \{0_1, 1_1, 0_5\},$$
$$\{0_5, 2_3, 2_0\}, \{0_4, 1_4, 2_0\}, \{0_4, 0_1, 0_0\}, \{0_5, 1_3, 2_2\}, \{0_5, 2_4, 0_2\}, \{\infty, 0_5, 0_3\},$$
$$\{0_3, 1_3, 2_4\}, \{0_3, 2_2, 2_1\}, \{0_4, 0_3, 1_1\}, \{0_3, 0_2, 2_0\}, \{0_3, 0_1, 1_0\}, \{0_5, 1_5, 1_4\}$$

$$\{0_5, 1_2, 1_1\}, \{0_5, 0_4, 2_2\}, \{0_5, 0_3, 1_0\}, \{0_3, 0_2, 1_1\}, \{0_2, 1_2, 2_0\}, \{0_1, 1_1, 1_5\}, \{\infty, 0_5, 0_2\},$$
$$\{0_5, 1_5, 2_3\}, \{0_3, 1_2, 0_1\}, \{0_3, 2_2, 2_0\}, \{0_4, 1_4, 1_2\}, \{0_4, 2_1, 1_0\}, \{0_5, 1_4, 0_0\},$$
$$\{0_4, 2_3, 1_1\}, \{\infty, 0_4, 0_1\}, \{0_5, 2_4, 2_0\}, \{0_0, 1_0, 0_1\}, \{\infty, 0_3, 0_0\}, \{0_3, 1_3, 0_4\}$$

c	$\Theta_{IB}\Theta_{IA}^{-1}(c)$	$\Theta_{IA}\Theta_{II}^{-1}(c)$	$\Theta_{IB}\Theta_{II}^{-1}(c)$
0_0	$1_3, 0_5, 0_3, 1_0, 2_3, 2_4, 0_1, 0_4, 1_2$	$\infty, 2_0, 0_1, 2_2, 1_5, 1_2, 0_3, 2_4, 0_5$	$1_5, 2_1, 0_3, 1_0, 2_2, 1_1, 2_0, 0_4, 1_4$
0_1	$2_2, 0_0, 2_4, 1_3, 0_3, 2_1, 1_5, 1_4, \infty$	$1_1, 1_0, 2_1, 1_2, 2_5, 0_4, 0_3, 2_2, 0_5$	$1_0, 0_3, 0_4, 0_2, 0_0, 1_5, \infty, 2_3, 2_1$
0_2	$0_3, 2_2, 2_5, 0_1, 2_1, 2_4, 1_5, 2_0, \infty$	$2_1, 0_5, 0_4, 2_2, 2_3, 0_3, 0_0, 1_2, 1_0$	$\infty, 2_0, 2_1, 0_5, 1_3, 1_1, 2_4, 0_4, 2_2$
0_3	$1_1, 0_4, 2_2, 1_2, 1_5, 2_3, 2_0, 1_0, \infty$	$0_0, 2_5, 2_4, 1_4, 1_2, 1_3, 1_1, 2_3, 0_1$	$0_4, 1_2, 2_3, 1_5, \infty, 2_5, 0_2, 0_0, 2_1$
0_4	$2_0, 0_5, 1_4, 0_1, 2_1, 1_5, 1_1, 1_2, 0_3$	$1_3, 0_1, 0_5, \infty, 0_0, 1_2, 2_0, 0_3, 2_4$	$1_4, 2_4, 1_5, 0_0, 1_3, 0_1, 2_5, 2_2, 0_2$
0_5	$1_4, 2_2, 1_0, 0_2, 1_5, 2_1, 2_3, 2_0, 1_2$	$2_3, \infty, 0_1, 2_5, 1_4, 0_2, 1_0, 2_4, 2_1$	$0_2, 2_4, 1_5, 1_3, 0_0, 2_5, 2_0, 0_1, 0_3$
∞	$0_0, 1_0, 2_0, 0_4, 1_4, 2_4, 0_5, 1_5, 2_5$	$0_0, 1_0, 2_0, 0_4, 1_4, 2_4, 0_5, 1_5, 2_5$	$0_3, 1_3, 2_3, 0_1, 1_1, 2_1, 0_2, 1_2, 2_2$

For $n = 21$, no success resulted from this same method using 4-rotational STS and an order two map on the orbits generating two of the three systems. Hill-climbing to a single STS orthogonal to each of its images under an order three shift of its orbits ((1)(234), say) has also produced no results. The frequency of partial designs encountered missing one base block seems to indicate no triple of pairwise orthogonal STS of order 21 exists having these additional properties.

4. Conclusion

In the first part of this paper we used finite fields to construct a rich collection of Steiner triple systems and then tested these for pairwise orthogonality. Using this method we found new lower bounds for $\sigma(q)$

n	$\sigma \geq$	Ref.	n	$\sigma \geq$	Ref.	n	$\sigma \geq$	Ref.	n	$\sigma \geq$	Ref.	n	$\sigma \geq$	Ref.
1	∞	--	45	3	11	91	3	15	181	7	9	343	9	6
3	1	--	49	4	8	97	4	16	193	3	8	349	9	9
7	2	--	51	3	11	103	8	6	199	9	6	361	5	8
9	1	[10]	55	3	15	109	6	9	211	14	6	367	11	6
13	2	[6]	57	3	11	115	3	15	223	12	6	373	9	9
15	2	[6]	61	5	8	121	5	8	229	7	9	379	11	6
19	3	17	63	3	11	127	6	[14]	241	4	8	397	10	9
21	2	[7]	67	8	6	133	3	15	271	9	6	409	6	8
25	4	10	69	3	11	139	8	[8]	277	8	9	421	10	9
27	3	11	73	4	16	145	3	15	283	11	6	433	4	8
31	6	[8]	75	3	11	151	14	6	289	3	8	439	14	6
33	3	11	79	4	6	157	8	[8]	307	10	6	457	6	8
37	3	15	81	3	11	163	9	6	313	5	8	463	15	6
39	3	11	85	4	16	169	4	8	331	11	6	487	16	6
43	4	[8]	87	2	[7]	175	3	15	337	4	8	499	17	6

for many prime powers $q < 500$. In the second part we employed several different variants of the hill-climbing algorithm for Steiner triple systems to get sets of orthogonal triple systems (in general) for non-prime power orders.

It is certainly the case that computing limitations, rather than nonexistence results, are the main obstacle in finding larger sets of mutually orthogonal STS. Without a doubt, higher quotient starters and hill-climbing can find still better bounds; however, there are memory constraints in storing orthogonailty graphs, and time constraints for probabilistic searching.

We conclude with a table of the best known lower bounds on $\sigma(n)$ for $n < 500$, with references (theorems, lemmas or tables in this paper or the original source). Entries in bold are exact. In the first two columns, all orders less than 90 are given. In the third column, only orders 1 (mod 6) are given. In the fourth and fifth columns, only prime-powers are listed.

Acknowledgments

The second author is grateful for the support of his NSERC PGS B award. The longer hill-climb searches were carried out on machines at the Center for Advanced Computing Research at the California Institute of Technology.

References

[1] C. J. Colbourn, P. B. Gibbons, R. Mathon, R. C. Mullin, and A. Rosa, *The spectrum of orthogonal Steiner triple systems*, Canad. J. Math. **46(2)** (1994), 239–252.

[2] J. H. Dinitz, *Room n-cubes of low order*, J. Austral. Math. Soc. (A) **36** (1984), 237–252.

[3] J. H. Dinitz, "Starters" in *The CRC Handbook of Combinatorial Designs*, (C. J. Colbourn and J. H. Dinitz, eds.) CRC Press, Inc., 1996, 467–473.

[4] J. H. Dinitz and D. R. Stinson, "Room squares and related designs," in *Contemporary Design Theory: A Collection of Surveys* (J. H. Dinitz and D. R. Stinson, eds) John Wiley & Sons, New York, 1992, 137–204.

[5] P. Dukes and E. Mendelsohn, *Skew-orthogonal Steiner triple systems*, J. Combin. Des. **7** (1999), 431-440.

[6] P. B. Gibbons, *A census of orthogonal Steiner triple systems of order 15*, Ann. Discrete Math. **26** (1985), 165–182.

[7] P. B. Gibbons and R. A. Mathon, *The use of hill-climbing to construct orthogonal Steiner triple systems*, J. Combin. Des. **1** (1993), 27–50.

[8] K. B. Gross, *On the maximal number of pairwise orthogonal Steiner triple systems*, J. Combin. Theory (A) **19** (1975), 256–263.

[9] R. C. Mullin and E. Nemeth, *On furnishing Room squares*, J. Combin. Theory **7** (1969), 266–272.

[10] R. C. Mullin and E. Nemeth, *On the nonexistence of orthogonal Steiner triple systems of order 9*, Canad. Math. Bull. **13** (1970), 131–134.

[11] C. D. O'Shaughnessey, *A Room design of order 14*, Canad. Math. Bull. **11** (1968), 191–194.

[12] S. Schreiber, *Cyclical Steiner triple systems orthogonal to their opposites*, Discrete Math. **77** (1989), 281–284.

[13] W. D. Wallis, *Combinatorial Designs*, Dekker **118**, New York, 1988.

[14] L. Zhu, *A construction for orthogonal Steiner triple systems*, Ars Combin. **9** (1980), 253–262.

Chapter 6

ON MINIMAL DEFINING SETS IN $AG(D, 3)^*$

Diane Donovan

Abdollah Khodkar

Anne Penfold Street

Centre for Discrete Mathematics and Computing
Department of Mathematics
The University of Queensland, 4072, Australia

Abstract   Many distinct objects may exemplify a particular combinatorial structure, for example, a block design or latin square. When studying such
objects with the same parameters, two questions arise naturally:

• Given two such objects, where and how do they differ?

• What portion of a particular object identifies it uniquely?

Here we consider triple systems and latin squares. The first question
leads to the ideas of a *trade* in a triple system and of a *latin trade* in a
latin square. The second question leads to the ideas of a *defining set*, in
particular a *minimal* defining set, in a triple system and of a *uniquely
completable set*, in particular a *critical* set, in a latin square. We study
the relationship between latin squares and triple systems, especially that
between the trades and defining sets of the triple system and the latin
trades and critical sets of the square.

We apply these ideas and construct new families of minimal defining
sets for triple systems associated with $AG(d, 3)$.

*Research supported by Australian Research Council Grants A49937047, A49802044

1.   Background

In [7] Gower identified sets of $2d$ hyperplanes in $AG(d, 3)$, $d \geq 2$, the lines of which uniquely determine the incidence structure of the affine geometry. In this paper we go beyond the specific examples given by Gower, and determine general conditions which, when satisfied, identify sets of hyperplanes for which the associated lines define the incidence structure. The Main Theorem of this paper is:

Theorem 1 *Let* **H** *be a set of hyperplanes of* $AG(d, 3)$, $d \geq 2$, *whose equations are:*

$$
\begin{cases}
a_{11}x_{d-1} & + & a_{12}x_{d-2} & + & \dots & + & a_{1d-1}x_1 & + & a_{1d}x_0 & = & b_1 \\
a_{21}x_{d-1} & + & a_{22}x_{d-2} & + & \dots & + & a_{2d-1}x_1 & + & a_{2d}x_0 & = & b_2 \\
a_{31}x_{d-1} & + & a_{32}x_{d-2} & + & \dots & + & a_{3d-1}x_1 & + & a_{3d}x_0 & = & b_3 \\
 & & & & \dots \\
 & & & & \dots \\
 & & & & \dots \\
a_{m1}x_{d-1} & + & a_{m2}x_{d-2} & + & \dots & + & a_{md-1}x_1 & + & a_{md}x_0 & = & b_m.
\end{cases}
$$

Let $\mathcal{H}$ *be the collection of blocks which lie within the hyperplanes of* **H**. *Let* P *be the partial Steiner latin square corresponding to* $\mathcal{H}$. *Suppose that:*

(1) *there is one point of* $AG(d, 3)$ *which is not incident with any hyperplane of* **H***;*

(2) P *is a uniquely completable set of order* 3^d*;*

(3) *for each block* $\{\bar{i}, \bar{j}, \bar{k}\}$ *of* $\mathcal{H}$ *there is a trade* T *of type one in* $AG(d, 3)$ *such that* $T \cap \mathcal{H} = \{\{\bar{i}, \bar{j}, \bar{k}\}\}$.

Let $\mathbf{H}^*$ *be the extension of* **H** *to an analogous set of hyperplanes in* $AG(d+1, 3)$, *and let* $\mathcal{H}^*$ *be the collection of blocks within the hyperplanes of* $\mathbf{H}^*$. *Finally, let* P^* *be the partial Steiner latin square corresponding to* $\mathcal{H}^*$. *Then:*

(1') *there is one point of* $AG(d + 1, 3)$ *which is not incident with any hyperplane of* $\mathbf{H}^*$*;*

(2') P^* *is a uniquely completable set of order* 3^{d+1}*;*

(3') *for each block* $\{\bar{i}, \bar{j}, \bar{k}\}$ *of* $\mathcal{H}^*$ *there is a trade* T^* *of type one in* $AG(d + 1, 3)$ *such that* $T^* \cap \mathcal{H}^* = \{\{\bar{i}, \bar{j}, \bar{k}\}\}$.

We then develop examples which show that this theory can be applied to determine new families of defining sets. The key to the determination of the results is the representation of the lines of $AG(d, 3)$ as blocks of a Steiner triple system (STS) or entries of an associated latin square. Through this approach the theory of direct products of latin squares is used to recursively develop sets of hyperplanes of $AG(d + 1, 3)$ from well chosen sets of hyperplanes of $AG(d, 3)$. In this way it is relatively straightforward to prove that the corresponding lines (blocks) form a minimal defining set in $AG(d + 1, 3)$ (the Steiner triple system).

The examples of minimal defining sets in the Steiner triple system are interesting in their own right. Let $\mathcal{M}$ be the collection of all minimal defining sets of a design D and define the spectrum to be $spec(D) = \{m | M$ is a minimal defining set of D and $|M| = m\}$. We say that the spectrum contains a *hole* if there exist $\ell < m < n$ such that $\ell, n \in spec(D)$, but $m \notin spec(D)$. Little is known about the spectra of designs or even whether there exist spectra with holes. As far as the authors are aware, the five Steiner triple systems set out in Table 1.1 are the only ones for which the spectra of minimal defining sets are fully known. However Ramsay [15] contains many partial results for the other 79 Steiner triple systems on 15 points. The results presented in this paper shed new light on the spectrum of minimal defining sets of the Steiner triple system corresponding to $AG(d, 3)$ and document new techniques which may be adapted to construct defining sets for general designs.

v	Type of STS	$spec(D)$	Reference
7		$\{3\}$	[9]
9		$\{4, 5\}$	[8]
13	cyclic	$\{9, 10, 11, 12, 13\}$	[12],[11]
	noncyclic	$\{8, 9, 10, 11, 12, 13\}$	[12],[11]
15	$PG(3, 2)$	$\{16, 17, 18, 19, 20, 21, 22\}$	[15]

Table 1.1 Spectra for minimal defining sets in some STSs of small order

2.   Introduction

We start with basic material which allows us to develop our Main Theorem (Theorem 1). In this section definitions and results are illustrated with carefully chosen examples. These examples will be used extensively in Sections 3, 4 and 5.

Definition 1 Let $\mathcal{F}$ be the Galois field of order q, denoted by $GF(q)$. An *affine d-dimensional space over* $\mathcal{F}$ is denoted by $AG(d, q)$ and defined as follows:

- points are vectors of $\mathcal{F}^d$;

- lines are cosets of 1-dimensional subspaces of $\mathcal{F}^d$;

- planes are cosets of 2-dimensional subspaces of $\mathcal{F}^d$;

- m-flats are cosets of m-dimensional subspaces of $\mathcal{F}^d$;

- hyperplanes are cosets of $(d-1)$-dimensional subspaces of $\mathcal{F}^d$.

Two m-flats are said to be *parallel* if they are both cosets of the same subspace, and the set of all the cosets of a particular subspace is known collectively as a *parallel class*.

Definition 2 Let $\bar{x} = (x_{d-1}, x_{d-2}, x_{d-3}, \ldots, x_1, x_0)$ be a point in $AG(d, 3)$. The integer

$$x = 3^{d-1}x_{d-1} + 3^{d-2}x_{d-2} + 3^{d-3}x_{d-3} + \ldots + 3x_1 + x_0$$

is called the *integer representation* of $\bar{x}$ and the vector $\bar{x}$ is called the *vector representation* of x.

Moreover, for $b \in \{0, 1, 2\}$, we define $b\bar{x}$ to be the point

$$(b, x_{d-1}, x_{d-2}, x_{d-3}, \ldots, x_1, x_0)$$

in $AG(d+1, 3)$. In this paper we sometimes assume that the points of $AG(d, 3)$ are integers x for $0 \leq x \leq 3^d - 1$ and the lines (blocks) of $AG(d, 3)$ are of the form $\{x, y, z\}$, where $0 \leq x, y, z \leq 3^d - 1$.

Note that there is a one-to-one correspondence between the integers x and the points of $AG(d, 3)$, where $0 \leq x \leq 3^d - 1$.

Example 2 When $d = 2$ the correspondence between vector and integer representations is that given in the following table.

Vector representations	Integer representations
$\bar{0} = (0,0)$	$0.3^1 + 0.3^0 = 0$
$\bar{1} = (0,1)$	$0.3^1 + 1.3^0 = 1$
$\bar{2} = (0,2)$	$0.3^1 + 2.3^0 = 2$
$\bar{3} = (1,0)$	$1.3^1 + 0.3^0 = 3$
$\bar{4} = (1,1)$	$1.3^1 + 1.3^0 = 4$
$\bar{5} = (1,2)$	$1.3^1 + 2.3^0 = 5$
$\bar{6} = (2,0)$	$2.3^1 + 0.3^0 = 6$
$\bar{7} = (2,1)$	$2.3^1 + 1.3^0 = 7$
$\bar{8} = (2,2)$	$2.3^1 + 2.3^0 = 8$

Since the vector space has nine vectors, the affine space has nine points. In the vector space, the equation $1x_1 + 0x_0 = 0$ defines a 1-dimensional subspace which contains the points with position vectors $(0,0), (0,1)$ and $(0,2)$. The corresponding line in the affine space is incident with the points 0, 1 and 2. The other subspaces in the vector space are determined by the equations $0x_1 + 1x_0 = 0$, $1x_1 + 2x_0 = 0$ and $1x_1 + 1x_0 = 0$ respectively, and the corresponding lines are incident with the points $\{0,3,6\}$, $\{0,4,8\}$, $\{0,5,7\}$.

These four subspaces have two cosets each. For example, the subspace with equation $1x_1 + 0x_0 = 0$ has cosets $1x_1 + 0x_0 = 1$ and $1x_1 + 0x_0 = 2$ which correspond to the lines incident with the points $\{3,4,5\}$ and $\{6,7,8\}$ respectively. Altogether there are 12 lines.

Definition 3 Let m be a positive integer and let $0 \leq a_{ij} \leq 2$ and $0 \leq b_i \leq 2$ for all $1 \leq i,j \leq m$. Let **H** be a set of m hyperplanes of $AG(d,3)$ whose equations are:

$$
\left\{
\begin{array}{ccccccccc}
a_{11}x_{d-1} & + & a_{12}x_{d-2} & + & \cdots & + & a_{1d-1}x_1 & + & a_{1d}x_0 & = & b_1 \\
a_{21}x_{d-1} & + & a_{22}x_{d-2} & + & \cdots & + & a_{2d-1}x_1 & + & a_{2d}x_0 & = & b_2 \\
a_{31}x_{d-1} & + & a_{32}x_{d-2} & + & \cdots & + & a_{3d-1}x_1 & + & a_{3d}x_0 & = & b_3 \\
& & & & \cdots & & & & & & \\
& & & & \cdots & & & & & & \\
& & & & \cdots & & & & & & \\
a_{m1}x_{d-1} & + & a_{m2}x_{d-2} & + & \cdots & + & a_{md-1}x_1 & + & a_{md}x_0 & = & b_m.
\end{array}
\right.
$$

By the *extension of* **H**, denoted **H***, we mean the following set of $m + 2$ hyperplanes of $AG(d + 1, 3)$.

$$
\begin{cases}
1x_d + & 0x_{d-1} + & 0x_{d-2} + \cdots + & 0x_1 + & 0x_0 = 0 \\
1x_d + & 0x_{d-1} + & 0x_{d-2} + \cdots + & 0x_1 + & 0x_0 = 1 \\
0x_d + & a_{11}x_{d-1} + & a_{12}x_{d-2} + \cdots + & a_{1d-1}x_1 + & a_{1d}x_0 = b_1 \\
0x_d + & a_{21}x_{d-1} + & a_{22}x_{d-2} + \cdots + & a_{2d-1}x_1 + & a_{2d}x_0 = b_2 \\
0x_d + & a_{31}x_{d-1} + & a_{32}x_{d-2} + \cdots + & a_{3d-1}x_1 + & a_{3d}x_0 = b_3 \\
& & \cdots & & \\
& & \cdots & & \\
& & \cdots & & \\
0x_d + & a_{m1}x_{d-1} + & a_{m2}x_{d-2} + \cdots + & a_{md-1}x_1 + & a_{md}x_0 = b_m.
\end{cases}
$$

Definition 4 A *triple system* is a pair $(V, \mathcal{B})$ where V is a v-set and $\mathcal{B}$ is a collection of b 3-subsets of V (*blocks* or *triples*) such that each element of V is contained in precisely r blocks and each 2-subset of V is contained in precisely λ blocks. The numbers v, b, r and λ are the *parameters* of the triple system and simple counting arguments show that $r = \lambda(v - 1)/2$ and $b = vr/3$. If $\lambda = 1$ the triple system is called a *Steiner* triple system $(STS(v))$ which has $r = (v - 1)/2$ and $b = v(v - 1)/6$.

A *partial* Steiner triple system is a pair $(V, \mathcal{B}')$ where $\mathcal{B}'$ is a collection of 3-subsets (blocks) of V such that each 2-subset of V is contained in at most one block.

Example 3 Let V and $\mathcal{B}$ respectively be the set of points and the set of lines of $AG(d, 3)$. Then the pair $(V, \mathcal{B})$ forms a Steiner triple system on 3^d points, denoted by $STS_A(3^d)$. In particular, if $d = 2$, then the set of points of the $STS(3^2)$ is $V = \{0, 1, 2, 3, 4, 5, 6, 7, 8\}$ and the set of blocks $\mathcal{B}$ are those of the following table, where each column shows the blocks of one parallel class.

$$
\begin{array}{cccc}
\{0, 1, 2\} & \{0, 3, 6\} & \{0, 4, 8\} & \{0, 5, 7\} \\
\{3, 4, 5\} & \{1, 4, 7\} & \{1, 5, 6\} & \{1, 3, 8\} \\
\{6, 7, 8\} & \{2, 5, 8\} & \{2, 3, 7\} & \{2, 4, 6\}
\end{array}
$$

Note that although every $STS(9)$ is isomorphic to $STS_A(3^2)$, there are in general many $STS(3^d)$ not isomorphic to $STS_A(3^d)$.

We now consider the idea of a defining set, introduced by Gray [9] in the more general context of balanced incomplete block designs. We need it only for Steiner triple systems.

Definition 5 A set of blocks which is a subset of a unique $STS(v)$ is said to be a *defining set* of that Steiner triple system. A *minimal* defining set is a defining set, no proper subset of which is a defining set. A *smallest* defining set is a defining set such that no other defining set has smaller cardinality.

Since $\mathcal{B}$ itself forms a defining set of the $STS(v)$ $(V, \mathcal{B})$, every Steiner triple system has a defining set. Since every defining set can be reduced, by deletion of blocks, to at least one minimal defining set, every Steiner triple system has at least one minimal defining set.

Example 4 In the $STS_A(3^2)$ of Example 3, the following set $\mathcal{S}$ of six blocks is a defining set.

$$\begin{array}{ccc} \{0,1,2\} & \{0,3,6\} & \{0,4,8\} \\ \{3,4,5\} & \{1,4,7\} & \\ & & \{1,3,8\} \end{array}$$

Its subsets $\mathcal{S}_5 = \mathcal{S} \setminus \{\{1,4,7\}\}$ and $\mathcal{S}_4 = \mathcal{S} \setminus \{\{0,4,8\}, \{1,3,8\}\}$ are minimal defining sets; $\mathcal{S}_4$ is a smallest defining set.

On the other hand, the following set of six blocks, $\mathcal{N}$,

$$\begin{array}{cccc} \{0,1,2\} & & & \\ \{3,4,5\} & & & \\ \{6,7,8\} & \{2,5,8\} & \{2,3,7\} & \{2,4,6\} \end{array}$$

is not a defining set since it can be completed to an $STS(9)$ by adjoining either the six remaining blocks of $\mathcal{B}$, or the six blocks of $\mathcal{F}$, given below.

$$\begin{array}{ccc} \{0,4,7\} & \{0,5,6\} & \{0,3,8\} \\ \{1,3,6\} & \{1,4,8\} & \{1,5,7\} \end{array}$$

Similarly, the following set $\mathcal{M}$ of six blocks, five of which lie in the set $\mathcal{B}$, is not a defining set of an $STS(9)$ since it cannot be completed to an $STS(9)$ at all.

$$\begin{array}{cccc} \{0,1,2\} & \{0,3,6\} & \{0,4,8\} & \{0,5,7\} \\ \{3,4,5\} & \{1,7,8\} & & \end{array}$$

Lemma 5 (Gray [9]) *Any defining set of an $STS(v)$ has at least $v - 1$ elements of V occurring in its blocks.*

Example 6 In the defining sets of $STS_A(3^2)$ given in Example 4, the blocks of S contain all elements of V, those of S_4 contain all except 8, and those of S_5 all except 7.

Definition 6 Let T_1 and T_2 be two collections of m 3-sets of elements of the v-set V. If each 2-set of elements of V occurs in the triples of T_1 with precisely the same multiplicity that it occurs in the triples of T_2, then the collections T_1 and T_2 are said to be *mutually balanced*. If for a collection of 3-sets T_1, there exists a collection of 3-sets T_2 such that T_1 and T_2 are mutually balanced and have no common triple, they are disjoint and we say T_1 is a *Steiner trade* of *volume* m. If the triples of T_1 are triples of a Steiner triple system $STS(v)$, then $STS(v)$ is said to *contain* the trade T_1.

Example 7 In the $STS_A(3^2)$ of Example 3 with the defining sets of Example 4, let the set of six blocks $T_1 = \mathcal{B} \setminus \mathcal{N}$. Let $T_2 = \mathcal{F}$. Then T_1 is a trade of volume six contained in $STS_A(3^2)$.

Lemma 8 (Gray [9]) *In an $STS(v)$, every defining set has at least one block in common with every trade.*

Example 9 In the $STS_A(3^2)$ of Examples 3 and 4, $S_4 \cap T_1 = \{\{0, 3, 6\}, \{1, 4, 7\}\}$ and $S_5 \cap T_1 = \{\{0, 3, 6\}; \{0, 4, 8\}; \{1, 3, 8\}\}$.

Definition 7 There are only two non-isomorphic Steiner trades of volume 6 and block size 3; see [14] for example. These are called a *trade of type one*, such as

$$\tau_1 = \{\{0, 2, 5\}, \{0, 3, 6\}, \{0, 4, 7\}, \{1, 2, 6\}, \{1, 3, 7\}, \{1, 4, 5\}\},$$

and a *trade of type two*, such as

$$\tau_2 = \{\{0, 3, 4\}, \{0, 5, 6\}, \{1, 3, 5\}, \{1, 4, 6\}, \{2, 3, 6\}, \{2, 4, 5\}\}.$$

In this paper only trades of type one occur. Note that the six blocks of a trade of type one consist of three disjoint pairs, whereas any two blocks of a trade of type two intersect each other.

Definition 8 A *latin square* L of order n is an $n \times n$ array with entries chosen from a set N, of size n, such that each element of N occurs precisely once in each row and column. Similarly, a *partial latin square*

P of order n is an $n \times n$ array with entries chosen from a set N, of size n, such that each element of N occurs at most once in each row and column. Thus P may contain a number of empty cells. For convenience, a (partial) latin square will sometimes be represented as a set of ordered triples $(i, j; k)$, which is read to mean that element k occurs in cell (i, j) of the (partial) latin square L. For a (partial) latin square of order n we label entries, rows and columns by $0, 1, 2, \ldots, n-1$. $|P|$ denotes the number of non-empty cells of the partial latin square P and is called the *size* of P, and the set of positions $S_P = \{(i, j) \mid (i, j; k) \text{ for some } k \in N\}$ is said to determine the *shape* of P.

Example 10 Define the partial latin squares R, S and W, and the latin square L, each of order 3, as follows.

Then R and S can each be completed to L but to no other latin square of order 3, and W cannot be completed to any latin square of order 3. Note that throughout this paper S stands for this particular partial latin square of order 3.

Definition 9 Let M and N be two latin squares of orders m and n, with entries chosen from the sets $\{0, 1, 2, \ldots, m-1\}$ and $\{0, 1, 2, \ldots, n-1\}$, respectively. Suppose that P is a (partial) latin square in M and Q is a (partial) latin square in N. Let P^r be the array obtained from P by adding rm to the entry in each non-empty cell of P, for $r = 0, 1, 2, \ldots, n-1$. Similarly, let M^r be the array obtained from M by adding rm to the entry in each cell of M, for $r = 0, 1, 2, \ldots, n-1$. Then we define the *completable product* of Q and P, with respect to M and N, to be the (partial) latin square T of order mn obtained by replacing each cell containing the entry r of Q with the array M^r and each cell of $N \setminus Q$ with the array P^r. The completable product of Q with P will be denoted $Q \otimes P$. If Q is a latin square then the completable product corresponds to the definition of the direct product of the latin squares Q and N, usually denoted $Q \times N$.

Example 11 Let R and S be the partial latin squares and L the latin square defined in Example 10. We consider three examples with $L = M = N$, and a fourth subsquare of the first which cannot be written as a completable product.

(i) $L = P = Q$.

$$L_9 = Q \otimes P = L \times L = \begin{array}{|c|c|c|} \hline L^0 & L^2 & L^1 \\ \hline L^2 & L^1 & L^0 \\ \hline L^1 & L^0 & L^2 \\ \hline \end{array} =$$

0	2	1	6	8	7	3	5	4
2	1	0	8	7	6	5	4	3
1	0	2	7	6	8	4	3	5
6	8	7	3	5	4	0	2	1
8	7	6	5	4	3	2	1	0
7	6	8	4	3	5	1	0	2
3	5	4	0	2	1	6	8	7
5	4	3	2	1	0	8	7	6
4	3	5	1	0	2	7	6	8

(ii) $S = P = Q$.

$$L_{9,4} = Q \otimes P = S \otimes S = \begin{array}{|c|c|c|} \hline L^0 & S^2 & S^1 \\ \hline S^2 & L^1 & S^0 \\ \hline S^1 & S^0 & S^2 \\ \hline \end{array} =$$

0	2	1	6			3		
2	1	0		7			4	
1	0	2						
6			3	5	4	0		
	7		5	4	3		1	
			4	3	5			
3			0			6		
	4			1			7	

(iii) $R = P = Q$.

$$Q \otimes P = R \otimes R = \begin{array}{|c|c|c|} \hline L^0 & L^2 & R^1 \\ \hline L^2 & R^1 & R^0 \\ \hline R^1 & R^0 & R^2 \\ \hline \end{array}$$

(iv) The following partial latin square, $L_{9,5}$, cannot be written as a completable product.

$$L_{9,5} =$$

0	2	1	6	8		3		4
2	1	0	8					3
1	0	2						
6	8		3	5	4	0		1
8			5	4	3			0
			4	3	5			
3			0			6		
4	3		1	0				8

Example 12 We also need some partial latin squares with special properties. We choose examples of order 5, where A belongs to the second main class of latin squares of order 5 (see [3], page 129) and each of the partial latin squares B, C and D can be completed to A.

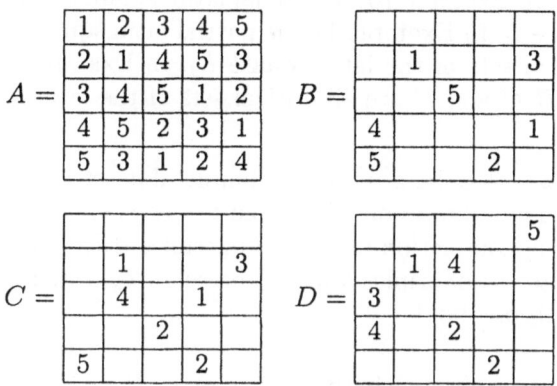

$$A = \begin{array}{|c|c|c|c|c|} \hline 1 & 2 & 3 & 4 & 5 \\ \hline 2 & 1 & 4 & 5 & 3 \\ \hline 3 & 4 & 5 & 1 & 2 \\ \hline 4 & 5 & 2 & 3 & 1 \\ \hline 5 & 3 & 1 & 2 & 4 \\ \hline \end{array} \qquad B = \begin{array}{|c|c|c|c|c|} \hline & & & & \\ \hline & 1 & & & 3 \\ \hline & & 5 & & \\ \hline 4 & & & & 1 \\ \hline 5 & & & 2 & \\ \hline \end{array}$$

$$C = \begin{array}{|c|c|c|c|c|} \hline & & & & \\ \hline & 1 & & & 3 \\ \hline & 4 & & 1 & \\ \hline & & 2 & & \\ \hline 5 & & & 2 & \\ \hline \end{array} \qquad D = \begin{array}{|c|c|c|c|c|} \hline & & & & 5 \\ \hline & 1 & 4 & & \\ \hline 3 & & & & \\ \hline 4 & & 2 & & \\ \hline & & & 2 & \\ \hline \end{array}$$

Definition 10 Let $(V, \mathcal{B})$ be a (partial) Steiner triple system of order v on the element set $V = \{0, 1, 2, \ldots, v-1\}$. We define the corresponding *(partial) Steiner latin square of order v* to be the array with entry k in the cell (i, j), $i \neq j$, if and only if $\{i, j, k\} \in \mathcal{B}$. Moreover, the cell (i, i) contains the entry i if and only if the element i occurs in a block. Note that some elements may not occur in any block of a partial Steiner triple system of order v.

Example 13 The Steiner triple system $STS_A(3^2)$ of Example 3 corresponds to the Steiner latin square L_9 given in Example 11, and its minimal defining sets S_5 and S_4 given in Example 4 to the partial squares $L_{9,5}$ and $L_{9,4}$ given in Example 11. Note that the element 7 missing from the blocks of S_5 does not appear on the diagonal of $L_{9,5}$, nor does the element 8 missing from the blocks of S_4 appear on the diagonal of $L_{9,4}$.

Definition 11 A *critical set* in a latin square L of order n is a set $C = \{(i, j; k) \mid i, j, k \in N\}$, such that both of the following conditions hold:

(1) L is the only latin square of order n with element k in cell (i, j) for each $(i, j; k) \in C$;

(2) no proper subset of C satisfies (1).

A *uniquely completable set* in a latin square L of order n is a partial latin square in L satisfying condition (1). Condition (2) guarantees that each entry of C is *necessary* for the completion to be unique.

Example 14 In Example 10 the partial latin square S is a critical set of the latin square L. In Example 11 the partial latin squares $L_{9,5}$ and $L_{9,4}$ are both critical sets of the latin square L_9. In Example 12 the partial latin squares B, C and D are all critical sets of the latin square A.

Definition 12 Let U be a uniquely completable set in the latin square L of order n. The adjunction of a triple $t = (r, c; s)$ is said to be *forced* (see [13]) during completion of a set P of triples ($|P| < n^2$, $U \subseteq P \subset L$) to the complete set of triples which represents L, if at least one of the following conditions holds:

(i) $\forall r' \neq r$, $\exists z \neq c$ such that $(r', z; s) \in P$ or $\exists z \neq s$ such that $(r', c; z) \in P$ (that is, in the partial completion F of L, each cell of column c except that in row r is either in a row of F which already contains the symbol s or is already filled with an element z distinct from s);

(ii) $\forall c' \neq c$, $\exists z \neq r$ such that $(z, c'; s) \in P$ or $\exists z \neq s$ such that $(r, c'; z) \in P$ (that is, in the partial completion F of L, each cell of row r except that in column c is either in a column of F which already contains the symbol s or is already filled with an element z distinct from s);

(iii) $\forall s' \neq s$, $\exists z \neq r$ such that $(z, c; s') \in P$ or $\exists z \neq c$ such that $(r, z; s') \in P$ (that is, in the partial completion F of L, every symbol except s already occurs either in column c or in row r of F).

The uniquely completable set U is called *strong* if we can define a sequence of sets of triples $U = F_1 \subset F_2 \subset F_3 \subset \ldots \subset F_r = L$ such that each triple $t \in F_{i+1} \setminus F_i$ is forced in F_i for $1 \leq i \leq r - 1$. If U is not strong, it is called *weak*. A completable set is *super-strong* if each triple in this sequence is forced by virtue of property (iii) alone. In particular, a critical set can be super-strong, strong but not super-strong, or weak.

Example 15 The critical sets $L_{9,5}$ and $L_{9,4}$ of Example 11 are both super-strong. In Example 12, the critical set B is super-strong, the critical set C is strong but not super-strong, and the critical set D is weak; see [1], [2].

Definition 13 Let P and P' be two partial latin squares of the same order, with the same size and shape. Then P and P' are said to be *mutually balanced* if the entries in each row (and column) of P are the same as those in the corresponding row (and column) of P'. They are said to be *disjoint* if no position in P' contains the same entry as the corresponding position in P. A *latin trade* (sometimes called a latin interchange) is a partial latin square I for which there exists another partial latin square I' of the same order, size and shape, with the property that I and I' are disjoint and mutually balanced. Then I' is called the *disjoint mate* of I. The smallest possible size for such partial latin squares is four, and such a latin trade is called an *intercalate*.

Example 16 The partial latin square I_1 is an example of an intercalate; that is a latin trade of size 4, with disjoint mate I_1'. The partial latin square I_2, is an example of a latin trade of order 3 and size 8, with disjoint mate I_2'.

$$
I_1 =
\begin{array}{|c|c|c|}
\hline
0 & & 1 \\
\hline
1 & & 0 \\
\hline
 & & \\
\hline
\end{array}
\quad
I_1' =
\begin{array}{|c|c|c|}
\hline
1 & & 0 \\
\hline
0 & & 1 \\
\hline
 & & \\
\hline
\end{array}
\quad
I_2 =
\begin{array}{|c|c|c|}
\hline
0 & & 2 \\
\hline
2 & 3 & 1 \\
\hline
3 & 1 & 0 \\
\hline
\end{array}
\quad
I_2' =
\begin{array}{|c|c|c|}
\hline
2 & & 0 \\
\hline
3 & 1 & 2 \\
\hline
0 & 3 & 1 \\
\hline
\end{array}
$$

In a given latin square L, every latin trade contained in L must intersect every critical set contained in L. More precisely, we have the following result.

Theorem 17 (Donovan and Howse [4]) *A partial latin square C of the latin square L is a critical set of L if and only if the following conditions hold:*

(1) *C intersects every latin trade that occurs in L;*

(2) *for each $(i, j; k) \in C$, there exists a latin trade I in L such that $I \cap C = \{(i, j; k)\}$.*

Lemma 18 *Consider τ_1 as given in Definition 7. Then τ_1 defines six disjoint latin trades each of size six.*

Proof. Define

$$I_1 = \{(0,2;5),(0,3;6),(0,4;7),(1,2;6),(1,3;7),(1,4;5)\}$$
$$I_2 = \{(2,0;5),(3,0;6),(4,0;7),(2,1;6),(3,1;7),(4,1;5)\}$$
$$I_3 = \{(0,5;2),(0,6;3),(0,7;4),(1,6;2),(1,7;3),(1,5;4)\}$$
$$I_4 = \{(5,0;2),(6,0;3),(7,0;4),(6,1;2),(7,1;3),(5,1;4)\}$$
$$I_5 = \{(5,2;0),(6,3;0),(7,4;0),(6,2;1),(7,3;1),(5,4;1)\}$$
$$I_6 = \{(2,5;0),(3,6;0),(4,7;0),(2,6;1),(3,7;1),(4,5;1)\}.$$

It is easy to see that I_r for $1 \le r \le 6$ is a latin trade. Moreover, $I_r \cap I_s = \emptyset$ for $1 \le r < s \le 6$. These latin trades are shown in the following table.

		5	6	7	2	3	4
		6	7	5	4	2	3
5	6				0	1	
6	7					0	1
7	5				1		0
2	4	0		1			
3	2	1	0				
4	3		1	0			

$\square$

In Section 3, we give some lemmas and constructions needed for the proof of the Main Theorem. The proof itself is given in Section 4.

3.   Further Preliminaries

Lemma 19 (Gower [6]) *Let P be a strongly completable set in the latin square M of order m and let Q be a (strongly) completable set in the latin square N of order n. Then $P \otimes Q$ is a (strongly) completable set of order mn in the latin square $M \times N$.*

Example 20 Let S and L be as defined in Example 11. Then S is a super-strong critical set in L. Now let $P = Q = S$ and $M = N = L$. Then $L_{9,4} = Q \otimes P = S \otimes S$ as in Example 11, and the fact that it is a strongly completable set of order 9 is confirmed by Lemma 19.

The following result states the relationship between a completable set and a defining set.

Lemma 21 *(Gower [5]) Let $(V, \mathcal{B})$ be a Steiner triple system of order v and let $\mathcal{B}' \subseteq \mathcal{B}$. Suppose that P is the partial Steiner latin square* *corresponding to* $(V, \mathcal{B}')$. *If* P *is a completable set of order* v *then* $\mathcal{B}'$ *is a defining set in* $(V, \mathcal{B})$.

Unfortunately the partial Steiner latin square corresponding to a defining set of an $STS(v)$ need not be uniquely completable. For an example of this, see the Appendix.

Example 22 Consider the latin square L_9 and the partial latin squares S and $L_{9,5}$ discussed in Example 10. Note that $L_{9,5}$ is a partial Steiner latin square. The partial Steiner triple system corresponding to $L_{9,5}$ consists of the blocks $\{0, 1, 2\}$, $\{0, 3, 6\}$, $\{0, 4, 8\}$, $\{1, 3, 8\}$ and $\{3, 4, 5\}$, which constitute the set S_5 given in Example 4 as a minimal defining set of the $STS(9)$ given in Example 3. Lemma 21 confirms that these blocks yield a defining set of order 9, and it is also easy to check that all the blocks are necessary for unique completion.

Now by Lemma 19, $R = S \otimes L_{9,5}$ is a strongly completable set of order 27. (Indeed, we see later that R is a critical set by Theorem 32, Section 6.) One can observe that R is a partial Steiner latin square. The partial Steiner triple system corresponding to R consists of the following 67 blocks.

$\{0, \ 1, \ 2\}$	$\{0, \ 3, \ 6\}$	$\{0, \ 4, \ 8\}$	$\{0, \ 5, \ 7\}$	$\{0, \ 9, 18\}$
$\{0, 10, 20\}$	$\{0, 11, 19\}$	$\{0, 12, 24\}$	$\{0, 13, 26\}$	$\{0, 15, 21\}$
$\{0, 17, 22\}$	$\{1, \ 3, \ 8\}$	$\{1, \ 4, \ 7\}$	$\{1, \ 5, \ 6\}$	$\{1, \ 9, 20\}$
$\{1, 10, 19\}$	$\{1, 11, 18\}$	$\{1, 12, 26\}$	$\{1, 17, 21\}$	$\{2, \ 3, \ 7\}$
$\{2, \ 4, \ 6\}$	$\{2, \ 5, \ 8\}$	$\{2, \ 9, 19\}$	$\{2, 10, 18\}$	$\{2, 11, 20\}$
$\{3, \ 4, \ 5\}$	$\{3, \ 9, 24\}$	$\{3, 10, 26\}$	$\{3, 12, 21\}$	$\{3, 13, 23\}$
$\{3, 14, 22\}$	$\{3, 15, 18\}$	$\{3, 17, 19\}$	$\{4, \ 9, 26\}$	$\{4, 12, 23\}$
$\{4, 13, 22\}$	$\{4, 14, 21\}$	$\{4, 17, 18\}$	$\{5, 12, 22\}$	$\{5, 13, 21\}$
$\{5, 14, 23\}$	$\{6, \ 7, \ 8\}$	$\{6, \ 9, 21\}$	$\{6, 12, 18\}$	$\{6, 15, 24\}$
$\{8, \ 9, 22\}$	$\{8, 10, 21\}$	$\{8, 12, 19\}$	$\{8, 13, 18\}$	$\{8, 17, 26\}$
$\{9, 10, 11\}$	$\{9, 12, 15\}$	$\{9, 13, 17\}$	$\{9, 14, 16\}$	$\{10, 12, 17\}$
$\{10, 13, 16\}$	$\{10, 14, 15\}$	$\{11, 12, 16\}$	$\{11, 13, 15\}$	$\{11, 14, 17\}$
$\{12, 13, 14\}$	$\{15, 16, 17\}$	$\{18, 19, 20\}$	$\{18, 21, 24\}$	$\{18, 22, 26\}$
$\{19, 21, 26\}$	$\{21, 22, 23\}$			

By Lemma 21, these blocks yield a defining set of order 27. (Indeed, these blocks yield a minimal defining set in $AG(3, 3)$ by Theorem 30, Section 5.)

Theorem 23 (Gower [7]) *Let* **H** *be the following 2d hyperplanes in* $AG(d, 3)$, $d \geq 2$.

$$
\begin{cases}
1x_{d-1} & + & 0x_{d-2} & + & \dots & + & 0x_1 & + & 0x_0 & = & 0 \\
1x_{d-1} & + & 0x_{d-2} & + & \dots & + & 0x_1 & + & 0x_0 & = & 1 \\
0x_{d-1} & + & 1x_{d-2} & + & \dots & + & 0x_1 & + & 0x_0 & = & 0 \\
0x_{d-1} & + & 1x_{d-2} & + & \dots & + & 0x_1 & + & 0x_0 & = & 1 \\
& & & & \dots & & & & & & \\
& & & & \dots & & & & & & \\
& & & & \dots & & & & & & \\
0x_{d-1} & + & 0x_{d-2} & + & \dots & + & 1x_1 & + & 0x_0 & = & 0 \\
0x_{d-1} & + & 0x_{d-2} & + & \dots & + & 1x_1 & + & 0x_0 & = & 1 \\
0x_{d-1} & + & 0x_{d-2} & + & \dots & + & 0x_1 & + & 1x_0 & = & 0 \\
0x_{d-1} & + & 0x_{d-2} & + & \dots & + & 0x_1 & + & 1x_0 & = & 1.
\end{cases}
$$

Then the lines contained within these hyperplanes correspond to blocks of the $PSTS(3^d)$ which are a minimal defining set for this Steiner triple system.

Theorem 24 (Gower [7]) *The hyperplanes of Theorem 23 can be replaced by another set of $2d$ hyperplanes of $AG(d, 3)$ provided these new hyperplanes are such that:*

 (1) they belong to d parallel classes with precisely two hyperplanes chosen from each of these parallel classes;

 (2) no line of $AG(d, 3)$ is contained within any d of them;

 (3) there is a point of $AG(d, 3)$ which is incident with none of these $2d$ hyperplanes.

By Lemma 5, it is necessary for a set of hyperplanes in $AG(d, 3)$ which provides a defining set in $AG(d, 3)$ to cover all points of $AG(d, 3)$ except possibly one. We state this fact in the following lemma and give a simple proof for it.

Lemma 25 *Let $\mathbf{H}$ be a set of hyperplanes in $AG(d, 3)$, where $d \geq 2$. Suppose that the points $\bar{i}$ and $\bar{j}$ of $AG(d, 3)$ are not incident with any hyperplane of $\mathbf{H}$. Then the lines within the hyperplanes of $\mathbf{H}$ do not form a defining set.*

Proof. Let $\bar{k} \neq -\bar{i} - \bar{j}$ be a point of $AG(d, 3)$. Let T consist of the following blocks:

$$
\{\bar{i},\ \bar{k},\ -\bar{i} - \bar{k}\},\quad \{\bar{i},\ -\bar{i} + \bar{j} + \bar{k},\ -\bar{j} - \bar{k}\},\quad \{\bar{i},\ \bar{i} - \bar{j} + \bar{k},\ \bar{i} + \bar{j} - \bar{k}\},
$$
$$
\{\bar{j},\ \bar{k},\ -\bar{j} - \bar{k}\},\quad \{\bar{j},\ -\bar{i} + \bar{j} + \bar{k},\ \bar{i} + \bar{j} - \bar{k}\},\quad \{\bar{j},\ \bar{i} - \bar{j} + \bar{k},\ -\bar{i} - \bar{k}\}.
$$

First we note that T is a type one trade. Secondly, for each $\{\bar{x}, \bar{y}, \bar{z}\}$ of T we have $\bar{x} + \bar{y} + \bar{z} = \bar{0}$. So T is a trade in $AG(d, 3)$. Thirdly, no block of T lies in any hyperplane of $\mathbf{H}$. Now the result follows. $\qquad \square$

Lemma 26 *Consider the set of hyperplanes* $\mathbf{H}$ *in* $AG(d, 3)$ *defined in Definition 3. Let L be the Steiner latin square corresponding to the blocks of $AG(d, 3)$. Let $\mathcal{H}$ be the collection of blocks contained within the hyperplanes of $\mathbf{H}$. Suppose that P is the partial Steiner latin square of order 3^d corresponding to $\mathcal{H}$. Let $\mathcal{H}^*$ be the set of blocks within the hyperplanes of $\mathbf{H}^*$, the extension of $\mathbf{H}$. Then the partial Steiner latin square P^* corresponding to $\mathcal{H}^*$ is*

$$
P^* = S \otimes P = \begin{array}{|c|c|c|}
\hline
L^0 & P^2 & P^1 \\
\hline
P^2 & L^1 & P^0 \\
\hline
P^1 & P^0 & P^2 \\
\hline
\end{array}
$$

where S is as in Example 10.

Proof. First for each block of $\mathcal{H}^*$ we find the corresponding elements of $S \otimes P$. Note that each block of $\mathcal{H}^*$ fills precisely six off-diagonal cells of P^*. Conversely, for each off-diagonal element of $S \otimes P$ we give its corresponding block of $\mathcal{H}^*$. Finally, for a diagonal element $(i, i; i)$ of $S \otimes P$ we prove that $\bar{i}$ occurs in a block of $\mathcal{H}^*$.

Let $\{\bar{i}, \bar{j}, \bar{k}\}$ be a block of $\mathcal{H}^*$ and let $\bar{i} = (i_d, i_{d-1}, i_{d-2}, \ldots, i_1, i_0)$, $\bar{j} = (j_d, j_{d-1}, j_{d-2}, \ldots, j_1, j_0)$, and $\bar{k} = (k_d, k_{d-1}, k_{d-2}, \ldots, k_1, k_0)$. Suppose that i, j and k are integer representations of $\bar{i}$, $\bar{j}$ and $\bar{k}$, respectively. Without loss of generality, we can assume $i < j < k$. This leads to

$$(i_d, j_d, k_d) \in \{(0, 0, 0), (1, 1, 1), (2, 2, 2), (0, 1, 2)\}.$$

Define $\bar{i}' = (i_{d-1}, i_{d-2}, \ldots, i_0)$, $\bar{j}' = (j_{d-1}, j_{d-2}, \ldots, j_0)$, and $\bar{k}' = (k_{d-1}, k_{d-2}, \ldots, k_0)$. Then $\bar{i}' + \bar{j}' + \bar{k}' = \bar{0}$ since $\bar{i} + \bar{j} + \bar{k} = \bar{0}$. So $\{\bar{i}', \bar{j}', \bar{k}'\}$ is a block of $AG(d, 3)$.

Case $(i_d, j_d, k_d) = (0, 0, 0)$: It is obvious that the points $\bar{i}$, $\bar{j}$ and $\bar{k}$ are in the first hyperplane of $\mathbf{H}^*$. So the block $\{\bar{i}, \bar{j}, \bar{k}\}$ lies in this hyperplane. On the other hand, the integer representations for $\bar{i}'$, $\bar{j}'$ and $\bar{k}'$ are i, j and k, respectively. Moreover, $(i_d, j_d, k_d) = (0, 0, 0)$ implies that $0 \leq i, j, k < 3^d$. So $(i, j; k) \in L^0$.

Case $(i_d, j_d, k_d) = (1, 1, 1)$: It is obvious that the points $\bar{i}$, $\bar{j}$ and $\bar{k}$ are in the second hyperplane of $\mathbf{H}^*$. So the block $\{\bar{i}, \bar{j}, \bar{k}\}$ lies in this hyperplane. On the other hand, the integer representations for $\bar{i}'$, $\bar{j}'$ and

$\bar{k}'$ are $i - 3^d$, $j - 3^d$ and $k - 3^d$, respectively. Moreover, $(i_d, j_d, k_d) = (1, 1, 1)$ implies that $3^d \leq i, j, k < 2.3^d$. So $(i - 3^d, j - 3^d; k) \in L^1$.

Case $(i_d, j_d, k_d) = (2, 2, 2)$: It is obvious that the points $\bar{i}$, $\bar{j}$ and $\bar{k}$ cannot be in the first or second hyperplane of $\mathbf{H}^*$. So the block $\{\bar{i}, \bar{j}, \bar{k}\}$ must be in one of the last m hyperplanes of $\mathbf{H}^*$. Let $\{\bar{i}, \bar{j}, \bar{k}\}$ be in the nth hyperplane of $\mathbf{H}^*$, where $3 \leq n \leq m + 2$. Then the points $\bar{i}'$, $\bar{j}'$ and $\bar{k}'$ are in the $(n - 2)$nd hyperplane of $\mathbf{H}$. So the block $\{\bar{i}', \bar{j}', \bar{k}'\}$ is in $\mathcal{H}$. Moreover, $(i_d, j_d, k_d) = (2, 2, 2)$ implies that $2.3^d \leq i, j, k < 3^{d+1}$. So $(i - 2.3^d, j - 2.3^d; k) \in P^2$.

Case $(i_d, j_d, k_d) = (0, 1, 2)$: It is obvious that the point $\bar{k}$ cannot be in the first or second hyperplane of $\mathbf{H}^*$. So the block $\{\bar{i}, \bar{j}, \bar{k}\}$ must be in one of the last m hyperplanes of $\mathbf{H}^*$. Let $\{\bar{i}, \bar{j}, \bar{k}\}$ be in the nth hyperplane of $\mathbf{H}^*$, where $3 \leq n \leq m + 2$. Then the points $\bar{i}'$, $\bar{j}'$ and $\bar{k}'$ are in the $(n - 2)$nd hyperplane of $\mathbf{H}$. Now if $i \not\equiv j \pmod{3^d}$ then $\bar{i}' \neq \bar{j}'$. So $\{\bar{i}', \bar{j}', \bar{k}'\}$ is a block in $\mathcal{H}$. Moreover, $(i_d, j_d, k_d) = (0, 1, 2)$ implies that $0 \leq i < 3^d$, $3^d \leq j < 2.3^d$, and $2.3^d \leq k < 3^{d+1}$. So $(i, j - 3^d, k) \in P^2$. Finally, if $i \equiv j \pmod{3^d}$ then $\bar{i}' = \bar{j}' = \bar{k}'$. So the cell $(i, j - 3^d)$ is a diagonal cell in P. Now by the definition of P^2 we have $(i, j - 3^d; k - 2.3^d) \in P$ if and only if $(i, j - 3^d; k) \in P^2$. Note that in this case the six off-diagonal elements corresponding to the block $\{\bar{i}, \bar{j}, \bar{k}\}$ are in the two P^0s, the two P^1s and the two (off-diagonal) P^2s in the latin square above.

Suppose that $(i, j; k)$ is an element of $S \otimes P$. We consider the following cases:

First, assume $i = j$. Then $(i, j; k)$ is a diagonal element of $S \otimes P$. So $\bar{i}'$ is incident with a hyperplane of $\mathbf{H}$ and $\bar{i}$ is incident with one of the last m hyperplanes of $\mathbf{H}^*$. Therefore $\bar{i}$ occurs in one of the blocks of $\mathcal{H}^*$.

Now we assume that $i \neq j$.

Case $0 \leq i, j, k < 3^d$: One can see that $\bar{i}$, $\bar{j}$ and $\bar{k}$ are incident with the first hyperplane of $\mathbf{H}^*$. Since $i \neq j$, $\{\bar{i}, \bar{j}, \bar{k}\}$ is a block of $\mathcal{H}^*$.

Case $3^d \leq i, j, k < 2.3^d$: One can see that $\bar{i}$, $\bar{j}$ and $\bar{k}$ are incident with the second hyperplane of $\mathbf{H}^*$. Since $i \neq j$, $\{\bar{i}, \bar{j}, \bar{k}\}$ is a block of $\mathcal{H}^*$.

Case $2.3^d \leq i, j, k < 3^{d+1}$: Suppose that $\bar{i} = (i_d, i_{d-1}, \ldots, i_1, i_0)$, $\bar{j} = (j_d, j_{d-1}, \ldots, j_1, j_0)$, and $\bar{k} = (k_d, k_{d-1}, \ldots, k_1, k_0)$. Then since $2.3^d \leq i, j, k < 3.3^d$ we have $i_d = j_d = k_d = 2$. Since $i \neq j$ and $(i - 2.3^d, j - 2.3^d; k) \in P^2$ we have $\{\bar{i}', \bar{j}', \bar{k}'\} \in \mathcal{H}$, where $\bar{i}' = (i_{d-1}, \ldots, i_1, i_0)$, $\bar{j}' = (j_{d-1}, \ldots, j_1, j_0)$, and $\bar{k}' = (k_{d-1}, \ldots, k_1, k_0)$. So the block $\{\bar{i}', \bar{j}', \bar{k}'\}$ lies in one of the hyperplanes of **H**. Therefore, the block $\{\bar{i},\bar{j},\bar{k}\}$ lies in one of the last m hyperplanes of **H***. Hence, $\{\bar{i},\bar{j},\bar{k}\} \in \mathcal{H}^*$.

The remaining six cases are as follows.

1	0	$\leq i <$	3^d	3^d	$\leq j <$	2.3^d	2.3^d	$\leq k <$	3^{d+1}		
2	0	$\leq i <$	3^d	2.3^d	$\leq j <$	3^{d+1}	3^d	$\leq k <$	2.3^d		
3	3^d	$\leq i <$	2.3^d	0	$\leq j <$	3^d	2.3^d	$\leq k <$	3^{d+1}		
4	3^d	$\leq i <$	2.3^d	2.3^d	$\leq j <$	3^{d+1}	0	$\leq k <$	3^d		
5	2.3^d	$\leq i <$	3^{d+1}	0	$\leq j <$	3^d	3^d	$\leq k <$	2.3^d		
6	2.3^d	$\leq i <$	3^{d+1}	3^d	$\leq j <$	2.3^d	0	$\leq k <$	3^d		

These cases are similar to the case $2.3^d \leq i,j,k < 3^{d+1}$ and we leave them to the reader. $\qquad\square$

4.   Proof of the Main Theorem

First we prove the following lemmas, in which we let **H**, **H***, $\mathcal{H}$ and $\mathcal{H}^*$ be as defined in Definition 3 and applied in Theorem 1.

Lemma 27 *Let $\{\bar{i},\bar{j},\bar{k}\}$ be a block of $\mathcal{H}$ and let*

$$T = \{\{\bar{i},\bar{j},\bar{k}\}, \{\bar{i},\bar{e},\bar{f}\}, \{\bar{i},\bar{g},\bar{h}\}, \{\bar{\ell},\bar{j},\bar{e}\}, \{\bar{\ell},\bar{k},\bar{g}\}, \{\bar{\ell},\bar{f},\bar{h}\}\}$$

be a trade in $AG(d,3)$ such that $T \cap \mathcal{H} = \{\{\bar{i},\bar{j},\bar{k}\}\}$. Then there exist trades T_0^* and T_1^*, of type one, in $AG(d+1,3)$ such that $T_0^* \cap \mathcal{H}^* = \{\{0\bar{i},0\bar{j},0\bar{k}\}\}$ and $T_1^* \cap \mathcal{H}^* = \{\{1\bar{i},1\bar{j},1\bar{k}\}\}$.

Proof. First define

$$\begin{aligned} T_0^* &= \{\{0\bar{i},0\bar{j},0\bar{k}\}, \{0\bar{i},1\bar{e},2\bar{f}\}, \{0\bar{i},1\bar{g},2\bar{h}\}, \\ &\quad \{2\bar{\ell},0\bar{j},1\bar{e}\}, \{2\bar{\ell},0\bar{k},1\bar{g}\}, \{2\bar{\ell},2\bar{f},2\bar{h}\}\}. \end{aligned}$$

Since for each block $\{\bar{r},\bar{s},\bar{t}\} \in T$ we have $\bar{r}+\bar{s}+\bar{t} = \bar{0}$ it follows that for each block $\{\bar{x},\bar{y},\bar{z}\} \in T_0^*$ we have $\bar{x}+\bar{y}+\bar{z} = \bar{0}$. Therefore, the blocks of T_0^* are in $AG(d+1,3)$. Using the fact that $T \cap \mathcal{H} = \{\{\bar{i},\bar{j},\bar{k}\}\}$, it is straightforward to see that $T_0^* \cap \mathcal{H}^* = \{\{0\bar{i},0\bar{j},0\bar{k}\}\}$.
Next define

$$\begin{aligned} T_1^* &= \{\{1\bar{i},1\bar{j},1\bar{k}\}, \{1\bar{i},0\bar{e},2\bar{f}\}, \{1\bar{i},0\bar{g},2\bar{h}\}, \\ &\quad \{2\bar{\ell},1\bar{j},0\bar{e}\}, \{2\bar{\ell},1\bar{k},0\bar{g}\}, \{2\bar{\ell},2\bar{f},2\bar{h}\}\}. \end{aligned}$$

Similarly, the blocks of T_1^* are in $AG(d+1,3)$ and $T_1^* \cap \mathcal{H}^* = \{\{1\bar{i},1\bar{j},1\bar{k}\}\}$.
$\square$

Lemma 28 *Let $\{\bar{i}, \bar{j}, \bar{k}\}$ be a block in $AG(d,3) \setminus \mathcal{H}$. Then there exist trades T_0^* and T_1^* of type one in $AG(d+1,3)$ such that $T_0^* \cap \mathcal{H}^* = \{\{0\bar{i}, 0\bar{j}, 0\bar{k}\}\}$ and $T_1^* \cap \mathcal{H}^* = \{\{1\bar{i}, 1\bar{j}, 1\bar{k}\}\}$.*

Proof. First define

$$T_0^* = \{\{0\bar{i}, 0\bar{j}, 0\bar{k}\}, \{0\bar{i}, 1\bar{j}, 2\bar{k}\}, \{0\bar{i}, 2\bar{j}, 1\bar{k}\},$$
$$\{2\bar{i}, 0\bar{j}, 1\bar{k}\}, \{2\bar{i}, 1\bar{j}, 0\bar{k}\}, \{2\bar{i}, 2\bar{j}, 2\bar{k}\}\}.$$

Obviously, any block of T_0^* is a block of $AG(d+1,3)$. Moreover, the fact that $\{\bar{i}, \bar{j}, \bar{k}\}$ is not a block of $\mathcal{H}$ leads to $T_0^* \cap \mathcal{H}^* = \{\{0\bar{i}, 0\bar{j}, 0\bar{k}\}\}$. Next define

$$T_1^* = \{\{1\bar{i}, 1\bar{j}, 1\bar{k}\}, \{1\bar{i}, 0\bar{j}, 2\bar{k}\}, \{1\bar{i}, 2\bar{j}, 0\bar{k}\},$$
$$\{2\bar{i}, 0\bar{j}, 1\bar{k}\}, \{2\bar{i}, 1\bar{j}, 0\bar{k}\}, \{2\bar{i}, 2\bar{j}, 2\bar{k}\}\}.$$

Similarly, any block of T_1^* is a block of $AG(d+1,3)$ and $T_1^* \cap \mathcal{H}^* = \{\{1\bar{i}, 1\bar{j}, 1\bar{k}\}\}$. $\square$

Lemma 29 *Let $\bar{\gamma} = (\gamma_{d-1}, \gamma_{d-2}, \ldots, \gamma_1, \gamma_0)$ be a point of $AG(d,3)$ which occurs in one of the blocks of $\mathcal{H}$. Then there exists a trade T^* of type one in $AG(d+1,3)$ such that $T^* \cap \mathcal{H}^* = \{\{0\bar{\gamma}, 1\bar{\gamma}, 2\bar{\gamma}\}\}$.*

Proof. Suppose that the point $\bar{\theta}$ of $AG(d,3)$ is not incident with any hyperplane of $\mathbf{H}$. So it does not occur in any block of $\mathcal{H}$. Now define

$$T^* = \{\{0\bar{\gamma}, 1\bar{\gamma}, 2\bar{\gamma}\}, \{0\bar{\gamma}, 1\bar{\theta}, 2\bar{\delta}\}, \{0\bar{\gamma}, 1\bar{\delta}, 2\bar{\theta}\},$$
$$\{0\bar{\theta}, 1\bar{\gamma}, 2\bar{\delta}\}, \{0\bar{\theta}, 1\bar{\delta}, 2\bar{\gamma}\}, \{0\bar{\theta}, 1\bar{\theta}, 2\bar{\theta}\}\}$$

where $\bar{\delta} = -\bar{\gamma} - \bar{\theta}$. First note that for each block $\{\bar{x}, \bar{y}, \bar{z}\}$ in T^* we have $\bar{x} + \bar{y} + \bar{z} = \bar{0}$. So T^* is a trade of type one in $AG(d+1,3)$. Secondly, it is easy to see that $T^* \cap \mathcal{H}^* = \{\{0\bar{\gamma}, 1\bar{\gamma}, 2\bar{\gamma}\}\}$. $\square$

Now we are ready to prove Theorem 1.

Proof of Theorem 1 ($1'$) Suppose $\bar{\theta}$ is the point which is not incident with any hyperplane of $\mathbf{H}$. Then $2\bar{\theta}$ is not incident with any hyperplane of $\mathbf{H}^*$.

($2'$) Let L be the Steiner latin square corresponding to the blocks of $AG(d,3)$. Then, by Lemma 26, P^* is the product of $S = \{(0,0;0),$

$(1,1;1)\}$ and P; that is:

$$P^* = S \otimes P = \begin{array}{|c|c|c|} \hline L^0 & P^2 & P^1 \\ \hline P^2 & L^1 & P^0 \\ \hline P^1 & P^0 & P^2 \\ \hline \end{array}$$

Since the partial latin square S is a super-strong critical set of order three, by Lemma 19, P^* is a completable set of order 3^{d+1}. So the blocks of $\mathcal{H}^*$ form a defining set in $AG(d+1,3)$.

$(3')$ Let $\{\bar{i}, \bar{j}, \bar{k}\}$ be a block of $\mathcal{H}^*$ and let $\bar{i} = (i_d, i_{d-1}, i_{d-2}, \ldots, i_1, i_0)$, $\bar{j} = (j_d, j_{d-1}, j_{d-2}, \ldots, j_1, j_0)$, and $\bar{k} = (k_d, k_{d-1}, k_{d-2}, \ldots, k_1, k_0)$. Suppose that i, j and k are integer representations of $\bar{i}$, $\bar{j}$ and $\bar{k}$, respectively. Without loss of generality, we can assume $i < j < k$. This leads to

$$(i_d, j_d, k_d) \in \{(0,0,0), (1,1,1), (2,2,2), (0,1,2)\}.$$

Define

$$\bar{i}' = (i_{d-1}, i_{d-2}, \ldots, i_0),$$
$$\bar{j}' = (j_{d-1}, j_{d-2}, \ldots, j_0),$$

and

$$\bar{k}' = (k_{d-1}, k_{d-2}, \ldots, k_0).$$

Then $\bar{i}' + \bar{j}' + \bar{k}' = \bar{0}$ since $\bar{i} + \bar{j} + \bar{k} = \bar{0}$. So $\{\bar{i}', \bar{j}', \bar{k}'\}$ is a block of $AG(d,3)$.

Case $(i_d, j_d, k_d) = (0,0,0)$: If $\{\bar{i}', \bar{j}', \bar{k}'\} \in \mathcal{H}$ we take T_0^* as defined in Lemma 27. If the block $\{\bar{i}', \bar{j}', \bar{k}'\}$ is not in $\mathcal{H}$ we take T_0^* as defined in Lemma 28. In both cases $T_0^* \cap \mathcal{H}^* = \{\{\bar{i}, \bar{j}, \bar{k}\}\}$.

Case $(i_d, j_d, k_d) = (1,1,1)$: This case is quite similar to the case $(i_d, j_d, k_d) = (0,0,0)$ and the proof is left to the reader.

Case $(i_d, j_d, k_d) = (2,2,2)$: By Lemma 26 the block $\{\bar{i}', \bar{j}', \bar{k}'\}$ is in $\mathcal{H}$. So there is a trade T of type one in $AG(d,3)$ such that $T \cap \mathcal{H} = \{\{\bar{i}', \bar{j}', \bar{k}'\}\}$. Now define

$$T^* = \{\{2\bar{x}, 2\bar{y}, 2\bar{z}\} \mid \{x, y, z\} \in T\}.$$

Then T^* is a trade of type one in $AG(d+1,3)$ and $T^* \cap \mathcal{H}^* = \{\{\bar{i}, \bar{j}, \bar{k}\}\}$.

Case $(i_d, j_d, k_d) = (0,1,2)$: First let $i \not\equiv j \pmod{3^d}$. Then by Lemma 26 the block $\{\bar{i}', \bar{j}', \bar{k}'\}$ is in $\mathcal{H}$. So there is a trade T of type one in $AG(3,d)$ such that $T \cap \mathcal{H} = \{\{\bar{i}', \bar{j}', \bar{k}'\}\}$. Now define

$$T^* = \{\{0\bar{x}, 1\bar{y}, 2\bar{z}\} \mid \{x, y, z\} \in T\}.$$

Then T^* is a trade of type one in $AG(d+1, 3)$ and $T^* \cap \mathcal{H}^* = \{\{\bar{i}, \bar{j}, \bar{k}\}\}$. Secondly, if $i \equiv j \pmod{3^d}$ then $\bar{i}' = \bar{j}' = \bar{k}'$ and $\bar{i}'$ occurs in a block of $\mathcal{H}$. So if T^* is the trade defined in Lemma 29 then $T^* \cap \mathcal{H}^* = \{\{\bar{i}, \bar{j}, \bar{k}\}\}$. This completes the proof.   □

Remark 4.1 Consider $\mathcal{H}$ as in Theorem 1. Then by Lemma 21 and Condition (2) in Theorem 1 we see that $\mathcal{H}$ is a defining set in $AG(d, 3)$. Now Condition (3) in Theorem 1 guarantees that each block of $\mathcal{H}$ is necessary for unique completion. Therefore $\mathcal{H}$ is a minimal defining set in $AG(d, 3)$.

5.   Sets of good hyperplanes in $AG(d, 3)$

In this section we show first that Theorem 23 is a simple corollary of Theorem 1. Then we introduce two sets of hyperplanes which provide minimal defining sets of different sizes in $AG(d, 3)$. These two sets are also different from the set introduced in Theorem 23. Note that for simplicity we use integer representation for the points of $AG(d, 3)$ in this section.

Proof of Theorem 23 If $d = 2$ then **H** consists of the following four hyperplanes of $AG(2, 3)$:

$$\begin{cases} 1x_1 & + & 0x_0 & = & 0 \\ 1x_1 & + & 0x_0 & = & 1 \\ 0x_1 & + & 1x_0 & = & 0 \\ 0x_1 & + & 1x_0 & = & 1. \end{cases}$$

One can see that the set of blocks which lie within these hyperplanes is

$$\mathcal{S}_4 = \{\{0, 1, 2\}, \{3, 4, 5\}; \{0, 3, 6\}, \{1, 4, 7\}\}.$$

Moreover, the partial Steiner latin square corresponding to $\mathcal{S}_4$ is the partial Steiner latin square $L_{9,4}$ given in Example 11. Now it is straightforward to check that:

(1) the point 8 ($\bar{8} = (2, 2)$) is not incident with any hyperplane of **H**;

(2) $L_{9,4}$ is a completable set of order 9;

(3) for each block $\{i, j, k\}$ of $\mathcal{H}$ there is a trade T of type one in $AG(2, 3)$ such that $T \cap \mathcal{H} = \{\{i, j, k\}\}$.

So by Remark 4.1 we see that $\mathcal{H}$ is a minimal defining set in $AG(2, 3)$. Now for $d \geq 3$ Theorem 23 follows by Theorem 1 and an induction on d. $\hfill\square$

Theorem 30 *Let* **H** *consist of the following* $2d+1$ *hyperplanes of* $AG(d, 3)$, *where* $d \geq 2$.

$$
\left\{
\begin{array}{ccccccccccc}
1x_{d-1} & + & 0x_{d-2} & + & 0x_{d-3} & + & \ldots & + & 0x_1 & + & 0x_0 & = & 0 \\
1x_{d-1} & + & 0x_{d-2} & + & 0x_{d-3} & + & \ldots & + & 0x_1 & + & 0x_0 & = & 1 \\
0x_{d-1} & + & 1x_{d-2} & + & 0x_{d-3} & + & \ldots & + & 0x_1 & + & 0x_0 & = & 0 \\
0x_{d-1} & + & 1x_{d-2} & + & 0x_{d-3} & + & \ldots & + & 0x_1 & + & 0x_0 & = & 1 \\
0x_{d-1} & + & 0x_{d-2} & + & 1x_{d-3} & + & \ldots & + & 0x_1 & + & 0x_0 & = & 0 \\
0x_{d-1} & + & 0x_{d-2} & + & 1x_{d-3} & + & \ldots & + & 0x_1 & + & 0x_0 & = & 1 \\
& & & & & & \cdots & & & & \\
& & & & & & \cdots & & & & \\
& & & & & & \cdots & & & & \\
0x_{d-1} & + & 0x_{d-2} & + & 0x_{d-3} & + & \ldots & + & 1x_1 & + & 0x_0 & = & 0 \\
0x_{d-1} & + & 0x_{d-2} & + & 0x_{d-3} & + & \ldots & + & 1x_1 & + & 0x_0 & = & 1 \\
0x_{d-1} & + & 0x_{d-2} & + & 0x_{d-3} & + & \ldots & + & 0x_1 & + & 1x_0 & = & 0 \\
0x_{d-1} & + & 0x_{d-2} & + & 0x_{d-3} & + & \ldots & + & 1x_1 & + & 2x_0 & = & 0 \\
0x_{d-1} & + & 0x_{d-2} & + & 0x_{d-3} & + & \ldots & + & 1x_1 & + & 1x_0 & = & 1 \\
\end{array}
\right.
$$

Then $\mathcal{H}$, *the collection of blocks contained within the hyperplanes of* **H**, *is a minimal defining set in* $AG(d, 3)$.

Proof. If $d = 2$ then **H** consists of the following five hyperplanes of $AG(2, 3)$:

$$
\left\{
\begin{array}{ccccc}
1x_1 & + & 0x_0 & = & 0 \\
1x_1 & + & 0x_0 & = & 1 \\
0x_1 & + & 1x_0 & = & 0 \\
1x_1 & + & 2x_0 & = & 0 \\
1x_1 & + & 1x_0 & = & 1.
\end{array}
\right.
$$

It is easy to see that the blocks which lie in these hyperplanes are

$$
S_5 = \{\{0, 1, 2\}, \{0, 3, 6\}, \{0, 4, 8\}, \{1, 3, 8\}, \{3, 4, 5\}\}.
$$

Moreover, the partial Steiner latin square corresponding to $\mathcal{H}$ is the partial Steiner latin square $L_{9,5}$ given in Example 11. Now it is straightforward to check that:

(1) the point 7 ($\bar{7} = (2, 1)$) is not incident with any hyperplane of **H**;

(2) $L_{9,5}$ is a completable set of order 9;

(3) for each block $\{i, j, k\}$ of $\mathcal{H}$ there is a trade T of type one in $AG(2,3)$ such that $T \cap \mathcal{H} = \{\{i, j, k\}\}$.

So by Remark 4.1 we see that $\mathcal{H}$ is a minimal defining set in $AG(2,3)$. Now for $d \geq 3$ the result follows by Theorem 1 and an induction on d. $\square$

Theorem 31 *Let* **H** *consist of the following* $2d+1$ *hyperplanes of* $AG(d, 3)$, *where* $d \geq 3$.

$$\begin{cases} 1x_{d-1} + 0x_{d-2} + \ldots + 0x_2 + 0x_1 + 0x_0 = 0 \\ 1x_{d-1} + 0x_{d-2} + \ldots + 0x_2 + 0x_1 + 0x_0 = 1 \\ 0x_{d-1} + 1x_{d-2} + \ldots + 0x_2 + 0x_1 + 0x_0 = 0 \\ 0x_{d-1} + 1x_{d-2} + \ldots + 0x_2 + 0x_1 + 0x_0 = 1 \\ 0x_{d-1} + 0x_{d-2} + \ldots + 0x_2 + 0x_1 + 0x_0 = 0 \\ 0x_{d-1} + 0x_{d-2} + \ldots + 0x_2 + 0x_1 + 0x_0 = 1 \\ \qquad\qquad \ldots \\ \qquad\qquad \ldots \\ \qquad\qquad \ldots \\ 0x_{d-1} + 0x_{d-2} + \ldots + 0x_2 + 1x_1 + 0x_0 = 0 \\ 0x_{d-1} + 0x_{d-2} + \ldots + 0x_2 + 1x_1 + 0x_0 = 1 \\ 0x_{d-1} + 0x_{d-2} + \ldots + 0x_2 + 0x_1 + 1x_0 = 0 \\ 0x_{d-1} + 0x_{d-2} + \ldots + 1x_2 + 1x_1 + 2x_0 = 0 \\ 0x_{d-1} + 0x_{d-2} + \ldots + 1x_2 + 1x_1 + 1x_0 = 1 \end{cases}$$

Then $\mathcal{H}$, *the collection of blocks contained within the hyperplanes of* **H**, *is a minimal defining set in* $AG(d, 3)$.

Proof. If $d = 3$ then **H** consists of the following seven hyperplanes of $AG(3,3)$:

$$\begin{cases} 1x_2 + 0x_1 + 0x_0 = 0 \\ 1x_2 + 0x_1 + 0x_0 = 1 \\ 0x_2 + 1x_1 + 0x_0 = 0 \\ 0x_2 + 1x_1 + 0x_0 = 1 \\ 0x_2 + 0x_1 + 1x_0 = 0 \\ 1x_2 + 1x_1 + 2x_0 = 0 \\ 1x_2 + 1x_1 + 1x_0 = 1 \end{cases}$$

It is straightforward to see that the blocks which lie in these hyperplanes are as follows.

{0, 1, 2}	{0, 3, 6}	{0, 4, 8}	{0, 5, 7}	{0, 9, 18}
{0, 10, 20}	{0, 11, 19}	{0, 12, 24}	{0, 14, 25}	{0, 15, 21}
{1, 3, 8}	{1, 4, 7}	{1, 5, 6}	{1, 9, 20}	{1, 10, 19}
{1, 11, 18}	{1, 14, 24}	{1, 16, 22}	{2, 3, 7}	{2, 4, 6}
{2, 5, 8}	{2, 9, 19}	{2, 10, 18}	{2, 11, 20}	{3, 4, 5}
{3, 9, 24}	{3, 12, 21}	{3, 13, 23}	{3, 14, 22}	{3, 15, 18}
{3, 16, 20}	{4, 10, 25}	{4, 12, 23}	{4, 13, 22}	{4, 14, 21}
{4, 15, 20}	{5, 12, 22}	{5, 13, 21}	{5, 14, 23}	{6, 7, 8}
{6, 9, 21}	{6, 12, 18}	{6, 15, 24}	{8, 9, 22}	{8, 10, 21}
{8, 14, 20}	{8, 15, 25}	{8, 16, 24}	{9, 10, 11}	{9, 12, 15}
{9, 13, 17}	{9, 14, 16}	{10, 12, 17}	{10, 13, 16}	{10, 14, 15}
{11, 12, 16}	{11, 13, 15}	{11, 14, 17}	{12, 13, 14}	{15, 16, 17}
{18, 19, 20}	{18, 21, 24}	{20, 21, 25}	{20, 22, 24}	{21, 22, 23}

Let P be the partial Steiner latin square corresponding to $\mathcal{H}$. Then:

(1) the point 26 ($\bar{26} = (2, 2, 2)$) is not incident with any hyperplane of **H**;

(2) a backtrack search shows that P is a completable set of order 27;

(3) a modification of the backtrack search given in [15] shows that for each block $\{i, j, k\}$ of $\mathcal{H}$ there is a trade T of type one in $AG(3, 3)$ such that $T \cap \mathcal{H} = \{\{i, j, k\}\}$.

So by Remark 4.1 we see that $\mathcal{H}$ is a minimal defining set in $AG(3, 3)$. Now for $d \geq 4$ the result follows by Theorem 1 and an induction on d. $\square$

6.   Related critical sets

In [7], Gower shows that the number of blocks in the minimal defining set in Theorem 23 is $\frac{1}{6}(3^d(3^d - 1) - 7^d + 1)$. In this section we calculate the number of blocks in minimal defining sets obtained in Theorems 30 and 31. Moreover, we prove that the partial Steiner latin squares corresponding to the minimal defining sets obtained in Theorems 23, 30 and 31 are critical sets.

Theorem 32 *Let* **H**, $\mathcal{H}$ *and* P *be as in Theorem 1. Then* P *is a critical set of order* 3^d.

Proof. By Condition (2) we only need to prove that each entry of P is necessary for completion. First consider the off-diagonal entry $(i, j; k)$ of P. By Condition (3) there is a trade T of type one in $AG(d, 3)$ such that $T \cap \mathcal{H} = \{\{i, j, k\}\}$. Now by an extension of Lemma 18 we see that the entry $(i, j; k)$ is necessary. Secondly, consider the entry $(i, i; i)$ of P. By Condition (1) there is a point, j say, of $AG(d, 3)$ which is not incident with any hyperplane of $\mathbf{H}$. So the cell (j, j) of P is empty. Let $\bar{k} = -\bar{i} - \bar{j}$, where $\bar{i}$ and $\bar{j}$ are vector representations of i and j respectively. Define the latin trade

$$I = \{(i, i; i), (i, k; j), (j, k; i), (j, j; j), (k, j; i), (k, i; j)\}.$$

Since $\{i, j, k\}$ is not a block of $\mathcal{H}$ and the cell (j, j) of P is empty it follows that $I \cap P = \{(i, i; i)\}$. So the entry $(i, i; i)$ of P is necessary. $\square$

Theorem 33 *For all $d \geq 2$ the partial Steiner latin square P corresponding to the minimal defining set in Theorem 23 is a critical set of size $9^d - 7^d$.*

Proof. First note that by Theorem 32 P is a critical set of order 3^d. Since each block of the minimal defining set fills exactly six off-diagonal cells of the partial Steiner latin square and there is exactly one point of $AG(d, 3)$ which is not incident with any hyperplanes in Theorem 23 it follows that the number of entries is

$$6(\frac{1}{6}(3^d(3^d - 1) - 7^d + 1)) + (3^d - 1)$$
$$= (3^d - 1)(3^d + 1) - 7^d + 1 = 9^d - 7^d. \qquad \square$$

Theorem 34 *For all $d \geq 2$ the partial Steiner latin square P in Theorem 30 is a critical set of size $9^d - 43.7^{d-2}$.*

Proof. First note that by Theorem 32 P is a critical set of order 3^d. If $d = 2$ the number of entries in the partial latin square P (see the partial latin square $L_{9,5}$ given in Example 11) is $38 = 3^4 - 43.7^0$. Now by an induction on d the proof for $d \geq 3$ follows. $\square$

Since any block $\{i, j, k\}$ of $\mathcal{H}$ corresponds to exactly six off-diagonal entries of P and there is exactly one point of $AG(d, 3)$ which is not in any block of $\mathcal{H}$ we have the following result on the number of blocks of $\mathcal{H}$.

Theorem 35 *Let* $\mathcal{H}$ *be as in Theorem 30. Then the number of blocks in* $\mathcal{H}$ *is*

$$\frac{1}{6}(3^d(3^d-1)-7^{d-2}+1)-7^{d-1}.$$

Theorem 36 *For all* $d \geq 3$ *the partial Steiner latin square* P *in Theorem 31 is a critical set of size* $9^d - 313.7^{d-3}$.

Proof. First note that by Theorem 32 P is a critical set of order 3^d. If $d = 3$ then P is the partial Steiner latin square corresponding to the blocks given in Theorem 31. The number of entries in P is $416 = 3^6 - 313$. Now by an induction on d the proof for $d \geq 4$ follows. $\qquad\square$

As for Theorem 35 we can prove the following result.

Theorem 37 *Let* $\mathcal{H}$ *be as in Theorem 31. Then the number of blocks in* $\mathcal{H}$ *is*

$$\frac{1}{6}(3^d(3^d-1)-7^{d-3}+1)-52.7^{d-3}.$$

7.   Appendix

The following 12 blocks form a defining set for the noncyclic $STS(13)$ on $\{0,1,2,\ldots,12\}$.

$$\{1,\ 7,\ 5\}\quad \{1,\ 6,\ 12\}\quad \{2,4,10\}\quad \{2,6,8\}\quad \{3,\ 4,\ 9\}$$
$$\{3,\ 5,\ 11\}\quad \{4,\ 7,\ 8\}\quad \{5,8,\ 9\}\quad \{6,7,9\}\quad \{7,\ 10,\ 11\}$$
$$\{8,\ 11,\ 12\}\quad \{9,\ 10,\ 12\}$$

Let P be the partial Steiner latin square corresponding to this defining set. Then P has precisely 9 completions. Here we present a completion which is not a Steiner latin square.

0	9	7	8	1	2	3	12	10	11	5	6	4
4	1	9	10	11	7	12	5	3	0	8	2	6
5	11	2	7	10	12	8	0	6	1	4	9	3
6	8	12	3	9	11	10	2	0	4	1	5	7
12	0	10	9	4	6	11	8	7	3	2	1	5
10	7	0	11	12	5	4	1	9	8	6	3	2
11	12	8	0	5	10	6	9	2	7	3	4	1
2	5	3	12	8	1	9	7	4	6	11	10	0
3	10	6	1	7	9	2	4	8	5	0	12	11
1	2	11	4	3	8	7	6	5	9	12	0	10
8	3	4	6	2	0	5	11	1	12	10	7	9
9	4	1	5	6	3	0	10	12	2	7	11	8
7	6	5	2	0	4	1	3	11	10	9	8	12

References

[1] Peter Adams, Richard Bean and Abdollah Khodkar, *A census of critical sets in the Latin squares of order at most six*, (submitted).

[2] Peter Adams and Abdollah Khodkar, *On the direct product of two weak uniquely completable partial latin squares*, Utilitas Mathematica **60** (2001), 249–253.

[3] J Denes and A D Keedwell, 'Latin squares and their applications', The English Universities' Press Ltd, London (1974).

[4] Diane Donovan and Adelle Howse, *Critical sets for latin squares of order 7*, Journal of Combinatorial Mathematics and Combinatorial Computing **28** (1998), 113–123.

[5] Rebecca A H Gower, *Minimal defining sets in a family of Steiner triple systems*, Australasian Journal of Combinatorics **8** (1993), 55–73.

[6] Rebecca A H Gower, *Critical sets in products of latin squares*, Ars Combinatoria, **55** (2000), 293–317.

[7] Rebecca A H Gower, *Defining sets for the Steiner triple systems from affine spaces*, Journal of Combinatorial Designs **5** (1997), 155–175.

[8] Ken Gray, *Further results on smallest defining sets of well-known designs*, Australasian Journal of Combinatorics, **1** (1990), 91–100.

[9] Ken Gray, *On the minimum number of blocks defining a design*, Bulletin of the Australian Mathematical Society **41** (1990), 97–112.

[10] Brenton Gray, Rudi Mathon, Tony Moran and Anne Penfold Street, *The spectrum of minimal defining sets of some Steiner systems*, Discrete Mathematics (to appear).

[11] Catherine S Greenhill, *An algorithm for finding smallest defining sets of t-designs*, Master's thesis, The University of Queensland, (1992).

[12] Catherine S Greenhill, *An algorithm for finding smallest defining sets of t-designs*, Journal of Combinatorial Mathematics and Combinatorial Computing, **14**, (1993), 39–60.

[13] A D Keedwell, *What is the size of the smallest latin square for which a weakly completable critical set of cells exists?* Ars Combinatoria **51** (1999), 97–104.

[14] G B Khosrovshahi and H R Maimani, *On $2 - (v, 3)$ Steiner trades of small volumes*, Ars Combinatoria **52** (1999), 199–220.

[15] Colin Ramsay, *Trades and defining sets: theoretical and computational results*, PhD thesis, The University of Queensland, Australia, 1998.

Chapter 7

HADAMARD MATRICES, ORTHOGONAL DESIGNS AND CONSTRUCTION ALGORITHMS

Stelios Georgiou

Department of Mathematics
National Technical University of Athens
Zografou 15773, Athens, Greece
email: sgeorg@math.ntua.gr

Christos Koukouvinos

Department of Mathematics
National Technical University of Athens
Zografou 15773, Athens, Greece
email: ckoukouv@math.ntua.gr

Jennifer Seberry

School of IT and Computer Science
University of Wollongong
Wollongong, NSW 2522, Australia
email: j.seberry@uow.edu.au

Abstract   We discuss algorithms for the construction of Hadamard matrices. We include discussion of construction using Williamson matrices, Legendre pairs and the discret Fourier transform and the two circulants construction.

Next we move to algorithms to determine the equivalence of Hadamard matrices using the profile and projections of Hadamard matrices.

A summary is then given which considers inequivalence of Hadamard matrices of orders up to 44.

The final two sections give algorithms for constructing orthogonal designs, short amicable and amicable sets for use in the Kharaghani array.

1. Algorithms for constructing Hadamard matrices

1.1 Hadamard matrices constructed from Williamson matrices

An Hadamard matrix H of order n has elements ± 1 and satisfies $HH^T = nI_n$. These matrices are used extensively in coding and communications (see Seberry and Yamada [75]). The order of an Hadamard matrix is 1, 2 or $n \equiv (0 \bmod 4)$. The first unsolved case is order 428. We use Williamson's construction as the basis of our algorithm to construct a distributed computer search for new Hadamard matrices. We briefly describe the theory of Williamson's construction below. Previous computer searches for Hadamard matrices using Williamson's condition are described in Section 1.1.1. The implementation of the search algorithm is presented in Section 1.1.2, and the results of the search are described in Section 1.1.3.

Theorem 1 (Williamson [88]) *Suppose there exist four* $(1,-1)$ *matrices A, B, C, D of order n which satisfy*

$$XY^T = YX^T, X, Y \in \{A, B, C, D\}$$

Further, suppose

$$AA^T + BB^T + CC^T + DD^T = 4nI_n \qquad (1.1)$$

Then

$$H = \begin{bmatrix} A & B & C & D \\ -B & A & -D & C \\ -C & D & A & -B \\ -D & -C & B & A \end{bmatrix} \qquad (1.2)$$

is an Hadamard matrix of order 4n constructed from a Williamson array.

Let the matrix T given below be called the shift matrix:

$$T = \begin{bmatrix} 0 & 1 & 0 & \cdots & 0 \\ 0 & 0 & 1 & \cdots & 0 \\ & \cdots & & \cdots & . \\ 0 & 0 & 0 & \cdots & 1 \\ 1 & 0 & 0 & \cdots & 0 \end{bmatrix} \tag{1.3}$$

and note

$$T^n = I, \ (T^i)^{\mathrm{T}} = T^{n-i} \tag{1.4}$$

If n is odd, T is the matrix representation of the nth root of unity ω, $\omega^n = 1$.

Let

$$\begin{cases} A = \sum_{i=0}^{n-1} a_i T^i, & a_i = \pm 1, a_{n-i} = a_i \\ B = \sum_{i=0}^{n-1} b_i T^i, & b_i = \pm 1, b_{n-i} = b_i \\ C = \sum_{i=0}^{n-1} c_i T^i, & c_i = \pm 1, c_{n-i} = c_i \\ D = \sum_{i=0}^{n-1} d_i T^i, & d_i = \pm 1, d_{n-i} = d_i \end{cases} \tag{1.5}$$

Then matrices A, B, C, D may be represented as polynomials. The requirement that $x_{n-i} = x_i, x \in \{a, b, c, d\}$ forces the matrices A, B, C, D to be symmetric.

Since A, B, C, D are symmetric, (1.1) becomes:

$$A^2 + B^2 + C^2 + D^2 = 4nI_n$$

and the relation $XY^{\mathrm{T}} = YX^{\mathrm{T}}$ becomes $XY = YX$ which is true for polynomials.

Definition 14 *Williamson matrices are* $(1, -1)$ *symmetric circulant matrices. As a consequence of being symmetric and circulant they commute in pairs.*

We use the following theorem of Williamson's as the motivator for our search algorithm:

Theorem 2 (Williamson [88]) *If there exist solutions to the equations*

$$\mu_i = 1 + 2 \sum_{j=1}^{s} t_{ij}(\omega^j + \omega^{n-j}), i = 1, 2, 3, 4 \tag{1.6}$$

where $s = \frac{1}{2}(n-1), \omega$ *is a nth root of unity, exactly one of* $t_{1j}, t_{2j}, t_{3j}, t_{4j}$ *is nonzero and equals* ± 1 *for each* $1 \leq j \leq s$, *and*

$$\mu_1^2 + \mu_2^2 + \mu_3^2 + \mu_4^2 = 4n$$

then there exist solutions to the equations:

$$\begin{cases} A = \sum_{i=0}^{n-1} a_i T^i, & a_0 = 1, a_i = a_{n-i} = \pm 1 \\ B = \sum_{i=0}^{n-1} b_i T^i, & b_0 = 1, b_i = b_{n-i} = \pm 1 \\ C = \sum_{i=0}^{n-1} c_i T^i, & c_0 = 1, c_i = c_{n-i} = \pm 1 \\ D = \sum_{i=0}^{n-1} d_i T^i, & d_0 = 1, d_i = d_{n-i} = \pm 1 \end{cases} \qquad (1.7)$$

That is, there exists an Hadamard matrix of order 4n.

In matrix form, $\omega^j + \omega^{n-j}$ is represented as $T^j + T^{n-j}$. Since these are symmetric, we write

$$\omega_j = \omega^j + \omega^{n-j}$$

Remark 1 The solutions for (1.6) are independent of the particular root ω, so if n as defined by (1.1) is prime, we can choose ω so that the first μ having any ω_j assigned has ω_1. Since the equations are true for all roots of unity ω, they are also true for $\omega = 1$.

Theorem 3 (Williamson [88]) *Let n be odd, and matrices A, B, C, D satisfy (1.1) and (1.5), suppose $a_0 = b_0 = c_0 = d_0$, then exactly three of $a_j, b_j, c_j, d_j, 1 \le j \le n-1$, have the same sign.*

1.1.1 Results from previous searches.
In many cases complete searches have been conducted for Hadamard matrices of Williamson type. Searches have also been conducted for special classes of Williamson type Hadamard matrices. Furthermore, an infinite class of such matrices is known and will also be discussed briefly.

- Baumert and Hall [4] report results of a complete search for orders $4t$, t odd and $3 \le t \le 23$. Some incomplete results for higher orders are also given.

- Sawade [72] reports results of a complete search for orders $4t$, $t = 25, 27$. The results for $t = 25$ were later demonstrated to be incomplete by Djokovic [10].

- Djokovic [8] reports results of a complete search for orders $4t$, $t = 29, 31$. Only a single non-equivalent solution was found for $t = 29$ and is equivalent to an earlier result due to Baumert [2].

- Koukouvinos and Kounias [52, 53] report results of a complete search for order $4t$, $t = 33$ and 39. These results were later demonstrated to be incomplete by Djokovic [9].

- Djokovic [9] reports results of a complete search for orders $4t$, $t = 33, 35, 39$.

- Djokovic [10] reports results of a complete search for orders $4t$, $t = 25, 37$. This extends results obtained by Sawade [72] for $t = 25$ and, for $t = 37$, by Williamson [88] and later Yamada [89] for a special class of matrices.

- Horton, Koukouvinos, and Seberry [41] report results of a complete search for orders $4t$, t odd and $25 \leq t \leq 37$. No new results were found, confirming existence results.

An infinite family of Hadamard matrices of Williamson type has been proved to exist under certain conditions [83, 87]:

Theorem 4 *If q is a prime power, $q \equiv 1 \pmod 4$, $q + 1 = 2t$, then there exists a Williamson matrix of order $4t$; we have $C = D$, and A and B differ only on the main diagonal.*

This theorem gives examples of Hadamard matrices of Williamson type for orders $4t$, $t = 31, 37, 41, 45, 49, 51, 55, \ldots$, for example.

Yamada [89] has searched for Hadamard matrices of Williamson type, with certain restrictions. These matrices are referred to as *Williamson type j matrices*. The Williamson equation for such matrices, of order $4n$ is:

$$
4n = \left(1 - 2\sum_{s \in A} c_s \omega_s\right)^2 + \left(1 - 2\sum_{s \in A} c_s \omega_{sj}\right)^2
$$
$$
+ \left(1 - 2\sum_{s \in B} d_s \omega_s\right)^2 + \left(1 - 2\sum_{s \in B} d_s \omega_{sj}\right)^2 \tag{1.8}
$$

where $c_s, d_s = \pm 1$, $\omega_s = \omega^s + \omega^{-s}$, $\omega^n = 1$, $j^2 \equiv -1 \pmod n$, A, B, jA, jB is a partition of $\{1, 2, \ldots, \frac{n-1}{2}\}$. Such a j exists if and only if all prime divisors of n are $\equiv 1 \pmod 4$. This led to some new results for $n = 29, 37, 41$.

1.1.2   Search method.

The search method to find Williamson matrices described in this section was given in [41].

Introduction.   The basic search method is to examine all possible combinations of $\omega_j, 1 \leq j \leq \frac{1}{2}(n-1)$ for each $\mu_i, i = 1, 2, 3, 4$, testing each set of μ so generated to see if it satisfies Williamson's condition and can be used to form an Hadamard matrix of order $4n$. This search method is documented in more detail in the following sections.

As a result of the large size of the search space, a distributed client/server approach was taken to the problem: the server breaks work up into smaller portions which are then processed by the clients; any results discovered are reported to the server by the client. Very little work is done by the server itself.

Using a distributed approach, we are able to perform large amounts of work in a fraction of the time required for a single computer to perform the same amount of work.

At various times during the performance of the searches, Macintosh computers and computers running some variety of UNIX have been available for use. To make best use of the available resources, and to eliminate any need to install software beyond that of the client program itself, all communication was performed using low-level networking APIs, sockets [78] on UNIX and Open Transport [1] on the Macintosh, rather than using a package such as PVM [14] or MPI [32] that in some cases can facilitate the construction of distributed programs.

Searches for Hadamard matrices of all orders up to and including order 148 have been performed using Williamson's method implemented by a client/server system. Towards the end of an initial search of order 148, 37 computers were involved, 20 270MHz Ultra 5 computers from Sun Microsystems, and 17 333MHz iMacs from Apple Computer. No computers not available on the local area network were employed in the initial search. However, a subsequent search performed to verify results utilized 35 350MHz Pentium-II computers at the University of Newcastle in addition to 30 local Ultra 5 computers.

The details of the implementation of Williamson's method within the framework of a client/server system are discussed in the following sections.

Decompose $4n$ into sum-of-squares representation.   The first step in performing a search is to decompose $4n$ into all possible sums-of-squares representations. Observing the form of (1.6), we see that when

$\omega = 1$ each μ_i satisfies:

$$\begin{aligned} |\mu_i| &\equiv 1 \bmod 4, \mu_i > 0; \text{ or} \\ |\mu_i| &\equiv 3 \bmod 4, \mu_i < 0. \end{aligned} \qquad (1.9)$$

For example, the possible decompositions for 148 are:

$$\begin{array}{cccc} 1, & 1, & 5, & 11 \\ 1, & 7, & 7, & 7 \\ 3, & 3, & 3, & 11 \\ 3, & 3, & 7, & 9 \\ 5, & 5, & 7, & 7 \end{array}$$

In the sections to follow, we write ω_{sub} to indicate some $\omega_k = \omega^k + \omega^{n-k}$ for $1 \le k \le \frac{1}{2}(n-1)$ when it is necessary to distinguish from an nth root of unity, ω.

Decide on the number of ω_{sub} assigned to each μ.   The next step is to assign a number of ω_{sub} to each μ. Using (1.9), we see that if $|\mu_i| \equiv 1 \bmod 4$, then of the ω_{sub} contributing to μ_i, the number being added to μ_i will always be $\frac{|\mu_i|-1}{4}$ greater than the number of ω_{sub} that are subtracted. A similar condition can be derived for $|\mu_i| \equiv 3 \bmod 4$. These ω_{sub} are termed "fixed"; others are "floating" and always occur in pairs, one added and the other subtracted. These conditions are enforced to help limit the size of the space to be searched.

All possible permutations of the number of floating ω_{sub} are assigned to each μ over the course of the search of a particular sum-of-squares representation, subject to certain restrictions that are useful for reducing the size of the space to be searched:

1 The number of ω_{sub} assigned to μ_i must be greater than or equal to the number of ω_{sub} assigned to μ_j where $j < i$ and μ_i and μ_j correspond to the same value in the sum-of-squares decomposition. We may apply this condition because for the purposes of testing the set of μ to see if Williamson's condition is satisfied, μ_i and μ_j are interchangeable, and it is desirable to perform the test only once rather than twice. This may be extended further if more than two μ have the same value in the sum-of-squares decomposition.

2 If n is prime, then we may always place ω_1 in the first μ to which any ω_{sub} are assigned. This corresponds to solving the set of μ for some nth root of unity, ω^j, such that ω_1 is present in the first μ to which any ω_{sub} are assigned. Furthermore, if there are ω_{sub}

both added and subtracted from this μ, we may either subtract or add ω_1; we do not need to check both. If this condition is in force, then condition 1 is not applied in the case of the μ to which ω_1 is assigned, but remains applicable for other μ corresponding to the same value from the sum-of-squares decomposition. Enforcing this condition can greatly reduce the size of the space to be searched: for example, applying this condition for searching for Hadamard matrices of size 148 reduces the size of the space to be searched to 37% of its size were this condition not to be enforced (reducing from about 32,387,862,644,280 to 12,062,406,963,464)

For each permutation of floating ω_{sub} that is generated, we must assign specific identities to each ω_{sub} and evaluate Williamson's condition.

Assign specific identities to each ω_{sub} .   We must now assign specific identities to each ω_{sub} so that Williamson's condition may be tested.

Let the number of ω_{sub} added to μ_i be represented by c_{2i-1} and the number of ω_{sub} subtracted from μ_i by c_{2i}. S_{2i-1} is the set of ω_{sub} added to μ_i and S_{2i} is the set of ω_{sub} subtracted from μ_i. That is, there are eight sets S, two for each μ. Some of these sets S may be empty.

$$\mu_i = 1 + 2 \sum_{\forall j \in S_{2i-1}} \omega_j - 2 \sum_{\forall j \in S_{2i}} \omega_j$$

Dividing ω_{sub} into two groups, one added to a μ and the other subtracted, helps to simplify the procedure for iterating over all possible combinations of ω_{sub} .

The sets S_i are formed by choosing c_i elements from the set of ω_{sub} not already allocated to an $S_j, j < i$. Recalling that $s = \frac{1}{2}(n-1)$, $S_{\mathrm{T},0}$ is defined as:

$$S_{\mathrm{T},0} = \{\omega_1, \omega_2, \omega_3, \ldots, \omega_s\}.$$

$S_{\mathrm{T},i}$ is defined as:

$$S_{\mathrm{T},i} = S_{\mathrm{T},i-1} - S_{i-1}, i = 1, \ldots, 8. \qquad (1.10)$$

For convenience, we say that:

$$S_0 = \emptyset$$

Williamson's condition may be tested once $S_1, \ldots, S_8$ have been generated. All possible combinations of c_i elements from $S_{\mathrm{T},i}$ are examined;

$j := 1$;
do
    for k **from** j **to** 8
        populate $S_{T,k}$ from $S_{T,k-1}$ and S_{k-1} using (1.10);
        generate combination S_k by choosing c_k elements from $S_{T,k}$;
        $\boxed{\text{Test Williamson Condition using } S_1, \ldots, S_8 \text{ to generate } \mu_1, \ldots, \mu_4;}$
    $j := 8$;
    $g := $ **false**;
    while $((j > 0)$ **and** $(g == $ **false**$))$
        generate new combination S_j using c_j elements from $S_{T,j}$
        if successful
            $g := $ **true**;
            $j := j + 1$;
        else
            $j := j - 1$;
while $(j > 0)$;

Figure 1.1. Segment of pseudocode illustrating generation of combinations for testing Williamson's condition.

once the combinations are exhausted, the next combination for S_{i-1} is generated. The process is illustrated by the small segment of pseudocode shown in Figure 1.1.

So it should be easy to see that the number of tests of Williamson's condition for a particular set of $c_1, \ldots, c_8$ can be calculated as follows:

$$\text{Evaluations} = \prod_{i=1}^{8} \binom{|S_{T,i}|}{c_i} \qquad (1.11)$$

Usually, however, the total number of evaluations performed will be less than this, for two reasons:

1 If condition 2 from Section 1.1.2 is applied, we choose one fewer ω_{sub} for the set S in which ω_1 is to appear.

2 If μ_i and $\mu_j, i < j$ correspond to the same value in the sum-of-squares decomposition of $4n$ and have the same number of ω_{sub} assigned, then we may require that if ω_x is the ω_{sub} of smallest subscript assigned to μ_i and ω_y has the smallest subscript assigned to μ_j, that $x < y$. Otherwise, work will be repeated when μ_i replicates a sequence that had previously occurred in μ_j. Enforcing this condition ensures that no repetition takes place and reduces the size of the search space slightly. The reduction is unfortunately not

as substantial as that for applying condition 2 from Section 1.1.2.

Dividing up the work for distribution.   The obvious manner
in which to reduce the amount of work performed by the clients to
a reasonable level was to make the server perform part of the work
described in Section 1.1.2. The server performs no evaluations itself, but
would choose sets $S_1, \ldots, S_i$, for some $i < 8$. The client would evaluate
all the possibilities for the choice of the remaining sets $S_{i+1}, \ldots, S_8$.

The server decides what value i should take by estimating the amount
of work involved in a subproblem using a modification of Equation (1.11).
Two constants $S_{\min}$ and $S_{\max}$ must be specified to the server: a subproblem is of acceptable size if its size lies between the two limits. Unfortunately, this does not yield subproblems with an even division of work:
there are some very large and very small subproblems. Very small subproblems can be solved quickly, and result in a large number of reports
of completed problems and requests for new problems being handled by
the server over a short period of time. This can cause congestion and is
not desirable.

The solution that was ultimately adopted was for the server to allocate
multiple small subproblems to a client looking for work. The server also
maintains a queue of pre-allocated subproblems ready for assignment to
clients, so that client requests can be satisfied as rapidly as possible.

1.1.3   Search results.

Lemma 5 *Let the Williamson decomposition into four squares be $s_1^2 +
s_2^2 + s_3^2 + s_4^2 = 4n$. Further, let the row sums of the four Williamson
matrices A, B, C, D be m_1, m_2, m_3, m_4. Let*

$$M = \frac{1}{2} \begin{bmatrix} -1 & 1 & 1 & 1 \\ 1 & -1 & 1 & 1 \\ 1 & 1 & -1 & 1 \\ 1 & 1 & 1 & -1 \end{bmatrix}, \; \underset{\sim}{s} = \begin{bmatrix} s_1 \\ s_2 \\ s_3 \\ s_4 \end{bmatrix}, \; \underset{\sim}{m} = \begin{bmatrix} m_1 \\ m_2 \\ m_3 \\ m_4 \end{bmatrix}$$

Then

$$s_1^2 + s_2^2 + s_3^2 + s_4^2 = 4n \Leftrightarrow m_1^2 + m_2^2 + m_3^2 + m_4^2 = 4n$$

and

$$M\underset{\sim}{s} = \underset{\sim}{m} \Leftrightarrow M\underset{\sim}{m} = \underset{\sim}{s}$$

Proof (1.6) gives, using the root $\omega = 1$, a decomposition with

$$s_i = \mu_i = 1 + 4\sum_{j=1}^{s} t_{ij}, \ i = 1, 2, 3, 4.$$

By Williamson's assumption condition,

$$s_1^2 + s_2^2 + s_3^2 + s_4^2 = 4n.$$

On the other hand,

$$
\begin{aligned}
m_1 &= \sum_{j=1}^{n} a_j \\
&= 1 - 2\sum_{j=1}^{\frac{n-1}{2}} t_{1j} + 2\sum_{j=1}^{\frac{n-1}{2}} t_{2j} + 2\sum_{j=1}^{\frac{n-1}{2}} t_{3j} + 2\sum_{j=1}^{\frac{n-1}{2}} t_{4j} \\
&= 1 - \frac{1}{2}(s_1 - 1) + \frac{1}{2}(s_2 - 1) + \frac{1}{2}(s_3 - 1) + \frac{1}{2}(s_4 - 1) \\
&= \frac{1}{2}(-s_1 + s_2 + s_3 + s_4)
\end{aligned}
$$

Similarly,

$$
\begin{aligned}
m_2 &= \frac{1}{2}(s_1 - s_2 + s_3 + s_4) \\
m_3 &= \frac{1}{2}(s_1 + s_2 - s_3 + s_4) \\
m_4 &= \frac{1}{2}(s_1 + s_2 + s_3 - s_4)
\end{aligned}
$$

and $M\underset{\sim}{s} = \underset{\sim}{m}$. Inverting we have, as $M^{-1} = M$, $M\underset{\sim}{m} = \underset{\sim}{s}$. It is easy to check that

$$m_1^2 + m_2^2 + m_3^2 + m_4^2 = s_1^2 + s_2^2 + s_3^2 + s_4^2 = 4n.$$

Unfortunately, no new matrices were found as a result of the searches run so far. However, we are able to provide independent verification of results from previous searches. This is considered of utility since some previous searches, such as that conducted by Sawade [72], for example, failed to reveal all solutions that are now known for the order searched, in that case, order 100. In particular, we provide verification of results reported by Djokovic [9, 10] for orders 100, 140 and 148. Results for order 100 are also verified by Christos Koukouvinos.

For reference purposes, tables of Hadamard matrices derived from Williamson matrices using circulant symmetric $(1, -1)$ matrices in the Williamson array for orders 100 through 180 are presented in Appendix 1 of [41]. A complete search of order 156 is claimed by Djokovic [9]. Results for orders 164, 172 and 180 are incomplete.

1.2   Hadamard matrices from Williamson matrices for non prime orders

An efficient algorithm to find Williamson matrices of order $n = p \cdot q$, i.e. n is not a prime has been described in [52]. This algorithm computes the solutions in groups of order p and q. In fact with the aim of this algorithm we can find all the inequivalent solutions which satisfy the Williamson equation in groups of orders p and q respectively. Then we can merge these solutions in order to find the solution in the group of order n. Of course this algorithm can also be used when n is prime power but it is not too efficient in this case. More details for this algorithm can be found in [52].

1.2.1   The method.
In this section we give the necessary tools needed for our algorithm. We want to construct the $(1, -1)$ circulant matrices:

$$A = (a_0, a_1, \ldots, a_{m-1}), \quad B = (b_0, b_1, \ldots, b_{m-1}),$$
$$C = (c_0, c_1, \ldots, c_{m-1}), \quad D = (d_0, d_1, \ldots, d_{m-1}),$$

such that

$$A^2 + B^2 + C^2 + D^2 = 4mI_m. \tag{1.12}$$

The symmetry requirement gives $v_i = v_{m-i}$, $i = 1, 2, \ldots, \frac{1}{2}(m-1)$, $v_i \in \{a_i, b_i, c_i, d_i\}$. Let $G_q^T = (I_p, I_p, \ldots, I_p)$ be a $p \times p \cdot q$ matrix, i.e., the unit matrix I_p of order p is repeated q times.

The following theorems have been proved in [52] and are essential tools for our algorithm.

Theorem 6 *If*

(1) $m = p \cdot q$, $p, q > 1$,

(2) $V = (v_0, v_1, \ldots, v_{m-1})$ *is circulant of order* m,

then

(a) $G_q^T \cdot V = U \cdot G_q^T$, where $U = (u_0, u_1, \ldots, u_{p-1})$ is circulant of order p with

$$u_j = \sum_{i \equiv j(mod\ p), i < m} v_i, \quad j = 0, 1, \ldots, p-1,$$

(b) U is symmetric if V is symmetric.

Now multiplying on the left A, B, C, D by G_q^T we obtain:

$$G_q^T A = X_p G_q^T, \ G_q^T B = Y_p G_q^T, \ G_q^T C = Z_p G_q^T, \ G_q^T D = W_p G_q^T$$

where

$$
\begin{aligned}
X_p &= (x_0, x_1, \ldots, x_{p-1}), &&\text{with } x_j = \sum_i a_i, \\
Y_p &= (y_0, y_1, \ldots, y_{p-1}), &&\text{with } y_j = \sum_i b_i, \\
Z_p &= (z_0, z_1, \ldots, z_{p-1}), &&\text{with } z_j = \sum_i c_i, \\
W_p &= (w_0, w_1, \ldots, w_{p-1}), &&\text{with } w_j = \sum_i d_i
\end{aligned}
\tag{1.13}
$$

and the summations are over all $i \equiv j(mod\ p)$, $i < m$.

If we multiply both members of (1.12), on the left by G_q^T and on the right by G_q we obtain in the symmetric case:

$$X_p^2 + Y_p^2 + Z_p^2 + W_p^2 = 4mI_p. \tag{1.14}$$

Of course we do not know A, B, C, D so we do not know X_p, Y_p, Z_p, W_p. However it is easier to find X_p, Y_p, Z_p, W_p satisfying (1.14) than A, B, C, D because p is much smaller than m. Now to construct X_p, Y_p, Z_p, W_p note that:

Theorem 7 *If*

(1) A, B, C, D are circulant and symmetric $(1, -1)$-matrices satisfying (1.12) with row (and hence column) sums a, b, c, d,

(2) X_p, Y_p, Z_p, W_p are as defined in (1.13),

then

(1) $\displaystyle\sum_{j=0}^{p-1} x_j = a, \quad \sum_{j=0}^{p-1} y_j = b, \quad \sum_{j=0}^{p-1} z_j = c, \quad \sum_{j=0}^{p-1} w_j = d,$

$$a^2 + b^2 + c^2 + d^2 = 4m, \quad -q \le x_j, y_j, z_j, w_j \le q, \quad x_j, y_j, z_j, w_j \text{ odd}, \quad (1.15)$$

$$x_j = x_{p-j}, \quad y_j = y_{p-j}, \quad z_j = z_{p-j}, \quad w_j = w_{p-j}, \quad j = 1, 2, \ldots, \tfrac{1}{2}(p-1),$$

(2) *If moreover* $a_0 + b_0 + c_0 + d_0 = 0, \pm 4,$ *then*

$$(x_0+y_0+z_0+w_0) - (a_0+b_0+c_0+d_0) = \begin{cases} 0(mod\ 8), & \text{if} \quad q \equiv 1(mod\ 4), \\ 4(mod\ 8), & \text{if} \quad q \equiv 3(mod\ 4), \end{cases}$$
$$(1.16)$$

$$x_j + y_j + z_j + w_j \equiv 2(mod\ 4), \quad j = 1, 2, \ldots, \tfrac{1}{2}(p-1).$$

1.2.2   The algorithm.

For a given decomposition $4m = a^2 + b^2 + c^2 + d^2$, with $m = p \cdot q$, $p < q$, the algorithm consists of four stages:

I)   1 Form all sequences $X_p = \{x_0, x_1, \ldots, x_{p-1}\}$ satisfying:

         (i) $\displaystyle\sum_{i=0}^{p-1} x_i = a$, (ii) $-q \le x_i \le q$ (iii) x_i odd,

         (iv) $x_i = x_{p-i}$, $\quad i = 1, 2, \ldots, \tfrac{1}{2}(p-1)$.

      2 Repeat the construction for Y_p, Z_p, W_p replacing a with b, c, d respectively.

      3 Examine which quadruples X_p, Y_p, Z_p, W_p satisfy $X_p^2 + Y_p^2 + Z_p^2 + W_p^2 = 4mI_p$.

II)   1 Repeat stage **I** interchanging p and q.

      2 Find all inequivalent solutions by applying the transformation $j \to j \cdot s(mod\ q)$ to each solution X_q, Y_q, Z_q, W_q, where $(s, m) = 1$ for every $s < q$.

III)   1 If there are h_1 solutions X_p, Y_p, Z_p, W_p, and h_2 inequivalent solutions $\hat{X}_q, \hat{Y}_q, \hat{Z}_q, \hat{W}_q$, form the $h_1 \cdot h2$ combined solutions $X_p, Y_p, Z_p, W_p, \hat{X}_q, \hat{Y}_q, \hat{Z}_q, \hat{W}_q$.

      2 Find $A = (a_0, a_1, \ldots, a_{m-1})$ from:

$$a_i = a_{m-i}, \quad i = 1, 2, \ldots, \frac{1}{2}(m-1),$$

$$\sum_{i \equiv j(mod\ p), i < m} a_i = x_j, \quad j = 0, 1, 2, \ldots, \frac{1}{2}(p-1),$$

$$\sum_{i \equiv j(mod\ q), i < m} a_i = \hat{x}_j, \quad j = 0, 1, 2, \ldots, \frac{1}{2}(q - 1),$$

where $X_p = (x_0, x_1, \ldots, x_{p-1})$, $\hat{X}_q = (\hat{x}_0, \hat{x}_1, \ldots, \hat{x}_{q-1})$.

3 Find B, C, D similarly.

IV) Examine which quadruples A, B, C, D satisfy $A^2 + B^2 + C^2 + D^2 = 4mI_m$.

Now repeat stages, **I, II, III, IV** for every decomposition of $4m$ as the sum of four odd squares.

If $p = q$ then the algorithm is:

1)   1 Perform steps 1, 2 ,3 of stage **I** of the previous algorithm.

     2 Find all inequivalent solutions by applying the transformation $j \rightarrow j \cdot s(mod\ p)$ to each solution X_p, Y_p, Z_p, W_p, where $(s, m) = 1$ for every $s < p$.

2)   1 Find $A = (a_0, a_1, \ldots, a_{m-1})$ from:

$$a_i = a_{m-i}, i = 1, 2, \ldots, \frac{1}{2}(m - 1),$$

$$\sum_{i \equiv j(mod\ p), i < m} a_i = x_j, j = 0, 1, 2, \ldots, \frac{1}{2}(p - 1),$$

where $X_p = (x_0, x_1, \ldots, x_{p-1})$.

     2 Find B, C, D similarly.

3) Examine which quadruples A, B, C, D satisfy $A^2 + B^2 + C^2 + D^2 = 4mI_m$.

Now repeat stages, **1, 2, 3,** for every decomposition of $4m$ as the sum of four odd squares.

This algorithm was used in [52, 53] for a complete search for orders $4t$, $t = 33, 39$. The same algorithm was used later by Djokovic [9] for orders $4t$, $t = 33, 35, 39$. He noted one more solution for $t = 33$ and $t = 39$ which was missing in [52, 53]. He also claimed the non existence results for $t = 35$.

1.3   Hadamard matrices from generalized Legendre pairs using the discrete Fourier transform

1.3.1   Definitions and notations.

Let U be a sequence of ℓ real numbers $u_0, u_1, ..., u_{\ell-1}$. The *periodic auto-correlation function, PAF, $P_U(j)$* of such a sequence is defined, reducing $i + j$ modulo ℓ, by:

$$P_U(j) = \sum_{i=0}^{\ell-1} u_i u_{i+j}, \quad j = 0, 1, ..., \ell - 1.$$

Two sequences U and V of identical length ℓ are said to be *compatible* if the sum of their periodic autocorrelations is a constant, say a, except for the 0-th term. That is,

$$P_U(j) + P_V(j) = a, \quad j \neq 0. \tag{1.17}$$

(Such pairs are said to have *constant periodic autocorrelation* even though it is the sum of the autocorrelations that is a constant.) If U and V are both ± 1 sequences, compatible and $a = -2$, then they are called a *generalized Legendre pair (or GL–pair)*.

In this section we are interested for compatible ± 1 sequences which are a GL–pair, and may be used as below to construct Hadamard matrices of order $2\ell + 2$. The Legendre or Jacobi symbol is written $(a|n)$ if n is prime or composite, respectively. When referring to the elements of a $-1, 0, 1$ sequence we often write '$-$' instead of -1 and '$+$' instead of 1.

The *discrete Fourier transform (DFT)* of a sequence U is given by

$$DFT_U(k) = \mu_k = \sum_{i=0}^{\ell-1} u_i \omega^{ik}, \quad k = 0, 1, ..., \ell - 1$$

where ω is a primitive ℓ-th root of unity $e^{\frac{2\pi i}{\ell}}$. If we take the squared magnitude of each term in the DFT of U, the resulting sequence is called the *power spectral density (PSD)* of U. Because we use them so often, the k-th terms in the PSDs of U and V will be denoted by $|\mu_k|^2$ and $|\nu_k|^2$, respectively.

Example 8 The PSD of the sequence 1 2 2 -2 0 0 0 is

$$49.000 \ 19.988 \ 13.220 \ 7.792 \ 7.792 \ 13.220 \ 19.988$$

If a sequence u is transformed by the operation of cyclically taking every d-th element, where $\gcd(d, \ell) = 1$, the sequence U is said to be *decimated* by d. That is, if $V = U$ decimated by d, then $v_i = u_{di \mod \ell}$.

Example 9

$$1111000 \text{ decimated by } 2 = 1100110$$
$$1111000 \text{ decimated by } 3 = 1101010$$

The set of all possible decimations of a sequence is called a *decimation class*. Since d is required to be relatively prime to ℓ, a sequence of length ℓ has $\phi(\ell)$ decimations, though sometimes they are not all distinct. We note that decimation by -1 is the same as reversing a sequence. Hence, by assuming that each sequence also represents its reverse, the maximum size of any decimation class is $\phi(\ell)/2$. Finally, we define compatibility between decimation classes. Two decimation classes are said to be compatible if and only if some sequence belonging to one class is compatible with some sequence in the other class.

1.3.2   Some preliminary results.

We make use of the following well-known theorem [70, Chapter 12], [82, Chapter 10].

Theorem 10 (Wiener–Khinchin Theorem) *The PSD of a sequence is equal to the DFT of its periodic autocorrelation function*

$$|\mu_k|^2 = \sum_{j=0}^{\ell-1} P_U(j)\omega^{jk}. \qquad (1.18)$$

The periodic autocorrelation function is equal to the inverse DFT of the sequence's PSD

$$P_U(j) = \frac{1}{\ell} \sum_{k=0}^{\ell-1} |\mu_k|^2 \omega^{-jk}. \qquad (1.19)$$

The next main theorem was proved in [13].

Theorem 11 *Two sequences are compatible if and only if their PSDs sum to a constant (i.e. $|\mu_k|^2 + |\nu_k|^2 = c$ iff $P_U(j) + P_V(j) = a$).*

Example 12 Two compatible sequences and their PSDs are shown below.

Sequences	PSD (terms 1 to 3)		
1 2 2 -2 0 0 0	19.988	13.220	7.792
2 1 -1 2 -1 0 0	5.012	11.780	17.208
	25.000	25.000	25.000

(hence $c = 25$)

In fact, the constant c depends only on the set of numbers comprising the sequences U and V. It is easily shown that

$$c = \frac{\ell \sum_{i=0}^{\ell-1} u_i^2 - \left(\sum_{i=0}^{\ell-1} u_i\right)^2}{\ell - 1} + \frac{\ell \sum_{i=0}^{\ell-1} v_i^2 - \left(\sum_{i=0}^{\ell-1} v_i\right)^2}{\ell - 1}. \qquad (1.20)$$

Hence, all permutations of the sequences yield the same constant. Theorem 11 is a generalization of results that have appeared in the literature in other forms, see for example Kounias, Koukouvinos, Nikolaou and Kakos [62].

The following useful relationships are easily proved by direct application of the definitions of decimation, autocorrelation and DFT.

- If a sequence is decimated by d, then its autocorrelation is likewise decimated by d, and its DFT and PSD are decimated by d^{-1} mod ℓ.

- It follows immediately that compatible sequences remain compatible if they are decimated by the same amount.

Remark 2 If U, V are $\pm 1, 0$–sequences then the above constant c is $c = w - a$, where w is the total number of non–zero entries and a is the constant from the periodic autocorrelation function of U and V.

1.3.3   Legendre sequences and modified Legendre sequences.

For the remainder of this section we consider only GL–pairs. The following is well known (see for example [38]) and is included for completeness only. Let p be an odd prime. The $-1, 0, 1$ sequence U of length p is called a *Legendre sequence L* if its elements $x_i = l_i$ satisfy

$$l_i = (i|p).$$

In other words, $l_0 = 0$ and for $i \neq 0$, $l_i = 1$ if i is a square modulo p and $l_i = -1$, otherwise. We call $(-1, L)$, $(0, L)$, or $(1, L)$ a *modified Legendre sequence*. The values of the modified Legendre sequence are exactly the same as those of the unmodified one except for l_0 which is set to -1, 0, or $+1$, respectively. $((0, L)$ is of course the original Legendre sequence but sometimes it is convenient to refer to it as an modified Legendre sequence.) Two sequences (e_1, L), (e_2, L) with $e_1, e_2 \in \{-1, 0, 1\}$ are called *modified Legendre sequences* and they are defined in the obvious manner.

Example 13 Let $p = 7$. The modified Legendre sequences $(0, L)$ and $(1, L)$ are given by

$$
\begin{aligned}
(0, L) &= 0 + + - + - - \\
(1, L) &= + + + - + - -
\end{aligned}
$$

The following two lemmas (see [13]) say that GL-pairs exist for lengths ℓ, where:

(i) ℓ is a prime (see for example [13]).

(ii) $2\ell + 1$ is a prime power (these arise from Szekeres difference sets, see for example [13] or [30]).

Lemma 14 *Let p be an odd prime then $(1, -L)$, $(1, L)$ is a GL-pair.*

This lemma shows the existence of a GL–pair for every odd prime p. We also note that

Lemma 15 *Let $p = 2\ell + 1$ be a prime power then there is a GL-pair.*

Theorem 16 *Suppose there is a GL-pair of length ℓ. Then there exists an Hadamard matrix of order $2\ell + 2$.*

Proof The sequences are used to make two circulant matrices A and B of order ℓ. Then the following matrix is the required Hadamard matrix.

$$
\left[
\begin{array}{cc|cc}
- & - & +\ \cdots\ + & +\ \cdots\ + \\
- & + & +\ \cdots\ + & -\ \cdots\ - \\
\hline
+ & + & & \\
\vdots & \vdots & A & B \\
+ & + & & \\
+ & - & & \\
\vdots & \vdots & B^T & -A^T \\
+ & - & & \\
\end{array}
\right]
$$

Corollary 17 *Suppose that there are* $2 - \{\ell; \frac{\ell+1}{2}, \frac{\ell+1}{2}; \frac{\ell+1}{2}\}$ *SDS. Then there exists an Hadamard matrix of order* $2\ell + 2$.

GL-pairs also exist for lengths ℓ, where:

(i) $\ell = 2^k - 1$, $k \geq 2$ (two Galois sequences are a GL-pair, see for example [71]).

(ii) $\ell = 49, 57$ (these have been found by a non-exhaustive computer search that uses generalized cyclotomy and master-switch techniques, see [30, 34]).

(iii) $\ell = 3, 5, \ldots, 45$ (these have been found and classified by exhaustive computer searches, see [13]).

(iv) $\ell = 47, 49$ and 51 (these have been found and classified by partial computer searches, see [13]).

(v) $\ell = 143$ (also verified the results for $\ell = 3, 5, 7, 11, 13, 15, 17, 19, 23, 25, 31, 35, 37, 41, 43, 53, 59, 61, 63$ see [18]).

GL-pairs do not exist for even lengths. It is indicated in [13] that the following lengths $\ell \leq 200$ are unresolved: $55, 77, 85, 87, 91, 93, 115, 117, 121, 123, 129, 133, 145, 147, 159, 161, 169, 171, 175, 177, 185, 187$ and 195.

We note here that a GL-pair for length $\ell = 143$ is constructed easily since $143 = 11 \cdot 13$ is a product of twin primes as indicated in Corollary 21.

1.3.4   The PSD test.

We suppose that the set of numbers comprising sequences U and V are fixed and that only permutations of these sequences will be considered. Now every term in a PSD is non–negative. Hence if the sequences U and V are compatible, then no term in their PSDs can exceed the constant c in Theorem 11. That is,

$$|\mu_k|^2 + |\nu_k|^2 = c \Longrightarrow |\mu_k|^2 \leq c.$$

Equivalently, if any term of a sequence's PSD exceeds c, then the sequence cannot be a member of a compatible pair and so maybe discarded from our search. This test can be generalized in a straightforward manner to any family of sequences over any alphabet that have constant periodic autocorrelation function. (Since, the nonperiodic autocorrelation function being constant implies that the periodic autocorrelation function is constant, the above test is also applicable for such candidate sequences.)

1.3.5   Empirical performance of the PSD test for binary sequences.

Exhaustive searches over the space of all binary $0, 1$–sequences were performed for various lengths and weights (number of ones) to see what fraction of sequences actually pass the PSD test. The lengths ℓ and weights w were chosen to correspond to supplementary difference sets used in the constructions of D–optimal designs [62] and Hadamard matrices (as described above) while c, the threshold for the PSD test, was determined by (1.20). The results are shown Table 1 of [13]. (The last three rows in this table are derived from a count of decimation classes rather than sequences, but the percentage reduction is approximately the same either way.) It is evident that very substantial reductions in the number of candidate sequences can be realized through the use of the PSD test.

The exhaustive search algorithm was divided into three steps. In the first step, all decimation classes of length ℓ and weight $w = \frac{\ell+1}{2}$ are exhaustively generated, and each one that passes the PSD test is saved in a list. In the second step, the list is sorted by offset. In this manner, pairs of classes with equal and opposite offsets can be quickly found, and the third step is to compute the autocorrelation functions of such pairs to confirm whether they are compatible or not.

The results from these three steps for $\ell = 15$ are illustrated in Table 2 of [13].

The results from the exhaustive searches for $\ell \leq 45$ are shown in Table 3 of [13].

1.4   Hadamard matrices from generalized Legendre pairs using supplementary difference sets

1.4.1   Some preliminary results.

We say that two sets of residues modulo ℓ, say P and Q, are $2-\{\ell;\, k_1,\, k_2;\, \lambda\}$ *supplementary difference sets* mod ℓ (abbreviated as *sds*) if $|P| = k_1$, $|Q| = k_2$, and for each non-zero residue $k(mod\ell)$ the congruences $i - j \equiv k;\quad i, j \in P,\ i - j \equiv k;\quad i, j \in Q$, have in total exactly λ solutions.

If P, Q are $2 - \{\ell;\, k_1,\, k_2;\, \lambda\}$ *sds*, then we construct the first row of the corresponding $(-1, 1)$ circulant incidence matrices $A = (a_{ij})$ and $B = (b_{ij})$, $i, j = 0, 1, \ldots, \ell - 1$, as follows:

$$a_{0j} = -1, \quad if \ \ j \in P \ \ and \ \ a_{0j} = 1, \quad otherwise,$$

and

$$b_{0j} = -1, \quad if \ \ j \in Q \ \ and \ \ b_{0j} = 1, \quad otherwise$$

We know (see [5] or [38]) that:

Theorem 18 *(i) If P, Q are supplementary difference sets $2-\{\ell;\, k_1,\, k_2;\, \lambda\}$ and A, B the corresponding $(-1, 1)$ incidence matrices, then*

$$AA^T + BB^T = 4(k_1 + k_2 - \lambda)I_\ell + 2(\ell - 2(k_1 + k_2 - \lambda))J_\ell \qquad (1.21)$$

(ii) Given two $\ell \times \ell$ circulant matrices A, B satisfying (1.21), then the corresponding sets P, Q are supplementary difference sets $2-\{\ell;\, k_1,\, k_2;\, \lambda\}$, where k_1, k_2 is the number of -1's in each row of A, B respectively.

We note that two compatible sequences may contain elements from any alphabet. If the elements of two compatible sequences are $-1, 1$ then they are described as $2 - \{\ell; k_1, k_2; \lambda\}$ *sds* as the previous theorem say. In this section we are interested in the particular case of

$2 - \{\ell; \frac{\ell+1}{2}, \frac{\ell+1}{2}; \frac{\ell+1}{2}\}$ since these give, compatible ± 1 sequences which are a GL-pair, and may be used to construct Hadamard matrices of order $2\ell + 2$.

In this particular case, relation (1.21) becomes

$$AA^T + BB^T = (2\ell + 2)I_\ell - 2J_\ell \qquad (1.22)$$

Multiplying on the left by e^T and on the right by e both sides of (1.22) we obtain:

$$(\ell - 2k_1)^2 + (\ell - 2k_2)^2 = 2 \qquad (1.23)$$

where e is the $\ell \times 1$ vector of one's. Since $k_1 = k_2 = (\ell + 1)/2$, we conclude that, the sum of the elements in each row and column of the circulant matrices A and B must be minus one. Since multiplication by -1 of the first row of A and/or B leaves relation (1.22) invariant, we deduce that the first element in the first rows of A and B will be $+1$ and from the remaining elements half will have positive sign and half negative one. Thus, a necessary condition for the existence of the $(-1, 1)$ circulant matrices A and B satisfying (1.22), or for the existence of the corresponding *sds* is that, ℓ should be odd.

Now we consider the first rows of A and B as two sequences of length ℓ. Using (1.19) it is easy to see that relation (1.22) is equivalent to

$$P_A(0) + P_B(0) = 2\ell \qquad (1.24)$$

$$P_A(s) + P_B(s) = -2, \quad for \ \ s = 1, 2, \ldots, \ell - 1 \qquad (1.25)$$

If a sequence A of length ℓ is transformed by the operation of cyclically taking every d-th element, where $(d, \ell) = 1$, the sequence A is said to be *decimated* by d. That is, if $A' = A$ decimated by d, then $a'_i = a_{di}$, reducing di modulo ℓ. The set of all possible decimations of a sequence is called a *decimation class*. Since d is required to be relatively prime to ℓ, a sequence of length ℓ has $\phi(\ell)$ decimations, though sometimes they are not all distinct. We note that decimation by -1 is the same as reversing a sequence. Hence, by assuming that each sequences also represents its reverse, the maximum size of any decimation class is $\phi(\ell)/2$. Any pair of sequences that can be transformed into another pair by exchanging the sequences, cyclically shifting or reversing either of the sequences, or decimating both by the same amount are considered equivalent. The corresponding *sds* are also considered equivalent. This notice of equivalent *sds* was also considered in [62].

Since in our case the parameters k_1 and k_2 of the *sds* are equal, we investigate multipliers of $2 - \{\ell; \frac{\ell+1}{2}, \frac{\ell+1}{2}; \frac{\ell+1}{2}\}$ *sds*. This efficient technique has already applied for some other families of *sds* in [15, 61]. In these cases the authors construct the set P and search for all possible w's prime to the ℓ, i.e. $(w, \ell) = 1$ such that $Q = wP \ (mod\ell)$, and P, Q constitute a *sds*, if such w's exist. They found many multipliers of the *sds* and constructed D-optimal designs for some orders.

In particular, Koukouvinos, Seberry, Whiteman, and Xia [61] used cyclotomy to prove the following theorem, where C_i are the cyclotomic classes in $GF(v)$ constructed by using a generator g of $GF(v) \setminus \{0\}$.

Theorem 19 (see [61]) *Let g be a generator of the cyclic group $GF(v) \setminus \{0\}$. Suppose*

(i) $v = 2q^2 + 2q + 1$ *is a prime power,*

(ii) *A and B are $2 - \{v; q^2, q^2; \lambda\}$ sds such that $2q+1$ is a multiplier ie $B = (2q + 1)A$, and $2q + 1 \in C_i$,*

(iii) *A and B are unions of cyclotomic classes.*

Then every $\alpha \in C_i$ or $\alpha \in C_i^{-1}$ is also a multiplier i.e. $B = \alpha A$.

1.4.2   Twin prime power construction.
For a comprehensive introduction to cyclotomy see [30] and [79].

Stanton and Sprott [77], Storer [79], and Whiteman [86], showed constructions of difference sets over $GF(p) \times GF(p+2)$, with p, $p+2$ both prime powers. Gysin and Seberry [35] constructed

$$\frac{p+1}{2} - \{p(p+2); \frac{p^2-1}{2}, 2, \ldots, 2; \frac{(p-1)^2}{4}\}$$

sds over $GF(p) \times GF(p+2)$, where p, $p+2$ are two prime powers, $p > 2$. In fact if x, y generate $GF(p)^*$, $GF(p + 2)^*$ respectively, they defined the following cyclotomic classes

$$C_i = \{(x^s, y^{s+i}) : s = 0, \ldots, f - 1\}$$

$$E_k = \{(x^{\frac{p-1}{2}s+k}, 0) : s = 0, 1\}$$

where $i = 0, 1$, $k = 0, \ldots, \frac{p-1}{2} - 1$, and $f = \frac{p^2-1}{2} = lcm(p - 1, p + 1)$.

Furthermore they defined $E = \{(x^s, 0) : s = 0, \ldots, p - 2\}$, $D = \{(0, y^s) : s = 0, \ldots, p\}$. Then using the classes C_0, E, and D they reproved the following theorem, which was originally proved by Stanton and Sprott [77], and Whiteman [86]. This is also included in [3].

Theorem 20 (Stanton-Sprott-Whiteman restated) *Let C_0, E be defined as above, then $\{C_0 \cup E \cup \{0\}\}$ is a*

$$\{p(p+2); \frac{p^2 - 1}{2} + p; \frac{(p+1)^2}{4} - 1\}$$

difference set over $GF(p) \times GF(p+2)$.

Gysin and Seberry [35] also noted the following corollary.

Corollary 21 *Let C_0, D be defined as above, then $\{C_0 \cup D\}$ is a*

$$\{p(p+2); \frac{(p+1)^2}{2}; \frac{(p+1)^2}{4}\}$$

difference set over $GF(p) \times GF(p+2)$.

Example 22 Let $p = 3$, $p + 2 = 5$, $(x, y) = (2, 2) = 2$. Now

$$
\begin{aligned}
C_0 &= \{1, 2, 4, 8\} \\
D &= \{3, 6, 12, 9\} \\
E = E_0 = E_1 &= \{5, 10\}
\end{aligned}
$$

in this case

$$\{C_0 \cup D\} = \{1, 2, 4, 8, 3, 6, 12, 9\},$$

is a $\{15; 8; 4\}$ difference set over $GF(3) \times GF(5) \simeq Z_{15}$.

Example 23 Let $p = 5$, $p + 2 = 7$, $(x, y) = (2, 3) = 17$. Now

$$
\begin{aligned}
C_0 &= \{1, 17, 9, 13, 11, 12, 29, 3, 16, 27, 4, 33\} \\
D &= \{15, 10, 30, 20, 25, 5\} \\
E &= \{21, 7, 14, 28\} \\
E_0 &= \{21, 14\} \\
E_1 &= \{7, 28\},
\end{aligned}
$$

In this case

$$\{C_0 \cup D\} = \{1, 17, 9, 13, 11, 12, 29, 3, 16, 27, 4, 33, 15, 10, 30, 20, 25, 5\},$$

is a $\{35; 18; 9\}$ difference set over $GF(5) \times GF(7) \simeq Z_{35}$.

We observe that the parameters of the difference sets constructed in corollary 2, are $\{\ell, \frac{\ell+1}{2}, \frac{\ell+1}{4}\}$. Hence, the above corollary motivate us to find $2 - \{\ell; \frac{\ell+1}{2}, \frac{\ell+1}{2}; \frac{\ell+1}{2}\}$ sds. Thus we have:

Theorem 24 *There exist* $2 - \{\ell; \frac{\ell+1}{2}, \frac{\ell+1}{2}; \frac{\ell+1}{2}\}$ *sds, where* $\ell = p(p+2)$ *and* $p, p+2$ *are two prime powers,* $p > 2$.

Proof Let D_1 be the $\{\ell, \frac{\ell+1}{2}, \frac{\ell+1}{4}\}$ difference set constructed in corollary 2. Then D_1 and $D_2 = D_1$ constitute a $2 - \{\ell; \frac{\ell+1}{2}, \frac{\ell+1}{2}; \frac{\ell+1}{2}\}$ sds.   □

Thus we conclude that:

Corollary 25 *Let* $\ell = p(p+2)$, *with* $p, p+2$ *both prime powers. Then there exist GL-pairs of length* ℓ.

1.4.3   The algorithm.

For the construction of $2 - \{\ell; \frac{\ell+1}{2}, \frac{\ell+1}{2}; \frac{\ell+1}{2}\}$ sds, we use the following algorithm, which is given in [18]. A modified version of this algorithm has been applied in [15]. This algorithm uses the idea of multipliers and is much faster than the algorithms that have been used in [5] and [33]. This algorithm provides the sds that can be constructed using multipliers and performs an exhaustive search for the multipliers of these sds. Not only the complexity of the algorithm is reduced but also using some powerful but elementary results from group theory the construction used in this algorithm give us a theoretical result on the multipliers of the corresponding sds. Modifications of the algorithm can be used for searching sds with same parameters $k_1 = k_2$ and their multipliers.

For a given ℓ odd

(i) Find positive integers k_1, k_2, λ satisfying:
   $k_1 = k_2 = \lambda = \frac{\ell+1}{2}$.

(ii) For an integer t, $1 \leq t < \ell$, $(t, \ell) = 1$, form all sets $\{a, at, \ldots, at^{m-1}\}$ with $at^m \equiv a \mod(\ell)$ for all $a = 0, 1, \ldots, \ell - 1$. Sort the sets by the smallest element and call them a_i, $i = 0, 1, \ldots, m$.

(iii) Find all possible multipliers using Lemmas 26 and 27. Try only one element from the groups a_i and a_i^{-1}, and do not try multipliers w, unless $(w, \ell) = 1$.

(iv) Form one set P with k_1 elements as union of sets found in step (ii).

(v) For each multiplier w found in step (iii), set $Q = wP$.

(vi) Examine if P, Q are supplementary difference sets $2 - \{\ell; k_1, k_2; \lambda\}$.

(vii) If the answer in (vi) is positive then save the set P and multiplier w.

(viii) If the multiplier that used was not the last, then go to step (v) and try the next multiplier.

(ix) Repeat steps (iv)-(viii) until all possible combinations of unions of sets P are examined.

(x) If the last possible union of sets P is reached, then go to step (ii) and use the next integer t to form the sets a_i.

(xi) Repeat steps (ii)-(x) until all values of t, $1 < t < \ell$, $(t, \ell) = 1$ are examined.

Next Lemmas which are essential in our search for multipliers of *sds* were proved in [18].

Lemma 26 *Let a_i, $i = 0, 1, \ldots, m$ be the subsets constructed in step (ii) of our algorithm and $P = a_{i_1} \cup a_{i_2} \cup \ldots \cup a_{i_n}$, $Q = w_1 P$, $w_1 \in a_j$, $j \in \{1, \ldots, m\}$ be $2 - \{v; k, k; \lambda\}$ supplementary difference set (we say that w_1 is a multiplier for the difference set). Then*

(i) *Every $w \in a_j$ is a multiplier for the supplementary difference set. That is $\forall w \in a_j$, P, $R = wP$ constitute a $2 - \{v; k, k; \lambda\}$ supplementary difference set.*

(ii) Every $w \in a_j^{-1}$ is also a multiplier.

Lemma 27 If $(w, \ell) > 1$ then w cannot be a multiplier.

The above algorithm can perform an exhaustive search for multipliers but only a partial search for the corresponding sds. If the sds can be constructed using multipliers then they will be easily found otherwise the sds cannot be constructed using multipliers but they may exist.

1.5   Hadamard matrices constructed from two circulant matrices

Let $A = \{A_j : A_j = \{a_{j1}, a_{j2}, ..., a_{jn}\}, \ j = 1, ..., \ell\}$, be a set of ℓ sequences of length n. The *non-periodic autocorrelation function (NPAF)* $N_A(s)$ of the above sequences is defined as

$$N_A(s) = \sum_{j=1}^{\ell} \sum_{i=1}^{n-s} a_{ji} a_{j,i+s}, \quad s = 0, 1, ..., n-1. \qquad (1.26)$$

If $A_j(z) = a_{j1} + a_{j2}z + ... + a_{jn}z^{n-1}$ is the associated polynomial of the sequence A_j, then

$$A(z)A(z^{-1}) = \sum_{j=1}^{\ell} \sum_{i=1}^{n} \sum_{k=1}^{n} a_{ji} a_{jk} z^{i-k} = N_A(0) + \sum_{j=1}^{\ell} \sum_{s=1}^{n-1} N_A(s)(z^s + z^{-s}).$$
$$(1.27)$$

It is clear that $P_A(s) = N_A(s) + N_A(n-s), \ s = 1, ..., n-1$. Therefore, if $N_A(s) = 0$ for all $s = 1, ..., n-1$, then $P_A(s) = 0$ for all $s = 1, ..., n-1$. But, $P_A(s)$ may equal zero for all $s = 1, ..., n-1$, even though the $N_A(s)$ are not.

Definition 15 (Golay sequences) Two sequences $A = \{a_1, a_2, ..., a_n\}$ and $B = \{b_1, b_2, ..., b_n\}$ of length n, with elements ± 1, are defined as *Golay sequences* of length n, if the following equations

$$N_A(s) + N_B(s) = 0 \ s \quad s = 1, 2, ..., n-1.$$

hold, where $N_A(s)$ is the nonperiodic autocorrelation function.

Example 28 The following binary sequences, with elements ± 1, are Golay sequences of length $n = 2, 10$ and 26 respectively.

(a) $n = 2$, $A = \{1, 1\}$, $B = \{1, -1\}$

(b) $n = 10$
$$A = \{1, -1, -1, 1, -1, 1, -1, -1, -1, 1\}$$
$$B = \{1, -1, -1, -1, -1, -1, -1, 1, 1, -1\}.$$

(c) $n = 26$

$$A = \{\ \ 1, 1, 1, -1, -1, 1, 1, 1, -1, 1, -1, -1, -1, -1, -1,$$
$$1, -1, 1, 1, -1, -1, 1, -1, -1, -1, -1\ \}$$
$$B = \{\ \ -1, -1, -1, 1, 1, -1, -1, -1, 1, -1, 1, 1, -1, 1, -1,$$
$$1, -1, 1, 1, -1, -1, 1, -1, -1, -1, -1\ \}.$$

Lemma 29 *If A and B are $n \times n$ circulant ± 1 matrices with first rows two Golay sequences $\{a_1, a_2, \ldots, a_n\}$, $\{b_1, b_2, \ldots, b_n\}$ of length n respectively, then*

$$AA^T + BB^T = \left(\sum_{i=1}^{n} (a_i^2 + b_i^2) \right) I_n = 2n I_n.$$

Lemma 30 *Let $A = \{a_1, a_2, \ldots, a_n\}$ and $B = \{b_1, b_2, \ldots, b_n\}$ are two Golay sequences of order n. Suppose that k_1 of the elements a_i are positive $(+1)$ and k_2 of the elements b_i are also positive $(+1)$. Then*

$$n = (k_1 + k_2 - n)^2 + (k_1 - k_2)^2$$

and n is even.

This condition is necessary but not sufficient for the existence of Golay sequences of order n.

Theorem 31 *If $A = \{a_1, a_2, \ldots, a_n\}$ and $B = \{b_1, b_2, \ldots, b_n\}$ are Golay sequences of length n and, $C = \{c_1, c_2, \ldots, c_m\}$ and $D = \{d_1, d_2, \ldots, d_m\}$ are Golay sequences of length m, then the sequences:*

$$X = A \times \left(\frac{C + D}{2} \right) + B \times \left(\frac{C - D}{2} \right)$$

$$Y = A \times \left(\frac{C^* - D^*}{2} \right) - B \times \left(\frac{C^* + D^*}{2} \right)$$

are Golay sequences of length nm.

So, as we know that Golay sequences of length $n = 2, 10, 26$ exist, then with the previous theorem we obtain that they exist in lengths $n = 2^a 10^b 26^c$, where a, b, c are non-negative integers. These results obtained by Golay [31] and Turyn [84], and these are the only known values of n that Golay sequences exist, These are the *Golay numbers*. It has been proved by Eliahou, Kervaire and Saffari [12] that Golay sequences do not exist for values $n = 34, 50, 58, 68$ and for every n that is divided by a prime number $p \equiv 3 \pmod 4$. The existence of Golay sequences of length n, if n, $n < 200$: $n = 74, 82, 106, 116, 122, 130, 136, 146, 148, 164, 170, 178, 194$, is an open problem.

The following theorem is analogous to Theorem 18 and can be used for the construction of Hadamard matrices, see [38] or [90].

Theorem 32 *If A, B are $v \times v$ (v even) circulant matrices with entries ± 1, satisfying:*

$$AA^T + BB^T = 2vI_v \qquad (1.28)$$

Then the matrix

$$H = \begin{bmatrix} A & B \\ -B^T & A^T \end{bmatrix}$$

is a Hadamard matrix of order $2v$.

Corollary 33 *If there are two $(1, -1)$ sequences of length n with zero PAF or NPAF then there exists a Hadamard matrix of order $2n$.*

Theorem 34 *There exist two sequences $(1, -1)$ with zero PAF for all lengths $n = 2^e \cdot 10^f \cdot 26^h \cdot 34$ for all non negative integers e, f, h.*

Proof There are Golay sequences X, Y of length $2^e \cdot 10^f \cdot 26^h$. The following sequences A and B of length 34 have zero PAF, and are given in [17].

$A= \{a, a, a, \bar{a}, \bar{a}, \bar{a}, \bar{b}, \bar{a}, \bar{b}, b, \bar{b}, b, a, \bar{b}, \bar{b}, b, b, a, b, \bar{b}, a, b, \bar{b}, b, b, a, a, \bar{a}, \\ b, \bar{b}, a, b, b, a\}$

$B= \{b, \bar{a}, \bar{a}, b, a, \bar{a}, \bar{b}, b, b, \bar{a}, \bar{a}, a, \bar{a}, b, a, \bar{a}, b, \bar{a}, \bar{a}, a, a, b, \bar{a}, a, \bar{a}, a, \bar{b}, a, \\ \bar{b}, \bar{b}, \bar{b}, b, b, b\}$

In these sequences we replace variables a, b by the sequences X, Y respectively to obtain the desired result. □

2.   On inequivalent Hadamard matrices

2.1   Basic definitions and preliminaries

A Hadamard matrix is said to be *normalized* if it has its first row and column all 1's. Thus we can normalize the Hadamard matrix by multiplying rows and columns by -1 where needed. In these matrices, n is necessarily 2 or a multiple of 4. Two Hadamard matrices H_1 and H_2 are called equivalent (or Hadamard equivalent, or H-equivalent) if one can be obtained from the other by a sequence of row negations, row permutations, column negations and columns permutations.

The discussion of Hadamard equivalence is quite difficult, principally because of the lack of a good canonical form. The exact results which have been discovered are as follows : Hadamard matrices of orders less than 16 are unique up to equivalence. There are precisely five equivalence classes at order 16, and three equivalence classes at order 20, see [36, 37]. There are precisely 60 equivalence classes at order 24, see [42, 47]. There are precisely 487 equivalence classes at order 28, see [48, 49]. The classification of Hadamard matrices of orders $n \geq 32$ is still remains an open and difficult problem since an algorithmic approach of an exhaustive search is an NP hard problem.

Given two Hadamard matrices of the same order, it can be quite difficult to decide whether or not they are equivalent.

The next two subsections discuss the use of the "profile" and "projections" of Hadamard matrices to determine inequivalence.

The following criterion (profile) was given in [6].

2.2   The profile criterion

Cooper, Milas and Wallis in [6] suggested the profile criterion to investigate the equivalence of Hadamard matrices. Later Lin, Wallis and Zhu in [64, 66, 67] proposed some modifications of this criterion. Suppose H is a Hadamard matrix of order $4n$ with typical entries h_{ij}. We write $P_{ijk\ell}$ for the absolute value of the generalized inner product of rows i, j, k and ℓ :

$$P_{ijk\ell} = |\sum_{x=1}^{4n} h_{ix}h_{jx}h_{kx}h_{\ell x}|$$

This criterion does not work in the case of Hadamard matrices of order $n = 20$ because it gives the same profile for all three equivalent classes of Hadamard matrices of this order.

Proposition 1 (see [6]) $P_{ijk\ell} \equiv 4n \ (mod \ 8)$.

We shall write $\pi(m)$ for the number of sets $\{i, j, k, \ell\}$ of four distinct rows such that $P_{ijk\ell} = m$. The definition and the above give that $\pi(m) = 0$ unless $m \geq 0$ and $m \equiv 4n \ (mod \ 8)$. We call $\pi(m)$ *the profile* (or 4-profile) of H.

The (unique) matrices of order $4, 8$ and 12 have profiles

$$\pi(4) = 1$$
$$\pi(0) = 56, \quad \pi(8) = 14$$
$$\pi(4) = 495, \quad \pi(12) = 0$$

respectively.

The five inequivalent classes of order 16 gave four distinct profiles.

class H_0 :	$\pi(0) = 1680,$	$\pi(8) = 0,$	$\pi(16) = 140$
class H_1 :	$\pi(0) = 1488,$	$\pi(8) = 256,$	$\pi(16) = 76$
class H_2 :	$\pi(0) = 1392,$	$\pi(8) = 484,$	$\pi(16) = 44$
class H_3 :	$\pi(0) = 1344,$	$\pi(8) = 448,$	$\pi(16) = 28$
class H_4 :	$\pi(0) = 1344,$	$\pi(8) = 448,$	$\pi(16) = 28$

The matrices of class H_4 are the transposes of the matrices of class H_3.

The three classes of order 20 all gave the same profile:

$$\pi(4) = 4560, \ \pi(12) = 285, \pi(20) = 0.$$

Similarly we can define a more general profile criterion based on more than 4 rows. For some modifications of the profile such as extended profile and generalized profile we refer the reader to [66]. We now give a modified version of the profile that was given in [6]. We observe that all the conditions which hold for the rows of a Hadamard matrix also hold for its columns.

We write $Q(m)$ for the absolute value of the generalized inner product of m columns, say $c_1, c_2, \ldots, c_m$ and we call this m-column profile.

$$Q(m) = |\sum_{x=1}^{4n} h_{xa_1} h_{xa_2} \cdots h_{xa_m}|$$

We shall write $q(s)$ for the number of sets $\{a_1, a_2, \ldots, a_m\}$ of m distinct rows such that $Q(m) = s$. The definition and the above give that $q(s) = 0$ unless $s \geq 0$. We call $q(s)$ the m-column profile (or m-cprofile) of H.

This criterion as well does not work in all case of Hadamard matrices.

Two more useful criterions to determine inequivalence of Hadamard matrices which are called "K-matrices" and "K-boxes" are also developed in [45, 46]. To save space we do not discuss these criteria here.

2.3 The projection and symmetric Hamming distance distribution algorithms

In this section we describe two new criteria, to test inequivalence in Hadamard matrices of order n, based on their projection properties and their symmetric Hamming distances.

Let H be a $n \times n$ Hadamard matrix. A $n \times k$ submatrix of H which consist of n rows and k columns is called a *projection* of H into k columns. The idea of the first criterion is that if two Hadamard matrices of order n are inequivalent then these matrices should have at least one different projection for some $k \leq n$ and vice versa (if there exist a $k \leq n$ such that the two Hadamard matrices give some different, inequivalent projections, then these Hadamard matrices are inequivalent).

Now we give in brief the description of our algorithm that can be used to determine all inequivalent projections for n and k.

First we give the definition of inequivalent projections of a Hadamard matrix of order n.

Definition 16 *Two projections, in k columns, of Hadamard matrices of order n are equivalent if one can be obtained from the other by one or more of the following transformations*

(a) *Sign changes in the columns (multiply one or more columns by -1).*

(b) *Sign changes in the rows (multiply one or more rows by -1).*

(c) *Permutations of the columns*

(d) *Permutations of the rows.*

The next algorithm gives us all the inequivalent projections of Hadamard matrices and through them the inequivalent Hadamard matrices.

The inequivalent projections algorithm:

(i) Set $k = 2$.

(ii) Find all projections for each Hadamard matrix of a given order n and k columns by taking all possible k columns of the entire $n \times n$ Hadamard matrix. These are $\binom{n}{k}$ projections in total.

(iii) From the projections found in step (ii) find the inequivalent ones using definition 16.

(iv) Check if the set of all projections of the first Hadamard matrix is different (non equivalent) with the set of all projections of the second Hadamard matrix.

(v) If the answer in step (iv) is true then stop and say that these two Hadamard matrices are inequivalent, otherwise increase k by 1.

(vi) If now $k \leq n$ then go to step (ii) and continue, otherwise stop and say that these Hadamard matrices are equivalent.

Lemma 35 *When we project a Hadamard matrix of order $4m$ into $k = 2$ columns we always obtain $\binom{4m}{2}$ equivalent projections. Each of these is equivalent to a projection which is consist of m times over the full 2^2 design.*

Proof A Hadamard matrix has its columns orthogonal to each other. Therefore, any two columns are equivalent with two columns for which each of the pairs $(1, 1), (1, -1), (-1, 1), (-1, -1)$ appear exactly m times. $\square$

Using the above lemma we can slightly improve this algorithm by not checking the projections in $k = 2$ columns, and starting the algorithm with $k = 3$.

In what follows by $log(x)$ we mean $log_a(x), \quad a > 1$.

Lemma 36 *Let h_k be a projection, in k columns, of a Hadamard matrix of order n. Then h_k cannot contains a full 2^k design if $k > \frac{log(n)}{log(2)}$.*

Proof A full 2^k experimental design has 2^k rows. A Hadamard matrix of order n has n rows. So if $2^k > n$ there cannot be a full 2^k design in a k column projection of this Hadamard matrix. We have that

$$2^k > n \implies k \cdot log(2) > log(n) \implies k > \frac{log(n)}{log(2)}.$$

Now if k is not an integer we take the next integer number. Thus, if k is not an integer we have that $k \geq \left\lceil \frac{log(n)}{log(2)} \right\rceil + 1$. □

Remark 3 If H_1, H_2 be two inequivalent Hadamard matrices of order n. The first Hadamard matrix H_1 will give at least one projection different (inequivalent) from all the projections of H_2 for some $k > \frac{log(n)}{log(2)}$.

Example 37 We give some orders of Hadamard matrices and the bound for k.

- For $n = 2^m$ we obtain $k \geq m + 1$.
- For $n = 12$ we obtain $k \geq 4$.
- For $n = 20$ we obtain $k \geq 5$.
- For $n = 24$ we obtain $k \geq 5$.
- For $n = 28$ we obtain $k \geq 5$.

Lemma 38 *For a Hadamard matrix of order n we have that if $2^m < n < 2^{m+1}$ then $k \geq m + 1$.*

Proof We know that $log_a(x)$ function is continuous, and increasing (since $a > 1$) function. Moreover, $\frac{log(n)}{log(2)} = log_2(n)$. Thus, since $log_2(2^m) = m$, we have that if $2^m < n < 2^{m+1}$ then $m < log_2(n) < m + 1$ and so $k \geq m + 1$. □

Theorem 39 *If two Hadamard are equivalent then their projections for all $k = 2, 3, \ldots, n - 1$ are equivalent as well.*

Proof Suppose that H_1 and H_2 are two equivalent Hadamard matrices of order n. Then, for a given k, both of them have $\binom{n}{k}$ projections in total. From the equivalence of the Hadamard matrices we have that each projection of the first Hadamard matrix is equivalent with one projection of the second Hadamard matrix and vice versa.   □

We will now discuss the complexity of the first new algorithm. First, we observe that all possible projections of a Hadamard matrix of order n in k columns are $\binom{n}{k}$. We note that the finding of inequivalent projections is computationally-intensive work, if we apply the definition of inequivalent projections. This is an NP hard problem when n and k increase. The sign changes in the columns (multiply one or more columns by -1) required 2^k possible multiplications and the sign changes in the rows (multiply one or more rows by -1) required 2^n possible multiplications. The permutations of the columns and rearrangements of the rows need $k!$ possible permutations. That is in total we have $2^{k+n} \cdot k! \cdot \binom{n}{k} = \frac{2^{k+n}n!}{(n-k)!}$ cases to check and that's a large complexity when k or n increases. So, if we are not interested in finding all inequivalent projections of Hadamard matrices we can apply the following algorithm which uses all projections and their symmetric Hamming distance distribution.

The symmetric Hamming distance of two $(1, -1)$ vectors of length n, is defined to be the smallest number of positions with the same entries and different entries. For example, the Hamming distance and symmetric Hamming distance of the two vectors $(1, 1, -1, 1, -1, -1, 1, -1)$ and $(1, -1, 1, -1, -1, 1, -1, 1)$ are 6 and 2 respectively. It is clear that if we have a Hadamard matrix H of order n, then the Hamming distance as well as the symmetric Hamming distance of any two distinct rows is $n/2$.

The *Hamming distance distribution* $(W(x))$ and the *symmetric Hamming distance distribution* $(SW(x))$, of a projection in k columns, is defined to be

$$W_k(x) = a_0 + a_1 x^1 + \ldots + a_k x^k \quad \text{and}$$

$$SW_k(x) = \begin{cases} \displaystyle\sum_{i=0}^{(k-1)/2} (a_i + a_{k-i}) x^i, & \text{when k is odd} \\ \displaystyle\sum_{i=0}^{(k-2)/2} (a_i + a_{k-i}) x^i + a_{\frac{k}{2}} x^{\frac{k}{2}}, & \text{when k is even} \end{cases}$$

respectively, where a_m is the number describing how many pairs of rows of the projection have distance m.

Example 40 Consider the projections for $k = 3$ and $n = 8$. A Hadamard matrix of order 8 is

$$\begin{array}{rrrrrrrr}
1 & 1 & 1 & 1 & 1 & 1 & 1 & 1 \\
1 & 1 & 1 & -1 & 1 & -1 & -1 & -1 \\
1 & 1 & -1 & -1 & -1 & 1 & 1 & -1 \\
1 & 1 & -1 & 1 & -1 & -1 & -1 & 1 \\
1 & -1 & 1 & 1 & -1 & 1 & -1 & -1 \\
1 & -1 & 1 & -1 & -1 & -1 & 1 & 1 \\
1 & -1 & -1 & 1 & 1 & -1 & 1 & -1 \\
1 & -1 & -1 & -1 & 1 & 1 & -1 & 1
\end{array}$$

Since $k = 3$ the projections are all possible 3-sets of columns. We will just illustrate with the sets of columns 2, 3, 4 and 2, 3, 5.

$$\begin{array}{rrr}
1 & 1 & 1 \\
1 & 1 & -1 \\
1 & -1 & -1 \\
1 & -1 & 1 \\
-1 & 1 & 1 \\
-1 & 1 & -1 \\
-1 & -1 & 1 \\
-1 & -1 & -1
\end{array} \quad \text{and} \quad
\begin{array}{rrr}
1 & 1 & 1 \\
1 & 1 & 1 \\
1 & -1 & -1 \\
1 & -1 & -1 \\
-1 & 1 & -1 \\
-1 & 1 & -1 \\
-1 & -1 & 1 \\
-1 & -1 & 1
\end{array}$$

We now consider the distance between all pairs of rows of these 8×3 matrices. The first set has distance 3 (4 times), 2 (12 times) and 1 (12 times) so its Hamming distance distribution and its symmetric Hamming distance distribution is

$$W_3(x) = 0 + 12x + 12x^2 + 4x^3, \quad SW_3(x) = 4 + 24x$$

respectively, while the second set has 0 (4 times) and 2 (24 times) so its Hamming distance distribution and its symmetric Hamming distance distribution is

$$W_3(x) = 4 + 24x^2, \quad SW_3(x) = 4 + 24x$$

respectively.   □

The Hamming distance distribution $W_k(x)$ is invariant only to permutations of columns or rows, or negations of columns while the symmetric Hamming distance distribution $SW_k(x)$ is invariant to permutations and negations of both rows and columns.

Lemma 41 *Two equivalent projections have the same symmetric Hamming distance distribution.*

Proof Let $P_a = \{a_1, a_2, \ldots, a_k\}, P_b = \{b_1, b_2, \ldots, b_k\}$ be two rows in a given projection in k columns. The result follows from the fact that the symmetric Hamming distance of these two rows is not affected if we apply some sing changes or permutations to both rows and columns. □

Lemma 42 *All projections of two Hadamard matrices H_1, H_2 of order n in $k = 1, 2$ columns are the same (actually these give only one inequivalent projection) even thought the Hadamard matrices are inequivalent.*

Proof Since any Hadamard matrix is equivalent to its normalized form, we can suppose that H_1, H_2 are normalized. Thus any column of H_1, H_2 have half $1's$ and half $-1's$. The assertion for $k = 1$ follows. For the case $k = 2$, since any two columns of a Hadamard matrix are orthogonal, it is easy to see that any projection in $k = 2$ columns is equivalent to a projection which is $n/4$ times the full 2^2 design.   □

Lemma 43 *Let H be a Hadamard matrix of order n. Any two rows of the Hadamard matrix have Hamming distace distribution and symmetric Hamming distace distribution $W_n(x) = SW_n(x) = x^{n/2}$.*

Proof Let $r = \{r_1, r_2, \ldots, r_n\}$ and $p = \{p_1, p_2, \ldots, p_n\}$ be the two rows of the Hadamard matrix. From the orthogonality of the rows we have that $\sum_{i=1}^{n} r_i p_i = 0$. This means that $n/2$ of the n pairs $(r_i, p_i) \in \{(1,1), (-1,-1)\}$ and the other $n/2$ pairs $(r_i, p_i) \in \{(-1,1), (1,-1)\}$, and thus the Hamming distace distribution and the symmetric Hamming distace distribution $W_n(x) = SW_n(x) = x^{n/2}$.   □

Definition 17 Let H be a Hadamard matrix of order n and P_k a set of k columns of H. We define the *complementary projection* of P_k to be the set of the columns of H which are not contained in P_k. Obviously the complementary projection of P_k consist of $n - k$ columns.

Remark 4 Let H_1, H_2 be two Hadamard matrices of order n. Suppose $r = \{r_1, r_2, \ldots, r_k\}$ and $p = \{p_1, p_2, \ldots, p_k\}$ be two rows of a projection of H_1 and $q = \{q_1, q_2, \ldots, q_k\}$ and $s = \{s_1, s_2, \ldots, s_k\}$ be two rows of a projection of H_2. Then $SW(x)$ of rows r, p is equal to $SW(x)$ of rows q, s if and only if the symmetric Hamming distance distribution of the corresponding rows of their complementary projections is equal.

Example 44 The complementary projections of the projections given in example 40 are

$$
\begin{array}{rrrrr}
1 & 1 & 1 & 1 & 1 \\
1 & 1 & -1 & -1 & -1 \\
1 & -1 & 1 & 1 & -1 \\
1 & -1 & -1 & -1 & 1 \\
1 & -1 & 1 & -1 & -1 \\
1 & -1 & -1 & 1 & 1 \\
1 & 1 & -1 & 1 & -1 \\
1 & 1 & 1 & -1 & 1
\end{array}
\quad and \quad
\begin{array}{rrrrr}
1 & 1 & 1 & 1 & 1 \\
1 & -1 & -1 & -1 & -1 \\
1 & -1 & 1 & 1 & -1 \\
1 & 1 & -1 & -1 & 1 \\
1 & 1 & 1 & -1 & -1 \\
1 & -1 & -1 & 1 & 1 \\
1 & 1 & -1 & 1 & -1 \\
1 & -1 & 1 & -1 & 1
\end{array}
$$

with symmetric Hamming distance distribution $SW_{8-3}(x) = SW_5(x) = 4 + 24x$.

From Lemmas 41, 42 and 43 it is obvious that:

Corollary 45 *All projections of two Hadamard matrices H_1, H_2 of order n in $k = 1, 2$ and $k = n$ columns have the same symmetric Hamming distance distribution.*

Using Remark 4 and the above lemmas we can conclude:

Corollary 46 *Let H_1, H_2 be two Hadamard matrices of order n. We need only to check the symmetric Hamming distance distribution of projections for $k = 3, 4, \ldots, n/2$ because if these have the same symmetric Hamming distance distribution, then the corresponding complementary projections will have the same symmetric Hamming distance distribution as well.*

In this way the modified algorithm (symmetric Hamming distance distribution algorithm) is much faster than the previous one but it might gives us an answer only to the question if the two Hadamard matrices are not equivalent and does not give us all inequivalent projections of the Hadamard matrices.

The Symmetric Hamming distance distribution algorithm:

(i) Set $k = 3$.

(ii) Find all projections for each Hadamard matrix of a given order n and k columns by taking all possible k columns of the entire $n \times n$ Hadamard matrix. These are $\binom{n}{k}$ projections in total.

(iii) In the projections found in step (ii) calculate the symmetric Hamming distance distributions for any two rows of the projection. These are $\binom{n}{2}$ symmetric Hamming distance distributions and save different symmetric Hamming distance distributions and how many times each of them appear.

(iv) Check if the set of all different symmetric Hamming distance distributions of the first Hadamard matrix is the same with the set of all different symmetric Hamming distance distribution of the second Hadamard matrix.

(v) If the answer in step (iv) is false, then stop and say that these two Hadamard matrices are inequivalent, otherwise increase k by 1.

(vi) If now $k \leq n/2$ then go to step (ii) and continue, otherwise stop and say that this algorithm cannot decide for the equivalence of these Hadamard matrices.

Let us discuss the complexity of the Hamming distance distribution algorithm. First, we observe again that all possible projections in k columns of a Hadamard matrix of order n is $\binom{n}{k}$. We note that finding the symmetric Hamming distance distribution of all projections is not computationally-intensive work, because it only needs $n(n-1)$ calculations. A calculation of the symmetric Hamming distance of two rows in a projection takes k comparisons and thus we have in total $\binom{n}{k}n(n-1)k$ multiplications, summations and comparisons. This is not an NP hard problem when n and k increase and it is much faster than the inequivalent projections algorithm.

2.4   Application of the new criterion in Hadamard matrices of small orders

In this section we apply our new algorithm in the cases of Hadamard matrices of small orders. When the Hadamard matrices are equivalent we have to check the symmetric Hamming distance distributions for all projections into $k = 3, \ldots, n/2$ columns. As an examples we give all symmetric Hamming distances of the unique Hadamard matrices of orders $n = 8, 12$ and for all $k = 3, 4, \ldots, n/2$. If the Hadamard matrices are inequivalent there exist $k \in \{2, 3, \ldots, n/2\}$ such that the symmetric Hamming distance distributions for the projections in k columns are different for each Hadamard matrix.

To save space, we give here the table with symmetric Hamming distance distribution only for orders 8 and 12. For larger orders the reader should consider [19].

2.4.1   Hadamard matrices of order $n = 8, 12$.
We know that there exist only one Hadamard matrix of these orders up to equivalence, see [7] for example. For the order $n = 4$ we have $k \leq n/2 = 2$ and thus all projection have the same symmetric Hamming distance distribution. The results of the application of the symmetric Hamming distance distribution algorithm for these orders $n \geq 8$ are given in Table 2.1. Since there is only one Hadamard matrix in each case we give all symmetric Hamming distance distributions for all projections into $k = 3, \ldots, n/2$ columns.

For example there are $\binom{8}{4} = 70$ projections in a Hadamard matrix of order $n = 8$ in $k = 4$ columns and $\binom{8}{2} = 28$ Hamming weights in each Hamming distance distribution of each projection.

When we say that the symmetric Hamming distance distribution is $0, 16, 12$ and times 56 that means that there are 0 pairs of rows in the projection with Hamming distance 0 or 4, 16 pairs of rows in the projection with Hamming distance 1 or 3 and 12 pairs of rows in the projection with Hamming distance 2. This distribution occurs for 56 of the 70 projections.

When we say that the symmetric Hamming distance distribution is $4, 0, 24$ and times 14 that means that there are 4 pairs of rows in the projection with Hamming distance 0 or 4, 0 pairs of rows in the projection with Hamming distance 1 or 3 and 24 pairs of rows in the projection with Hamming distance 2. This distribution occurs for 14 of the 70 projections.

As you can see the total number of symmetric Hamming distance (the sum of all symmetric Hamming distances in the symmetric Hamming distance distribution) is $\binom{8}{2} = 28$ and the total number of times each distribution occurs (the sum of all different symmetric Hamming distance distributions) is $\binom{8}{4} = 70$.

H_{name}	n	k	Symmetric Hamming distance	times
H_8	8	3	4,24	56
H_8	8	4	0,16,12	56
H_8	8	4	4,0,24	14
H_{12}	12	3	12,54	220
H_{12}	12	4	4,32,30	495
H_{12}	12	5	1,15,50	792
H_{12}	12	6	0,6,30,30	792
H_{12}	12	6	1,0,45,20	132

Table 2.1. Application of the symmetric Hamming distance distribution algorithm for $n = 8$ and 12

2.4.2 Hadamard matrices of order $n = 16$.

We know that there are exactly five inequivalent Hadamard matrices of this order, see [36]. The results of the application of the symmetric Hamming distance distribution algorithm for this order are given in [19]. Observe that for $k = 3$ the symmetric Hamming distance distributions of all projections of all five matrices are exactly the same. For $k = 4, 5$ and 6 we have four different symmetric Hamming distance distributions (thus four inequivalent Hadamard matrices) and we have to go up to $k = 7$ to obtain all five of them.

2.4.3 Hadamard matrices of order $n = 20$.

We know that there are exactly three inequivalent Hadamard matrices of this order, see [37]. The results of the application of the symmetric Hamming distance distribution algorithm for this order are given in [19]. Observe that for $k = 3, 4$ and 5, the symmetric Hamming distance distributions of all projections of all three matrices are exactly the same. For $k = 6$ all three have different symmetric Hamming distance distributions and thus we obtain all three inequivalent Hadamard matrices.

2.4.4 Hadamard matrices of order $n = 24$.

We know that there are exactly 60 inequivalent Hadamard matrices of this order, see [42, 47]. For Hadamard matrices of order 24 it is not convenient to give all different symmetric Hamming distance distributions for all k. We shall only discuss the results our algorithm gives. For $k = 3$ all sixty matrices give the same symmetric Hamming distance distributions thus we obtain only one of the sixty inequivalent Hadamard matrices. For $k = 4$ and $k = 5$ the algorithm finds 35 different symmetric Hamming distance distributions and thus 35 of the sixty inequivalent Hadamard matrices. Finally for $k = 6$ we obtain 60 different symmetric Hamming distance distributions and thus all 60 inequivalent Hadamard matrices. For more details in this order the reader should consider [19].

2.4.5 Hadamard matrices of order $n = 28$.

In the case $n = 28$ there are 487 inequivalent Hadamard matrices, see [48, 49]. If we apply our algorithm to this case we obtain the following results. For $k = 3$ all 487 matrices give the same symmetric Hamming distance distributions thus we obtain only one of the 487 inequivalent Hadamard matrices. The algorithm moves to $k = 4$ and finds 60 different symmetric Hamming distance distributions and thus 60 of the 487 inequivalent Hadamard matrices. Also for $k = 5$ we obtain 60 different symmetric Hamming distance distributions and thus 60 of the 487 inequivalent Hadamard matrices. Finally for $k = 6$ we obtain 487 different symmetric Hamming distance distributions, and thus all 487 inequivalent Hadamard matrices. For more details in this order the reader should consider [19].

2.4.6 Hadamard matrices of order 32.

The classification of Hadamard matrices of orders $n \geq 32$ is still remains an open and difficult problem since an algorithmic approach using an exhaustive search is an NP hard problem. In particular, in this case, Lin, Wallis and Zhu [65] found 66104 inequivalent Hadamard matrices of order 32. Extensive results appear in [68] and [69]. Thus the lower bound for inequivalent Hadamard matrices of order 32 is 66104.

2.4.7 Hadamard matrices of order 36.

There are at least 1036 inequivalent Hadamard matrices of order 36. In fact this number is obtained as follows: Seberry's home page "http://www.uow.edu.au/~ jennie" gives 192 inequivalent Hadamard matrices of order 36. These are supplied by E. Spence (180 matrices) see

[76], Z. Janko, (1 matrix of Bush-type) see [43] and V. D. Tonchev (11 matrices) see [80]. Using an efficient algorithm and the Magma software Georgiou and Koukouvinos [23] found that 172 of their transposes, are inequivalent to these. They also in [23] improved further this bound to 1036 by constructing 672 new Hadamard matrices of order 36.

2.4.8   Hadamard matrices of order 40.

Lam, Lam and Tonchev [63] showed that the lower bound for inequivalent Hadamard matrices of order 40 is 3.66×10^{11}.

2.4.9   Hadamard matrices of order 44.

Recently Topalova [81] classified the Hadamard matrices of order 44 with an automorphism of order 7, and found 384 inequivalent Hadamard matrices of this order. Using an efficient algorithm and the Magma software Georgiou and Koukouvinos [25] found that 6 of their transposes, are inequivalent to these. Two more Hadamard matrices were given in Sloane's web page ("http://www.research.att.com/~njas/hadamard/"). Georgiou and Koukouvinos in [25] showed that the transposes of these two matrices are inequivalent to all known Hadamard matrices of order 44. Also, in [25], they further improved this lower bound to 6018 by constructing 5624 new Hadamard matrices.

3.   Algorithms for constructing orthogonal designs

3.1   Basic definitions and preliminaries

An *orthogonal design* of order n and type $(s_1, s_2, \ldots, s_u)$ $(s_i > 0)$, denoted $OD(n; s_1, s_2, \ldots, s_u)$, on the commuting variables $x_1, x_2, \ldots, x_u$ is an $n \times n$ matrix A with entries from $\{0, \pm x_1, \pm x_2, \ldots, \pm x_u\}$ such that

$$AA^T = (\sum_{i=1}^{u} s_i x_i^2) I_n.$$

Alternatively, the rows of A are formally orthogonal and each row has precisely s_i entries of the type $\pm x_i$. In [29], where this was first defined, it was mentioned that

$$A^T A = (\sum_{i=1}^{u} s_i x_i^2) I_n$$

and so our alternative description of A applies equally well to the columns of A. It was also shown in [29] that $u \leq \rho(n)$, where $\rho(n)$ (Radon's function) is defined by $\rho(n) = 8c + 2^d$, when $n = 2^a b$, b odd, $a = 4c + d$, $0 \leq d < 4$.

Some small orthogonal designs are given in the following example, see [73].

Example 47 Some small orthogonal designs.

$$\begin{bmatrix} x & y \\ y & -x \end{bmatrix}, \quad \begin{bmatrix} a & -b & -c & -d \\ b & a & -d & c \\ c & d & a & -b \\ d & -c & b & a \end{bmatrix}, \quad \begin{bmatrix} a & b & b & d \\ -b & a & d & -b \\ -b & -d & a & b \\ -d & b & -b & a \end{bmatrix}, \quad \begin{bmatrix} a & 0 & -c & 0 \\ 0 & a & 0 & c \\ c & 0 & a & 0 \\ 0 & -c & 0 & a \end{bmatrix}$$

$$OD(2; 1, 1) \quad OD(4; 1, 1, 1, 1) \qquad OD(4; 1, 1, 2) \qquad\quad OD(4; 1, 1)$$

$OD(4; 1, 1, 1, 1)$ is the Williamson array. $\qquad\qquad\qquad\qquad\qquad\qquad\square$

A weighing matrix $W = W(n, k)$ is a square matrix with entries $0, \pm 1$ having k non-zero entries per row and column and inner product of distinct rows zero. Hence W satisfies $WW^T = kI_n$, and W is equivalent to an orthogonal design $OD(n; k)$. The number k is called the *weight* of W.

We make extensive use of the book of Geramita and Seberry [30]. We quote the following theorems, giving their reference from the aforementioned book, that we use:

Lemma 48 [30, Lemma 4.11, The Doubling Lemma] *If there exists an orthogonal design $OD(n; s_1, s_2, \ldots, s_u)$ then there exists an orthogonal design $OD(2n; s_1, s_1, es_2, \ldots, es_u)$ where $e = 1$ or 2.* $\qquad\square$

Lemma 49 [30, Lemma 4.4, The Equating and Killing Lemma] *If A is an orthogonal design $OD(n; s_1, s_2, \ldots, s_u)$ on the commuting variables $\{0, \pm x_1, \pm x_2, \ldots, \pm x_u\}$ then there is an orthogonal design $OD(n; s_1, s_2, \ldots, s_i + s_j, \ldots, s_u)$ and $OD(n; s_1, s_2, \ldots, s_{j-1}, s_{j+1}, \ldots, s_u)$ on the $u - 1$ commuting variables $\{0, \pm x_1, \pm x_2, \ldots, \pm x_{j-1}, \pm x_{j+1}, \ldots, \pm x_u\}$.* $\square$

Theorem 50 [30, Theorems 2.19 and 2.20] *Suppose $n \equiv 0 (mod\ 4)$. Then the existence of a $W(n, n-1)$ implies the existence of a skew-symmetric $W(n, n-1)$. The existence of a skew-symmetric $W(n, k)$ is equivalent to the existence of an $OD(n; 1, k)$.* $\qquad\square$

Theorem 51 [30, Proposition 3.54 and Theorem 2.20] *An orthogonal design* $OD(n; 1, k)$ *can only exist in order* $n \equiv 4(mod\ 8)$ *if* k *is the sum of three squares. An orthogonal design* $OD(n; 1, n - 2)$ *can only exist in order* $n \equiv 4(mod\ 8)$ *if* $n - 2$ *is the sum of two squares.* $\square$

Theorem 52 [30, Theorem 4.49] *Suppose there exist four circulant matrices* A, B, C, D *of order* n *satisfying*

$$AA^T + BB^T + CC^T + DD^T = fI_n$$

Let R *be the back diagonal matrix. Then*

$$GS = \begin{pmatrix} A & BR & CR & DR \\ -BR & A & D^T R & -C^T R \\ -CR & -D^T R & A & B^T R \\ -DR & C^T R & -B^T R & A \end{pmatrix}$$

is a $W(4n, f)$ *when* A, B, C, D *are* $(0, 1, -1)$ *matrices, and an orthogonal design* $OD(4n; s_1, s_2, \ldots, s_u)$ *on* $x_1, x_2, \ldots, x_u$ *when* A, B, C, D *have entries from* $\{0, \pm x_1, \ldots, \pm x_u\}$ *and* $f = \sum_{j=1}^{u}(s_j x_j^2)$. $\square$

Corollary 53 *If there are four sequences* A, B, C, D *of length* n *with entries from* $\{0, \pm x_1, \pm x_2, \pm x_3, \pm x_4\}$ *with zero periodic or non-periodic autocorrelation function, then these sequences can be used as the first rows of circulant matrices which can be used in the Goethals-Seidel array to form an* $OD(4n; s_1, s_2, s_3, s_4)$. *We note that if there are sequences of length* n *with zero non-periodic autocorrelation function, then there are sequences of length* $n + m$ *for all* $m \geq 0$. $\square$

3.2   Construction algorithms

In this section we are interested in the construction of orthogonal designs using four circulant matrices in the Goethals-Seidel array. Specifically, for positive integers $s_1, s_2, \ldots, s_u$ and odd n, the method searches for four circulant matrices A_1, A_2, A_3, A_4 of order n with entries from $\{0, \pm x_1, \pm x_2, \ldots, \pm x_u\}$, $u \leq 4$, such that

$$A_1 A_1^T + A_2 A_2^T + A_3 A_3^T + A_4 A_4^T = \left(\sum_{i=1}^{u} s_i x_i^2\right) I_n. \qquad (3.29)$$

In the remainder of this section, when four circulant (or group circulant) matrices of order n, with entries from the set $\{0, \pm x_1, \pm x_2, \ldots, \pm x_u\}$, satisfy equation (3.29) will be said that these matrices satisfy the *additive property*.

3.2.1 The matrix based algorithm.

Suppose the row and column sum of A_i is

$$r_i = p_{1i}x_1 + p_{2i}x_2 + p_{3i}x_3 + p_{4i}x_4, \quad i = 1, 2, 3, 4$$

Let e^T be the $1 \times n$ vector of 1's, then $e^T A_i = r_i e^T$. Multiplying on the left of (3.29) by e^T and the right of (3.29) by e we have

$$\sum_{i=1}^{4} (e^T A_i)(e^T A_i)^T = n \sum_{i=1}^{4} s_i x_i^2 \quad \text{or}$$

$$\sum_{i=1}^{4} (r_i e^T)(r_i e^T)^T = n \sum_{i=1}^{4} r_i^2 = n \sum_{i=1}^{4} s_i x_i^2$$

Thus we have

$$s_1 x_1^2 + s_2 x_2^2 + s_3 x_3^2 + s_4 x_4^2 = x_1^2 \sum_{i=1}^{4} p_{1i}^2 + x_2^2 \sum_{i=1}^{4} p_{2i}^2 + x_3^2 \sum_{i=1}^{4} p_{3i}^2$$

$$+ x_4^2 \sum_{i=1}^{4} p_{4i}^2 + 2x_1 x_2 \sum_{i=1}^{4} p_{1i} p_{2i}$$

$$+ 2x_1 x_3 \sum_{i=1}^{4} p_{1i} p_{3i} + 2x_1 x_4 \sum_{i=1}^{4} p_{1i} p_{4i}$$

$$+ 2x_2 x_3 \sum_{i=1}^{4} p_{2i} p_{3i} + 2x_2 x_4 \sum_{i=1}^{4} p_{2i} p_{4i}$$

$$+ 2x_3 x_4 \sum_{i=1}^{4} p_{3i} p_{4i}$$

Hence we have four integer vectors $p_1^T = (p_{11}, p_{12}, p_{13}, p_{14})$, $p_2^T = (p_{21}, p_{22}, p_{23}, p_{24})$, $p_3^T = (p_{31}, p_{32}, p_{33}, p_{34})$, $p_4^T = (p_{41}, p_{42}, p_{43}, p_{44})$, which are pairwise orthogonal. Also $|p_1^T|^2 = s_1$, $|p_2^T|^2 = s_2$, $|p_3^T|^2 = s_3$, $|p_4^T|^2 = s_4$.

Form these vectors into an orthogonal integer matrix P with $P^T = (p_1, p_2, p_3, p_4)$. Then $PP^T = diag(s_1, s_2, s_3, s_4)$ and $det P = \sqrt{s_1 s_2 s_3 s_4}$. But P is integer so $s_1 s_2 s_3 s_4$ is a square. Thus we have

Lemma 54 *The Goethals-Seidel construction for an orthogonal design* $OD(4n; s_1, s_2, s_3, s_4)$ *can only be used if*

(i) *there is an integer matrix P satisfying* $PP^T = diag\,(s_1, s_2, s_3, s_4)$
and hence

(ii) $s_1 s_2 s_3 s_4$ *is a square.*   □

Since the row sum of A_j is $\displaystyle\sum_{i=1}^{4} p_{ij} x_i$ for $1 \leq j \leq 4$, the 4×4 matrix
$P = (p_{ij})$ is called the *sum matrix* of A_1, A_2, A_3, A_4.

In this section we are interested in the construction of orthogonal designs using four circulant matrices in the Gorthals-Seidel array. Specifically, for positive integers $s_1, s_2, \ldots, s_u$ and odd n, the method searches for four circulant matrices A_1, A_2, A_3, A_4 of order n with entries from $\{0, \pm x_1, \pm x_2, \ldots, \pm x_u\}$ that satisfy equation (3.29).

Definition 18 If A_1, A_2, A_3, A_4 are $n \times n$ circulant matrices with entries from $\{0, \pm x_1, \pm x_2, \ldots, \pm x_u\}$ and the first row of A_j has m_{ij} entries of the kind $\pm x_i$, then the $u \times 4$ matrix $M = (m_{ij})$ is called the *entry matrix* of (A_1, A_2, A_3, A_4).   □

The elements of the entry matrices satisfy the following conditions.

$$\sum_{j=1}^{4} m_{ij} = s_i \quad \text{for} \ \ 1 \leq i \leq u$$
$$\sum_{i=1}^{u} m_{ij} \leq n \quad \text{for} \ \ 1 \leq j \leq 4 \tag{3.30}$$

Thus the rows of the entry matrices refer to the variables x_i and the columns to the circulant matrices A_1, A_2, A_3, A_4 which are constructed from four sequences of length n as described in Corollary 53.

Definition 19 Suppose that the row sum of A_j is $\displaystyle\sum_{i=1}^{u} p_{ij} x_i$ for $1 \leq j \leq 4$.
Then the $u \times 4$ integral matrix $P = (p_{ij})$ is called the *sum matrix* of (A_1, A_2, A_3, A_4). The *fill matrix* of (A_1, A_2, A_3, A_4) is $M - abs(P)$, where $abs(P)$ denotes the matrix having as elements the absolute values of elements of P. The content of A_i is determined by the i-th columns of the sum and fill matrices.   □

The following theorem may be used to find the sum matrix of a solution of (3.29).

Theorem 55 (Eades Sum Matrix Theorem) The sum matrix P of a solution of (3.29) satisfies $PP^T = diag(s_1, s_2, \ldots, s_u)$. □

The algorithm

Step 1. Find all sum matrices P of the desired orthogonal design using theorem 55.

Step 2. Select the first sum matrix.

Step 3. For the selected sum matrix P find all entry matrices M and the corresponding fill matrices (Q=M-abs(P)) using equations given by (3.30).

Step 4. Select the first entry matrix M and the corresponding fill matrix Q.

Step 5. Using P, M and Q write down the elements of sequences A_j, $j = 1, 2, 3, 4$.

Step 6. Construct all possible sequences A_j with entries we found in Step 5 and their corresponding PAF.

Step 7a. Combine the lists find in Step 6 and check if a combination gives zero PAF and if so save these sequences into PAF solution file.

Step 7b. If a zero PAF solution exist then search if some permutation of these sequences have zero NPAF and if so save these sequences into NPAF solution file.

Step 8. If there are more entry matrices then select the next entry matrix M and the corresponding fill matrix Q and go to Step 5.

Step 9. If there are more sum matrices then select the next sum matrix P and go to Step 3.

For more details about the construction of orthogonal designs which uses entry matrices, see [30].

3.2.2   The extension algorithm.

This algorithm extents already known orthogonal designs on t variables into new orthogonal designs on $t + 1$ variables. The algorithm is given briefly in the next steps.

Step 1. Input the sequences of the known orthogonal design $OD(4n; s_1, \ldots, s_t)$ on t variables $(a_1, a_2, \ldots, a_t)$, you wish to extent.

Step 2. In these sequences replace all zeros with variables x_i (a deferent variable on each zero).

Step 3. Using the new sequences and the equation

$$P_{A_1}(s) + P_{A_2}(s) + P_{A_3}(s) + P_{A_4}(s) = 0, \quad s = 1, 2, \ldots, \frac{(n-1)}{2}$$

create a system of equations.

Step 4. Solve this system ofequations and find all possible values x_i, where $x_i \in \{-1, 0, 1\}$, that satisfy equations given in Step 3.

Step 5. For all solutions, diferent from the zero solution, (of weight $k \neq 0$) replace ± 1 by $\pm a_{t+1}$ respectively and obtain the $OD(4n; s_1, \ldots, s_t, k)$ on $t+1$ variables $(a_1, a_2, \ldots, a_t, a_{t+1})$.

Then next example illustrates how this algorithm works.

Example 56 Start with the four sequences of length 9 and type $(5, 9)$ with $NPAF = 0$ (Step 1).

$$
\begin{array}{ccccccccc}
b & 0 & -b & 0 & 0 & 0 & 0 & 0 & 0 \\
b & a & -b & 0 & 0 & 0 & 0 & 0 & 0 \\
b & a & 0 & a & -b & 0 & 0 & 0 & 0 \\
b & a & b & -a & b & 0 & 0 & 0 & 0
\end{array}
$$

Now fill each zero position with one of the 22 variables $x_1, x_2, \ldots, x_{22}$ (Step 2). Thus we obtain

$$
\begin{array}{ccccccccc}
b & x_1 & -b & x_2 & x_3 & x_4 & x_5 & x_6 & x_7 \\
b & a & -b & x_8 & x_9 & x_{10} & x_{11} & x_{12} & x_{13} \\
b & a & x_{14} & a & -b & x_{15} & x_{16} & x_{17} & x_{18} \\
b & a & b & -a & b & x_{19} & x_{20} & x_{21} & x_{22}
\end{array}
$$

Using relations

$$P_{A_1}(s) + P_{A_2}(s) + P_{A_3}(s) + P_{A_4}(s) = 0, \quad s = 1, 2, \ldots, \frac{(n-1)}{2}$$

we construct the following twelve equations (Step 3):

$$2x_{14} = 0$$
$$x_7 - x_2 + x_{13} - x_8 + x_{18} - x_{15} + x_{22} + x_{19} = 0$$
$$x_3x_2 + x_4x_3 + x_5x_4 + x_6x_5 + x_7x_6 + x_9x_8 + x_{10}x_9$$
$$+x_{11}x_{10} + x_{12}x_{11} + x_{13}x_{12} + x_{16}x_{15} + x_{17}x_{16} + x_{18}x_{17}$$
$$+x_{20}x_{19} + x_{21}x_{20} + x_{22}x_{21} = 0$$
$$x_{13} + x_8 + x_{18} + x_{15} + x_{22} - x_{19} = 0$$
$$x_6 - x_3 + x_{12} - x_9 + x_{17} - x_{16} + x_{21} + x_{20} = 0$$
$$x_1x_7 + x_2x_1 + x_4x_2 + x_5x_3 + x_6x_4 + x_7x_5 + x_{10}x_8$$
$$+x_{11}x_9 + x_{12}x_{10} + x_{13}x_{11} + x_{17}x_{15} + x_{18}x_{16}$$
$$+x_{21}x_{19} + x_{22}x_{20} = 0$$
$$x_{12} + x_9 + x_{17} + x_{16} + x_{21} - x_{20} = 0$$
$$x_5 - x_7 + x_2 - x_4 + x_{11} - x_{13} + x_8 - x_{10}$$
$$+x_{16} - x_{17} + x_{20} + x_{22} + x_{19} + x_{21} = 0$$
$$x_1x_6 + x_3x_1 + x_5x_2 + x_6x_3 + x_7x_4 + x_{11}x_8 + x_{12}x_9$$
$$+x_{13}x_{10} + x_{14}x_{18} + x_{15}x_{14} + x_{18}x_{15} + x_{22}x_{19} = 0$$
$$x_{11} + x_{10} + x_{16} + x_{18} + x_{15} + x_{17}$$
$$+x_{20} - x_{22} + x_{19} - x_{21} = 0$$
$$x_4 - x_6 + x_3 - x_5 + x_{10} - x_{12} + x_9 - x_{11}$$
$$+x_{15} - x_{18} + x_{19} + x_{21} + x_{20} + x_{22} = 0$$
$$x_1x_5 + x_2x_7 + x_4x_1 + x_6x_2 + x_7x_3 + x_8x_{13}$$
$$+x_{12}x_8 + x_{13}x_9 + x_{14}x_{17} + x_{16}x_{14} = 0$$

By solving this system of equations (Step 4) we find, among others, the following solutions of weight $9, 14, 16$:

x_1	x_2	x_3	x_4	x_5	x_6	x_7	x_8	x_9	x_{10}	x_{11}	x_{12}	x_{13}
0	−1	0	0	0	0	−1	−1	−1	0	−1	1	0
0	−1	−1	−1	0	−1	1	0	−1	0	1	−1	−1
0	−1	−1	−1	−1	−1	1	1	−1	0	0	1	−1

x_{14}	x_{15}	x_{16}	x_{17}	x_{18}	x_{19}	x_{20}	x_{21}	x_{22}
0	1	1	−1	0	0	0	0	0
0	1	0	1	−1	0	−1	0	1
0	1	0	0	−1	1	−1	−1	1

The first one gives the orthogonal design of order 36 on three variables, $OD(36; 5, 9, 9)$ (Step 5, by replacing ± 1 by $\pm c$ respectively).

$$
\begin{array}{ccccccccc}
 & & & & (5,9,9) & & & & \\
b & 0 & -b & -c & 0 & 0 & 0 & 0 & -c \\
b & a & -b & -c & -c & 0 & -c & c & 0 \\
b & a & 0 & a & -b & c & c & -c & 0 \\
b & a & b & -a & b & 0 & 0 & 0 & 0 \\
\end{array}
$$

The second one gives the orthogonal design of order 36 on three variables, $OD(36; 5, 9, 14)$ (Step 5, by replacing ± 1 by $\pm c$ respectively).

$$
\begin{array}{ccccccccc}
 & & & & (5,9,14) & & & & \\
b & 0 & -b & -c & -c & -c & 0 & -c & c \\
b & a & -b & 0 & -c & 0 & c & -c & -c \\
b & a & 0 & a & -b & c & 0 & c & -c \\
b & a & b & -a & b & 0 & -c & 0 & c \\
\end{array}
$$

The third one gives the orthogonal design of order 36 on three variables, $OD(36; 5, 9, 16)$ (Step 5, by replacing ± 1 by $\pm c$ respectively).

$$
\begin{array}{ccccccccc}
 & & & & (5,9,16) & & & & \\
b & 0 & -b & -c & -c & -c & -c & -c & c \\
b & a & -b & c & -c & 0 & 0 & c & -c \\
b & a & 0 & a & -b & c & 0 & 0 & -c \\
b & a & b & -a & b & c & -c & -c & c \\
\end{array}
$$

3.2.3   The merge algorithm.

This algorithm relies on the two previously mentioned algorithms (the matrix based algorithm and the extension algorithm) given in [11, 30, 55] and in [22, 54] respectively.

The merge algorithm combines features of both algorithms with a new result given here to obtain a new, much faster, algorithm. It is an exhaustive search algorithm (i.e. if the orthogonal design exists it will be found otherwise it does not exist constructed from four sequences).

Notation 1 For the remainder of this section we use the following notations.

1 $\mathcal{N}$ denotes the set of non negative integers.

2 $\mathcal{N}^k$ denotes the space $\mathcal{N}^k = \underbrace{\mathcal{N} \times \mathcal{N} \times \cdots \times \mathcal{N}}_{k \; times}$ with elements

$$\mathbf{v} \in \mathcal{N}^k, \; \mathbf{v}^{\mathbf{T}} = [v_1, v_2, \ldots, v_k], \; v_i \in \mathcal{N}, \; i = 1, 2, \ldots, k.$$

3 $\mathcal{N}^{k \times \ell}$ will be the matrix space with dimension $k \times \ell$ and elements from $\mathcal{N}$. That is if $M \in \mathcal{N}^{k \times \ell}$ then

$$M = \begin{bmatrix} m_{11} & m_{12} & \cdots & m_{1\ell} \\ m_{21} & m_{22} & \cdots & m_{2\ell} \\ \vdots & \vdots & & \vdots \\ m_{k1} & m_{k2} & \cdots & m_{k\ell} \end{bmatrix} = \begin{bmatrix} \mathbf{m_1^T} \\ \mathbf{m_2^T} \\ \vdots \\ \mathbf{m_k^T} \end{bmatrix}$$

with $m_{ij} \in \mathcal{N}$, $\mathbf{m_i} \in \mathcal{N}^\ell$, $i = 1, 2, \ldots, k$, $j = 1, 2, \ldots, \ell$.   □

Let D be an $OD(4n; u_1, u_2, \ldots, u_t)$ with entries from the set $\{0, \pm x_1, \pm x_2, \ldots, \pm x_t\}$ where $x_1, x_2, \ldots, x_t$ are commuting variables. Using the terminology of [30], the symbols M_i represent the non-isomorphic *entry matrices* of the orthogonal design.

From the above construction of the sequences, we observe that we can permute rows and/or columns of the sum matrix P and the entry matrix M without obtaining an essentially different sum or entry matrix. It would be as though we interchanged the variables and/or the sequences of the orthogonal design. When we form the content of the sequences, we should take into account that the row and column order of the sum and the entry matrices must agree. That is to say that the same permutations of rows and/or columns should be operated to both these matrices. In the same way, we can multiply by -1 any rows and/or columns of the sum matrix P without obtaining an essentially different sum matrix.

Herein (because we use many non-isomorphic entry matrices from different orthogonal designs) we will use the *type* of the orthogonal design in the symbol of the entry matrices, so that seeing the entry matrix we can tell from which orthogonal design it comes. For D we will write $M_{(u_1, u_2, \ldots, u_t), i}$ for its non-isomorphic entry matrices. Then we can write the entry matrices using their rows as follows

$$M_{(u_1, u_2, \ldots, u_t), i} = \begin{bmatrix} \mathbf{v_1^T} \\ \mathbf{v_2^T} \\ \vdots \\ \mathbf{v_t^T} \end{bmatrix} \in \mathcal{N}^{t \times 4}, \; \mathbf{v_j} \in \mathcal{N}^4, \; j = 1, 2, \ldots, t.$$

Let $\mathcal{D}_{(u_1,u_2,...,u_t)}$ be the set of all non isomorphic entry matrices of the orthogonal design $OD(4n; u_1, u_2, \ldots, u_t)$. We write $M_{(u_1,u_2,...,u_t),i}|_{\mathcal{D}_{u_k,u_j}}$ for the entry matrix $M_{(u_1,u_2,...,u_t),i}$ after we eliminate all rows except from rows k and j. That is

$$M_{(u_1,u_2,...,u_t),i}|_{\mathcal{D}_{u_k,u_j}} = \begin{bmatrix} \mathbf{v}_k^T \\ \mathbf{v}_j^T \end{bmatrix} \in \mathcal{N}^{2\times 4}.$$

In order to illustrate the above notations and definitions we give the following example.

Example 57 Suppose we are searching for the $OD(4n; u_1, u_2, u_3, u_4) = OD(20; 2, 3, 6, 9)$. There is up to isomorphism only one sum matrix

$$P = \begin{bmatrix} 1 & 1 & 0 & 0 \\ 1 & -1 & 1 & 0 \\ -1 & 1 & 2 & 0 \\ 0 & 0 & 0 & 3 \end{bmatrix}$$

satisfying $PP^T = diag(2,3,6,9)$ as described in Theorem 55. From this matrix P we obtain the following three non-isomorphic entry matrices.

$$M_1 = \begin{bmatrix} 1 & 1 & 0 & 0 \\ 1 & 1 & 1 & 0 \\ 3 & 1 & 2 & 0 \\ 0 & 2 & 2 & 5 \end{bmatrix}, \ M_2 = \begin{bmatrix} 1 & 1 & 0 & 0 \\ 1 & 1 & 1 & 0 \\ 1 & 1 & 4 & 0 \\ 2 & 2 & 0 & 5 \end{bmatrix}, \ M_3 = \begin{bmatrix} 1 & 1 & 0 & 0 \\ 1 & 1 & 1 & 0 \\ 1 & 1 & 2 & 2 \\ 2 & 2 & 2 & 3 \end{bmatrix}.$$

Using our terminology these are:

$$M_{(u_1,u_2,u_3,u_4),1} = \begin{bmatrix} 1 & 1 & 0 & 0 \\ 1 & 1 & 1 & 0 \\ 3 & 1 & 2 & 0 \\ 0 & 2 & 2 & 5 \end{bmatrix}, \ M_{(u_1,u_2,u_3,u_4),2} = \begin{bmatrix} 1 & 1 & 0 & 0 \\ 1 & 1 & 1 & 0 \\ 1 & 1 & 4 & 0 \\ 2 & 2 & 0 & 5 \end{bmatrix},$$

$$M_{(u_1,u_2,u_3,u_4),3} = \begin{bmatrix} 1 & 1 & 0 & 0 \\ 1 & 1 & 1 & 0 \\ 1 & 1 & 2 & 2 \\ 2 & 2 & 2 & 3 \end{bmatrix}.$$

With this terminology we can easily see that by setting the first variable equal to zero (i.e. eliminating the first row $\mathbf{v}_1^T$) in the above entry matrices, we obtain the following entry matrices of an orthogonal design

$OD(20; 3, 6, 9)$:

$$M_{(u_2,u_3,u_4),1} = \begin{bmatrix} 1 & 1 & 1 & 0 \\ 3 & 1 & 2 & 0 \\ 0 & 2 & 2 & 5 \end{bmatrix}, M_{(u_2,u_3,u_4),2} = \begin{bmatrix} 1 & 1 & 1 & 0 \\ 1 & 1 & 4 & 0 \\ 2 & 2 & 0 & 5 \end{bmatrix},$$

$$M_{(u_2,u_3,u_4),3} = \begin{bmatrix} 1 & 1 & 1 & 0 \\ 1 & 1 & 2 & 2 \\ 2 & 2 & 2 & 3 \end{bmatrix}.$$

Similarly the entry matrices of an orthogonal design $OD(20; 5, 6, 9)$ obtained by setting first and second variable be the same symbol (i.e. replacing rows $\mathbf{v_1^T}, \mathbf{v_2^T}$ by row $\mathbf{v_1^T} + \mathbf{v_2^T}$) are

$$M_{(u_1+u_2,u_3,u_4),1} = \begin{bmatrix} 2 & 2 & 1 & 0 \\ 3 & 1 & 2 & 0 \\ 0 & 2 & 2 & 5 \end{bmatrix}, M_{(u_1+u_2,u_3,u_4),2} = \begin{bmatrix} 1 & 1 & 4 & 0 \\ 2 & 2 & 0 & 5 \end{bmatrix},$$

$$M_{(u_1+u_2,u_3,u_4),3} = \begin{bmatrix} 2 & 2 & 1 & 0 \\ 1 & 1 & 2 & 2 \\ 2 & 2 & 2 & 3 \end{bmatrix}.$$

□

Now from [30] we have that from an orthogonal design over t variables we can obtain an orthogonal design over $t - 1$ variables by "killing" one variable (i.e. setting one variable equal to zero) or "equating" two variables (i.e. setting two variables be the same symbol). If we do these many times we obtain the following lemma:

Lemma 58 *If an orthogonal design $OD(4n; u_1, u_2, \ldots, u_t)$ exist then the following orthogonal designs exist:*

i) *All orthogonal designs $OD(4n; u_{i_1}, u_{i_2}, \ldots, u_{i_k})$ for all $k = 1, 2, \ldots, t$, over k variables and for all $\{i_1, i_2, \ldots, i_k\} \subseteq \{1, 2, \ldots, t\}$.*

ii) *All orthogonal designs*

$$OD\left(4n; \sum_{j=k_0=1}^{k_1} u_{i_j}, \sum_{j=k_1+1}^{k_2} u_{i_j}, \ldots, \sum_{j=k_{m-1}+1}^{k_m} u_{i_j}\right)$$

over m variables where $1 \leq m \leq t$, $1 \leq k_i \leq t$, $\forall\, i = 1, 2, \ldots m$, $k_1 \leq k_2 \leq \ldots \leq k_m$, $u_{i_j} \neq u_{i_\ell}$, $\forall\, j, \ell = 1, 2, \ldots, k_m$ and $i \neq \ell$, $\bigcup_{j=1}^{k_m} u_{i_j} \subseteq \{u_1, u_2, \ldots, u_t\}$.

Proof By equating and killing variables we obtain the desired result.
$\square$

From the above lemma it is obvious that

Corollary 59 *If there exist* k : $1 \leq k \leq t$ *and* $\{i_1, i_2, \ldots, i_k\} \subseteq \{1, 2, \ldots, t\}$ *such that an orthogonal design* $OD(4n; u_{i_1}, u_{i_2}, \ldots, u_{i_k})$ *does not exist then an orthogonal design* $OD(4n; u_1, u_2, \ldots, u_t)$ *cannot exist.*

Our method relies on searching for $OD(4n; u_k, u_j)$, $1 \leq k, j \leq t$, in two variables, which is much faster, rather than using the matrix based algorithm, described in [30] for $OD(4n; u_1, u_2, \ldots, u_t)$, in t variables, which is much slower. Then we use the extension algorithm to construct the orthogonal design we want.

Moreover we do not have to check all non-isomorphic entry matrices $M_{(u_k, u_j), i}$ but only a few of them. We also can select the k, j in such way that we minimize the set of $M_{(u_k, u_j), i}$ we have to search.

Let D be the orthogonal design $OD(4n; u_1, u_2, \ldots, u_t)$. The steps of our algorithm are:

Step 0: Find all non-isomorphic entry matrices $M_{(s_1, s_2, \ldots, s_u), i}$ for D as it is described in [30].

Step 1: For $k, j \in \{1, 2, \ldots, u\}, k < j$ find all non-isomorphic entry matrices $M_{(s_k, s_j), i}$ for the orthogonal design $OD(4n; s_k, s_j)$.

Step 2: For all the above $\binom{u}{2}$ combinations check if $M_{(s_1, s_2, \ldots, s_u), i}|_{\mathcal{D}_{(s_k, s_j)}}$ is equal with any $M_{(s_k, s_j), \ell} \in \mathcal{D}_{(s_k, s_j)}$. Ignore similar matrices $M_{(s_1, s_2, \ldots, s_u), i}|_{\mathcal{D}_{(s_k, s_j)}}$ produced after using the two rows of $M_{(s_1, s_2, \ldots, s_u), i}$ and eliminate all others rows. These are the matrices that can be extended to $M_{(s_1, s_2, \ldots, s_u), i}$ and thus these might produce the orthogonal design D.

Step 3: Select the k, j which give the smallest number of entry matrices $M_{(s_1, s_2, \ldots, s_u), i}|_{\mathcal{D}_{(s_k, s_j)}}$.

Step 4: Apply first algorithm (matrix based algorithm) to the selected entry matrices specified in Step 3, and find all $OD(4n; s_k, s_j)$.

Step 5:   For each $OD(4n; s_k, s_j)$ found in Step 4, apply the second algorithm (extension algorithm), by replacing each zero by a unique variable x_p, $p = 1, 2, \ldots, 4n - (s_k + s_j)$.

Step 6:   Exhaustively search all possibilities then if the solution exists, it will be found, otherwise an $OD(4n; s_1, s_2, \ldots, s_u)$ does not exist constructed by four sequences.

Example 60 We will apply our algorithm to search for an orthogonal design $D = OD(36; u_1, u_2, u_3) = OD(36; 6, 7, 21)$.

Step 0: The following ten matrices are all the non-isomorphic entry matrices $M_{(u_1, u_2, u_3), i}$ for D as it is described in [30]:

$$
1) \begin{bmatrix} 3 & 1 & 2 & 0 \\ 3 & 1 & 1 & 2 \\ 2 & 6 & 6 & 7 \end{bmatrix}, \quad
2) \begin{bmatrix} 3 & 1 & 2 & 0 \\ 1 & 3 & 1 & 2 \\ 4 & 4 & 6 & 7 \end{bmatrix}, \quad
3) \begin{bmatrix} 3 & 1 & 2 & 0 \\ 1 & 1 & 1 & 4 \\ 4 & 6 & 6 & 5 \end{bmatrix},
$$

$$
4) \begin{bmatrix} 3 & 1 & 2 & 0 \\ 1 & 1 & 3 & 2 \\ 4 & 6 & 4 & 7 \end{bmatrix}, \quad
5) \begin{bmatrix} 1 & 1 & 4 & 0 \\ 3 & 1 & 1 & 2 \\ 4 & 6 & 4 & 7 \end{bmatrix}, \quad
6) \begin{bmatrix} 1 & 1 & 4 & 0 \\ 1 & 1 & 3 & 2 \\ 6 & 6 & 2 & 7 \end{bmatrix},
$$

$$
7) \begin{bmatrix} 1 & 1 & 4 & 0 \\ 1 & 1 & 1 & 4 \\ 6 & 6 & 4 & 5 \end{bmatrix}, \quad
8) \begin{bmatrix} 1 & 1 & 2 & 2 \\ 3 & 1 & 1 & 2 \\ 4 & 6 & 6 & 5 \end{bmatrix}, \quad
9) \begin{bmatrix} 1 & 1 & 2 & 2 \\ 1 & 1 & 3 & 2 \\ 6 & 6 & 4 & 5 \end{bmatrix},
$$

$$
10) \begin{bmatrix} 1 & 1 & 2 & 2 \\ 1 & 1 & 1 & 4 \\ 6 & 6 & 6 & 3 \end{bmatrix} \qquad \square
$$

Step 1: We have that

$$
|\mathcal{D}_{(u_1, u_2)}| = 10, \quad |\mathcal{D}_{(u_1, u_3)}| = 53, \quad |\mathcal{D}_{(u_2, u_3)}| = 21
$$

Step 2: By setting the first variable equal to zero (i.e. eliminating the first row $\mathbf{v}_1^T$) we only get 5 non-isomorphic entry matrices $M_{(u_1, u_2, u_3), i}|_{\mathcal{D}_{(u_2, u_3)}}$ from the 21 entry matrices of the orthogonal design $OD(36; 7, 21)$. Those come from the matrices $M_{(u_1, u_2, u_3), i}$ numbered i=1,2,3,8, and 10 above by deleting the first row.

By setting the second variable equal to zero we get 10 non-isomorphic entry matrices $M_{(u_1, u_2, u_3), i}|_{\mathcal{D}_{(u_1, u_3)}}$ from the 53 entry matrices of the orthogonal design $OD(36; 6, 21)$. Those come from the matrices $M_{(u_1, u_2, u_3), i}$ numbered $i = 1, 2, \ldots, 10$ above by deleting the second row.

By setting the third variable equal to zero we get only 10 non-isomorphic entry matrices $M_{(u_1, u_2, u_3), i}|_{\mathcal{D}_{(u_1, u_2)}}$ from the 10 entry matrices of the orthogonal design $OD(36; 6, 7)$. Those come from the matrices $M_{(u_1, u_2, u_3), i}$ numbered $i = 1, 2, \ldots, 10$ above by deleting the third row.

Step 3: Clearly in the case $k = 2$ and $j = 3$ we have fewer entry matrices to check than in any of the other cases, i.e five.

Step 4: Now we get all the quadruples of sequences with PAF=0 or NPAF=0, which can be used for the construction of $OD(36; 7, 21)$, via the Goethals-Seidel Array. This is applied to all five entry matrices described in steps 2 and 3.

Step 5: For each $OD(4n; u_k, u_j) = OD(36; 7, 21)$ found in Step 4, apply the second algorithm (extension algorithm), by replacing the zero of the sequences by the unique variables x_p, $p = 1, 2, \ldots, 8$.

We want to make clear that if an $OD(36; 6, 7, 21)$ existed it would have been found. We did not find any solutions by step 5 and thus, since our search is exhaustive for the orthogonal design $OD(36; 6, 7, 21)$, this design does not exist using four sequences.   □

Example 61 Applying our algorithm we try to find the $OD(36; 6, 8, 19)$ and the $OD(36; 7, 8, 19)$. There are 22 non-isomorphic entry matrices $M_{(6,8,19), i}$ corresponding to the orthogonal design $OD(36; u_1, u_2, u_3) = OD(36; 6, 8, 19)$ and 22 for the second orthogonal design $OD(36; u_4, u_2, u_3) = OD(36; 7, 8, 19)$.

By setting the first variable equal to zero we get only 17 non-isomorphic entry matrices $M_{(6,8,19), i}|_{\mathcal{D}_{(u_2, u_3)}}$ for the $OD(36; 8, 19)$.

We observe that the matrices $M_{(6,8,19), i}|_{\mathcal{D}_{(u_2, u_3)}}$ are exactly the same as the matrices $M_{(7,8,19), i}|_{\mathcal{D}_{(u_2, u_3)}}$ for the second orthogonal design.

Thus by searching those 17 non-isomorphic entry matrices we can perform an exhaustive search for both orthogonal designs. Using the matrix based algorithm we would have had to check 44 entry matrices using three variables for both designs.

Applying our algorithm and following the same process as in the previous example we find, among others, the following solutions, which have PAF=0:

$$OD(36; 6, 8, 19)$$

```
b -c 0 b b b a c -a
b b -b b c -a -b c a
c b -b -b -a -b b -a 0
b -b -b -c b -a b -a 0
```

$$OD(36; 7, 8, 19)$$

```
a -b -b -b c -a -c -b -c
b -a a b -c -b b -b -c
b -b a a b b -b 0 -c
a -b -b -b b a b 0 c
```

□

The interesting reader can find more on this algorithm in [21].

Remark 5 Using the above algorithms, cases where $n \equiv 0(\bmod\ 4)$, have been studied. In particular all orthogonal designs of orders $4n$, $n = 1, 3, 5, 7, 9$ had been completely studied, (see [20, 21, 50, 51, 55, 57]).

3.3   Amicable sets of matrices and constructions of orthogonal designs using the Kharaghani array

A pair of matrices A, B is said to be amicable (anti-amicable) if $AB^T - BA^T = 0$ ($AB^T + BA^T = 0$). Following [7] a set $\{A_1, A_2, \ldots, A_{2n}\}$ of square real matrices is said to be *amicable* if

$$\sum_{i=1}^{n} \left(A_{\sigma(2i-1)} A_{\sigma(2i)}^T - A_{\sigma(2i)} A_{\sigma(2i-1)}^T \right) = 0 \tag{3.31}$$

for some permutation σ of the set $\{1, 2, \ldots, 2n\}$. For simplicity, we will always take $\sigma(i) = i$ unless otherwise specified. So

$$\sum_{i=1}^{n} \left(A_{2i-1} A_{2i}^T - A_{2i} A_{2i-1}^T \right) = 0. \tag{3.32}$$

Clearly a set of mutually amicable matrices is amicable, but the converse is not true in general. Throughout the section R_k denotes the back diagonal identity matrix of order k. A set of matrices $\{B_1, B_2, \ldots, B_n\}$ of order m with entries in $\{0, \pm x_1, \pm x_2, \ldots, \pm x_u\}$ is said to satisfy an additive property of type $(s_1, s_2, \ldots, s_u)$ if

$$\sum_{i=1}^{n} B_i B_i^T = \sum_{i=1}^{u} (s_i x_i^2) I_m. \tag{3.33}$$

Let $\{A_i\}_{i=1}^{8}$ be an amicable set of circulant matrices (or type 1) of type $(s_1, s_2, \ldots, s_u)$ of order t. Then the Kharaghani array from [7]

$$H = \begin{pmatrix} A_1 & A_2 & A_4 R_n & A_3 R_n & A_6 R_n & A_5 R_n & A_8 R_n & A_7 R_n \\ -A_2 & A_1 & A_3 R_n & -A_4 R_n & A_5 R_n & -A_6 R_n & A_7 R_n & -A_8 R_n \\ -A_4 R_n & -A_3 R_n & A_1 & A_2 & -A_8^T R_n & A_7^T R_n & A_6^T R_n & -A_5^T R_n \\ -A_3 R_n & A_4 R_n & -A_2 & A_1 & A_7^T R_n & A_8^T R_n & -A_5^T R_n & -A_6^T R_n \\ -A_6 R_n & -A_5 R_n & A_8^T R_n & -A_7^T R_n & A_1 & A_2 & -A_4^T R_n & A_3^T R_n \\ -A_5 R_n & A_6 R_n & -A_7^T R_n & -A_8^T R_n & -A_2 & A_1 & A_3^T R_n & A_4^T R_n \\ -A_8 R_n & -A_7 R_n & -A_6^T R_n & A_5^T R_n & A_4^T R_n & -A_3^T R_n & A_1 & A_2 \\ -A_7 R_n & A_8 R_n & A_5^T R_n & A_6^T R_n & -A_3^T R_n & -A_4^T R_n & -A_2 & A_1 \end{pmatrix} \tag{3.34}$$

is a Kharaghani type orthogonal design $OD(8m; s_1, s_2, \ldots, s_u)$.

We present an algorithm which uses the known sets of four circulant matrices to construct an amicable set of eight matrices suitable for the array given by (3.34).

The algorithm

Step 1 Find four circulants matrices A, B, C, D of order n with variables a, b, c, d satisfying

$$AA^T + BB^T + CC^T + DD^T = (r_1 a^2 + r_2 b^2 + r_3 c^2 + r_4 d^2) I_n$$

for some integers r_i, by using any of the above algorithms.

Step 2 Form four new circulant matrices E, F, G, H from A, B, C, D just by replacing a, b, c, d with e, f, g, h respectively. Obviously the new matrices satisfy the previous conditions but on variables e, f, g, h.

Step 3 Search the set $\{A, B, C, D, E, F, G, H\}$ for a combination suitable to form an amicable set of eight matrices.

Step 4 If we find such a set, we replace the matrices in the array given by (3.34).

Notation 2 With the expression $circ(a, b, c, \ldots, z)$ we will denote the circulant matrix with first row the sequence in the brackets.

Example 62 Let $A = circ(a, b, c)$, $B = circ(d, -a, b)$, $C = circ(-c, d, a)$ and $D = circ(-b, c, d)$. Then $AA^t + BB^t + CC^t + DD^t = 3(a^2 + b^2 + c^2 + d^2)I_3$. We form the matrices $E = circ(e, f, g)$, $F = circ(h - e, f)$, $G = circ(-g, h, e)$ and $H = circ(-f, g, h)$. Then obviously we have that $EE^T + FF^T + GG^T + HH^T = 3(e^2 + f^2 + g^2 + h^2)I_3$. A computer search finds that

$$AH^T - HA^T + BG^T - GB^T + CF^T - FC^T + DE^T - ED^T = 0$$

So, we have found an amicable set of eight circulant matrices, the {A,H,B,G,C,F,D,E}. If we substitute these matrices in the array of the corollary, we get an $OD(24; 3, 3, 3, 3, 3, 3, 3, 3)$.

Example 63 Let $A = circ(a, b, b, d, -d)$, $B = circ(-b, a, a, c, -c)$, $C = circ(d, c, c, -a, a)$, $D = circ(-c, d, d, -b, b)$. Then $AA^T + BB^T + CC^T + DD^T = 5(a^2 + b^2 + c^2 + d^2)I_5$. We form the matrices $E = circ(e, f, f, h, -h)$, $F = circ(-f, e, e, g, -g)$, $G = circ(h, g, g, -e, e)$, $H = circ(-g, h, h, -f, f)$ just by substituting the variables a,b,c,d for e,f,g,h respectively. Then we have $EE^T + FF^T + GG^T + HH^T = 5(e^2 + f^2 + g^2 + h^2)I_5$. A computer search finds the amicable set

$$AE^T - EA^T + BH^T - HB^T + GC^T - CG^T + DF^T - FD^T = 0$$

So, we have the $\{A, E, B, H, G, C, D, F\}$ amicable set of matrices. If we substitute these matrices in Kharaghani array we obtain the $OD(40; 5, 5, 5, 5, 5, 5, 5, 5)$.

Remark 6 Using the above algorithm, and the Kharaghani array many new orthogonal designs of orders $8n$ are constructed, (see [16, 24, 26, 27, 38, 39, 7, 58, 60]).

4. Short amicable sets and Kharaghani type orthogonal designs

4.1 Preliminary results and basic definitions

Short amicable set were defined in [28] as a set of matrices $\{A_i\}_{i=1}^4$ of order m and type (u_1, u_2, u_3, u_4), abbreviated as $4 - SAS(m; u_1, u_2, u_3,$

$u_4; G)$, if (3.32) and (3.33) are satisfied for $n = 4$ and $u \leq 4$. $4 - SAS(m; u_1, u_2, u_3, u_4; G)$ can be used in either the Goethals-Seidel array or the *short Kharaghani array*

$$\begin{bmatrix} A & B & CR & DR \\ -B & A & DR & -CR \\ -CR & -DR & A & B \\ -DR & CR & -B & A \end{bmatrix}$$

to form an $OD(4m; u_1, u_2, u_3, u_4)$. In all cases, the group G of the matrices in the *amicable set* is such that the extension by Seberry and Whiteman [74] of the group from circulant to type 1 allows the same extension to R.

In general a set of $2n$ matrices of order m and type $(s_1, s_2, \ldots, s_u)$ that satisfy equations (3.32) and (3.33) will be denoted as $2n - SAS(m; s_1, s_2, \ldots, s_u; G)$. Moreover if these matrices are circulant they will be denoted as $2n - SCAS(m; s_1, s_2, \ldots, s_u; Z_m)$.

In [28] where short amicable sets were first defined, it was mentioned that:

Remark 7   1 If there exists a $2 - SAS(n; s_1, s_2; G)$ and a $2 - SAS(n; s_3, s_4; G)$ then there exists a $4 - SAS(n; s_1, s_2, s_3, s_4; G)$.

2 If there exists a $2 - SAS(n; s_1, s_2; G)$, $2 - SAS(n; s_3, s_4; G)$, $2 - SAS(n; s_5, s_6; G)$ and a $2 - SAS(n; s_7, s_8; G)$ there exists an $8 - AS(n; s_1, s_2, s_3, s_4, s_5, s_6, s_7, s_8; G)$.

3 If there exists a $4 - SAS(n; s_1, s_2, s_3, s_4; G)$ and a $4 - SAS(n; s_5, s_6, s_7, s_8; G)$ there exists an $8 - AS(n; s_1, s_2, s_3, s_4, s_5, s_6, s_7, s_8; G)$.

Thus we can obtain many classes of $4 - SAS(n; s_1, s_2, s_3, s_4; G)$ combining together two pairs of the given $2 - SAS(n; s_1, s_2; G)$ and $2 - SAS(n; s_3, s_4; G)$. Moreover, in Table 4.3, we give some $4 - SAS(m; u_1, u_2, u_3, u_4; Z_m)$ that cannot be constructed by this method.

Generally, unless we have other information regarding the structure, we are unable to ensure that the matrix R with the desired properties for the Kharaghani, Goethals-Seidel or short Kharaghani arrays exists unless the amicable sets have been group generated (circulant or type 1) or constructed from blocks of these kinds. Thus is we have the required matrix R_i for the group G_i, $i = 1, 2$ then $R_G = R_1 \times R_2$ will be the required matrix for $G = G_1 \times G_2$, (see [74]).

Let A_1 and A_2 be matrices of order m. We define $circ(A_1, A_2) = \begin{bmatrix} A_1 & A_2 \\ A_2 & A_1 \end{bmatrix}$. Amicable sets made from $2n$ such block circulant matrices will be called *block amicable sets*, *short block amicable sets* or *2-short block amicable sets*, $2n - SBAS(2m; s_1, s_2, \ldots, s_u; G)$, $n = 1, 2, 4$, where, using R_t for the back-diagonal matrix of order t, $G = Z_2 \times Z_m$ and $R_G = R_2 \times R_m$. Here, if A_1 and A_2 are circulant, then we use the back-diagonal matrix of the same order for R ensuring $A_i(A_j R)^T = A_j R A_i^T$. The required $R_G = R_2 \times R$.

We denote the product $Z_p \times Z_p \times \cdots \times Z_p (r \ times)$ by $EA(p^r)$ the Elementary Abelian group. Moreover $-a$ is denoted by $\bar{a}$.

Throughought this section we use the symbol 0_m to denote the sequence of length m with all elements zero and the symbol O_t to denote the $t \times t$ matrix with all entries zero.

For the undefined terms we refer the reader to the book by Geramita and Seberry [30].

4.2   Constructions

Theorem 64 *Write 0_s for the sequence of s zeros, and let a, b, c and d be commuting variables. Use the matrices A_1, A_2, A_3 and A_4 given by*

$$A_1 = circ(0_s b a \bar{b} 0_s), \qquad A_2 = circ(0_s c 0 c 0_s),$$
$$A_3 = circ(0_s \bar{c} d c 0_s), \qquad A_4 = circ(0_s b 0 b 0_s),$$

can be used in the Goethals-Seidel array to obtain an $OD(8s+12; 1, 1, 4, 4)$.

Proof Observe that

$$A_1 A_1^T + A_2 A_2^T + A_3 A_3^T + A_4 A_4^T = (a^2 + d^2 + 4b^2 + 4c^2) I_n$$

and

$$A_1 A_2^T - A_2 A_1^T + A_3 A_4^T - A_4 A_3^T = 0.$$

Thus A_1, A_2, A_3, A_4 are a short amicable set and satisfy the additive property (3.33), so they can be used in the Goethals-Seidel array to obtain an $OD(8s + 12; 1, 1, 4, 4)$. $\qquad\square$

The Melding Construction
Suppose the matrices A_1, A_2, A_3 and A_4 are are short amicable sets, on the set of commuting variables $\{0, \pm x_1, \pm x_2, \cdots, \pm x_u\}$ or from $\{0, \pm 1\}$,

and satisfy the additive property

$$\sum_{i=1}^{4} \left(A_i A_i^T \right) = \sum_{j=1}^{u} p_j x_j^2 I_n, \qquad (4.35)$$

and the matrices A_5, A_6, A_7 and A_8 are also short amicable sets, on the set of commuting variables $\{0, \pm y_1, \pm y_2, \cdots, \pm y_v\}$ or from $\{0, \pm 1\}$, and satisfy the additive property

$$\sum_{i=5}^{8} \left(A_i A_i^T \right) = \sum_{j=1}^{v} q_j y_j^2 I_n. \qquad (4.36)$$

Then the eight matrices will form an amicable set so we can use the two together in the Kharaghani array to obtain an $OD(8n; p_1, p_2, \cdots, p_u, q_1, q_2, \cdots, q_v)$. $\qquad \square$

order	type	group	order	type	group
n	1, 1	Z_n	$6n$	4, 4	Z_{6n}
$2n$	2, 2	Z_{2n}	$6n$	5, 5	Z_{6n}
$4n$	1, 4	Z_{4n}	$7n$	4, 4	Z_{7n}
$4n$	4, 4	Z_{4n}	$8n$	8, 8	Z_{8n}

order	type	group	order	type	group
$10n$	4, 4	Z_{10n}	$14n$	8, 8	Z_{14n}
$10n$	9, 9	Z_{10n}	$14n$	10, 10	Z_{14n}
$12n$	8, 8	Z_{12n}	$14n$	13, 13	Z_{14n}
$13n$	9, 9	Z_{13n}			

Table 4.2. Order and type for small 2-short amicable sets for all $n \geq 1$.

Using Table 4.2, Remark 7 and the above Melding Construction we obtain many 4-short amicable sets and 8-amicable sets.

Type	A_1 A_2	A_3 A_4	ZERO
(1,1,1,1)	a c	b d	NPAF n
(1,1,1,4)	0 -d a d 0 d 0 d	0 b 0 0 0 c 0 0	NPAF $4n$
(1,1,2,2)	a 0 b 0	c d c -d	NPAF $2n$
(1,1,2,8)	0 -c a c 0 c b c	0 -c b -c 0 -c d c	NPAF $4n$
(1,1,4,4)	a b -a c 0 c	a 0 a c d -c	NPAF $3n$
(1,1,5)	-a a a c 0 0	a 0 a 0 b 0	NPAF $4n$
(1,1,5,5)	-c a c 0 c -d c 0	-d b d 0 d c d 0	NPAF $4n$
(1,1,8,8)	0 -c -d a d c 0 c d 0 d c	0 c -d 0 -d c 0 -c d b -d c	NPAF $6n$
(1,2,2,4)	0 -d a d 0 d 0 d	c 0 b 0 c 0 -b 0	NPAF $4n$
(1,4,4,4)	0 -b a b 0 b 0 b	d c -d c -c d c d	NPAF $4n$
(2,2,2,2)	a b c d	a -b c -d	NPAF $2n$
(2,2,4,4)	a 0 b 0 a 0 -b 0	d c -d c -c d c d	NPAF $4n$
(2,2,5,5)	0 a 0 0 b 0 0 a 0 0 -b 0	c -d 0 -d c d d c 0 c d -c	NPAF $6n$
(2,2,8,8)	-d c a c d 0 -d -c a -c d 0	d -c b c d 0 -d -c b c -d 0	NPAF $6n$
(3,3)	a b a 0	b -a b 0	NPAF $2n$
(4,4,4,4)	a a b -b d d -c c	b b -a a c c d -d	NPAF $4n$
(4,4,8,8)	d a -c c a -d -d -b c c b -d	d b c -c b -d d -a c c a d	NPAF $6n$
(5,5)	a a -a b b -b	a 0 a b 0 b	NPAF $3n$
(5,5,5,5)	-a b a 0 a b b a -b 0 -b a	-c d c 0 c d d c -d 0 -d c	NPAF $6n$
(6,6)	a -b a b a b	a a -a b b -b	NPAF $3n$
(6,6,12)	c a c b -c a -c b -c -a c b	c a c -a c -a -c b c -b -c -b	NPAF $6n$
(8,8)	a a a -a b b b -b	b b -b b a a -a a	NPAF $4n$

Table 4.3. Short amicable sets.

(8,8,8,8)	a a a-a b b-b b c c c-c d d-d d	b b b-b a a-a a d d d-d c c-c c	NPAF 8n
(10,10,10,10)	disjoint	from Golay	NPAF n ≥ 10
(13,13)	c 0 -c c -c 0 0 c c g 0 -g g -g 0 0 g g	c c -c c c c 0 0 -c g g -g g g g 0 0 -g	NPAF 9n
(13,13,13,13)	from disjoint length 18	sequences weight 13	NPAF n ≥ 18
(16,16,16,16)	disjoint	from Golay	NPAF n ≥ 16
(17,17,17,17)	from disjoint length 26	sequences weight 17	NPAF n ≥ 26
(20,20,20,20)	dijoint	from Golay	NPAF n ≥ 20
(25,25,25,25)	from disjoint length 36	sequences weight 25	NPAF n ≥ 36
(26,26,26,26)	disjoint	from Golay	NPAF n ≥ 26
(14,14)	a b -b -b b a a b -a a a -a b b	-b a -b a -b b b a b a b a -a -a	NPAF 7n
(17,17)	a -a a a a a -a a 0 c -c c c c c -c c 0	c -c -c c c c -c -c a -a -a a a a a -a -a	PAF 9n

Table 4.3: (continued).

4.3 Some general results

We now consider the use of sequences with zero non-periodic auto-correlation function to make an amicable set of matrices. We refer the reader to [73, 75] for any undefined terms.

The next theorem was proved in [59].

Theorem 65 (General construction) *Let* X, Y *be two disjoint* $(0, \pm 1)$ *sequences with zero non-periodic autocorrelation function of length* n *and weight* k, *Let* a, b, c, d *be commuting variables and write* aV, bW *for the circulant (type 1) matrices of order* n *formed by using the first rows with the elements of* X *multiplied by* a *and the elements of* Y *multiplied by* b *respectively.*

Let A_i *be the circulant matrices of order* n *given by*

$$A_1 = aV + bW \quad A_2 = cV + dW \quad A_3 = dV - cW \quad A_4 = bV - aW$$
$$(4.37)$$

then $\{A_i\}_{i=1}^4$ is a short amicable set satisfying

$$\sum_{i=1}^{2} \left(A_{2i-1} A_{2i}^T - A_{2i} A_{2i-1}^T \right) = 0, \qquad (4.38)$$

and the additive property

$$\sum_{i=1}^{4} \left(A_i A_i^T \right) = k(a^2 + b^2 + c^2 + d^2) I_n. \qquad (4.39)$$

Corollary 66 *Let X, Y be a pair of disjoint $(0, \pm 1)$ sequences with zero non-periodic autocorrelation function of length n and weight k. Then there exists a short amicable set which can be used to form an $OD(4n; k, k, k, k)$.*

For $\alpha, \beta, \gamma, \delta, \epsilon, \phi, \psi, \mu, \nu$ non-negative integers, Koukouvinos and Se-berry [56, p. 160] show that there exist two disjoint $(0, \pm 1)$ sequences, with zero non-periodic autocorrelation function, of length $\geq n$, $n \in N = \{2 \times 2^\alpha 6^\beta 10^\gamma 9^\delta 14^\epsilon 18^\phi 26^\psi 24^\mu 34^\nu\}$ and weight k, $k \in K = \{2^\alpha 5^\beta 10^\gamma 13^\delta 17^\epsilon 25^\phi 26^\psi 34^\mu 50^\nu\}$. These give the results presented in Table 4.4.

Type	ZERO	Type	ZERO
(1,1,1,1)	NPAF $n \geq 1$	(2,2,2,2)	NPAF $n \geq 2$
(4,4,4,4)	NPAF $n \geq 4$	(5,5,5,5)	NPAF $n \geq 6$
(8,8,8,8)	NPAF $n \geq 8$	(10,10,10,10)	NPAF $n \geq 10$
(13,13,13,13)	NPAF $n \geq 18$	(16,16,16,16)	NPAF $n \geq 16$
(17,17,17,17)	NPAF $n \geq 26$	(20,20,20,20)	NPAF $n \geq 20$
(25,25,25,25)	NPAF $n \geq 36$	(26,26,26,26)	NPAF $n \geq 26$

Table 4.4. Short amicable sets from corollary 66

For more details about short amicable sets and their use in the con-struction of Kharaghani type orthogonal designs the interesting reader is refer to [28, 59].

References

[1] Apple Computer. Inside Macintosh: Networking with Open Transport, 1997. Available from http://developer.apple.com/tech-pubs/mac/pdf/Networking0T.pdf.

[2] L. D. Baumert, Hadamard matrices of orders 116 and 232, *Bull. Amer. Math. Soc.*, 72 (1966), 237.

[3] L. D. Baumert, *Cyclic Difference Sets*, Lecture Notes in Mathematics, Vol. 182, Springer-Verlag, Berlin-Heidelberg-New York, 1971.

[4] L. D. Baumert, and M. Hall Jr., Hadamard matrices of the Williamson type, *Math. Comput.*, 19 (1965), 442-447.

[5] T. Chadjipantelis and S. Kounias, Supplementary difference sets and D-optimal designs for $n \equiv 2(mod\,4)$, *Discrete Math.*, 57 (1985), 211-216.

[6] J. Cooper, J. Milas, and W. D. Wallis, Hadamard equivalence, in *Combinatorial Mathematics*, Lecture Notes in Mathematics, Vol. 686, Springer-Verlag, Berlin, Heidelberg, New York, 1978, 126–135.

[7] R. Craigen, *Hadamard Matrices and Designs*, The CRC Handbook of Combinatorial Designs, eds. C. J. Colbourn and J. H. Dinitz, CRC Press, Boca Raton, Fla., (1996), 370–377.

[8] D. Z. Djokovic, Williamson matrices of orders $4 \cdot 29$ and $4 \cdot 31$, *J. Combin. Theory Ser. A*, 59 (1992), 442–447.

[9] D. Z. Djokovic, Williamson matrices of orders 4n for n=33,35,39, *Discrete Math.*, 115 (1993), 267–271.

[10] D. Z. Djokovic, Note on Williamson matrices of orders 25 and 37, *J. Combin. Math. Combin. Comput.*, 18 (1995), 171–175.

[11] P. Eades and J. Seberry Wallis, An infinite family of skew-weighing matrices, *Combinatorial Mathematics IV*, in Lecture Notes in Mathematics, Vol. 560, Springer-Verlag, Berlin-Heidelberg-New York, pp. 27-40, 1976.

[12] S. Eliahou, M. Kervaire, and B. Saffari, A new restriction on the lengths of Golay complementary sequences, *J. Combin. Theory Ser. A*, 55 (1990), 49-59.

[13] R. J. Fletcher, M. Gysin, and J. Seberry, Application of the discrete Fourier transform to the search for generalized Legendre pairs and Hadamard matrices, *Australas. J. Combin.*, 23 (2001), 75-86.

[14] Al. Geist, A. Beguelin, J. Dongarra, W. Jiang, R. Manchek, and V. Sunderam. PVM:Parallel Virtual Machine–A User's Guide and Tutorial for Networked Parallel Computing. MIT Press, 1994. Available as Postscript from http://www.netlib.org/pvm3/book/pvm-book. ps.

[15] S. Georgiou and C. Koukouvinos, On multipliers of supplementary difference sets and D-optimal designs for $n \equiv 2mod4$, *Utilitas Math.*, 56 (1999), 127-136.

[16] S. Georgiou and C. Koukouvinos, On amicable sets of matrices and orthogonal designs, *Int. J. Appl. Math.*, 4 (2000), 211-224.

[17] S. Georgiou and C. Koukouvinos, On sequences with zero auto-correlation and orthogonal designs, *J Combin. Theory Ser. A*, 94 (2001), 15-33.

[18] S. Georgiou and C. Koukouvinos, On generalized Legendre pairs and multipliers of the corresponding supplementary difference sets, *Utilitas Math.*, (to appear).

[19] S. Georgiou and C. Koukouvinos, On equivalence of Hadamard matrices and projection properties, *Ars Combin.*, (to appear).

[20] S. Georgiou, C. Koukouvinos, M. Mitrouli, and J. Seberry, Necessary and Sufficient Conditions for two variable orthogonal designs in order 44:Addendum, *J. Combin. Math. Combin. Comput.*, 34 (2000), 59-64.

[21] S. Georgiou, C. Koukouvinos, M. Mitrouli and J. Seberry, A new algorithm for computer searches for orthogonal designs, *J. Combin. Math. Combin. Comput.*, 39 (2001), 49-63.

[22] S. Georgiou, C. Koukouvinos, M. Mitrouli and J. Seberry, Necessary and sufficient conditions for three and four variable orthogonal designs in order 36, *J. Statist. Plann. Inference*, (to appear).

[23] S. Georgiou and C. Koukouvinos, On inequivalent Hadamard matrices of order 36, *Ars Combin.*, (to appear).

[24] S. Georgiou, C. Koukouvinos, and J. Seberry, On full orthogonal designs in order 72, *J. Combin. Math. Combin. Comput.*, (to appear).

[25] S. Georgiou and C. Koukouvinos, On inequivalent Hadamard matrices of order 44, *Ars Combin.*, (to appear).

[26] S. Georgiou, C. Koukouvinos, and J. Seberry, On full orthogonal designs in order 56, *Ars. Combin.*, (to appear).

[27] S. Georgiou, C. Koukouvinos and J. Seberry, Some results on Kharaghani type orthogonal designs, *Utilitas Math.*, (to appear).

[28] S. Georgiou, C. Koukouvinos and J. Seberry, Short amicable sets, (submitted).

[29] A. V. Geramita, J. M. Geramita, and J. Seberry Wallis, Orthogonal designs, *Linear and Multilinear Algebra*, 3 (1976), 281-306.

[30] A. V. Geramita, and J. Seberry, *Orthogonal designs: Quadratic forms and Hadamard matrices*, Marcel Dekker, New York-Basel, 1979.

[31] M. J. E. Golay, Complementary sequences, *IRE Trans. Inform. Theory*, 7 (1961), 82-87.

[32] W. Gropp, L. Ewing, and A. Skjellum, Using MPI: Portable Parallel Programming with the Message-Passing Interface, MIT Press, 1994.

[33] M. Gysin, New D-optimal designs via cyclotomy and generalized cyclotomy, *Australas. J. Combin.*, 15 (1997), 247-255.

[34] M. Gysin and J. Seberry, An experimental search and new combinatorial designs via a generalization of cyclotomy, *J. Combin. Math. Combin. Comput.*, 27 (1998), 143-160.

[35] M. Gysin and J. Seberry, On new families of supplementary difference sets over rings with short orbits, *J. Combin. Math. Combin. Comput.*, 28 (1998), 161-186.

[36] M. Hall Jr., Hadamard matrices of order 16, *JPL Research Summary* No. 36-10, Vol. 1 (1961), 21–26.

[37] M. Hall Jr., Hadamard matrices of order 20, *JPL Technical Report* No. 32-76, Vol.1 (1965).

[38] W.H. Holzmann, and H. Kharaghani, On the Plotkin arrays, *Australas. J. Combin.*, 22 (2000), 287-299.

[39] W.H. Holzmann, and H. Kharaghani, On the orthogonal designs of order 24, *Discrete Appl. Math.*, 102 (2000), 103-114.

[40] W.H. Holzmann, and H. Kharaghani, On the orthogonal designs of order 40, *J. Statist. Plann. Inference*, 96 (2001), 415-429.

[41] J. Horton, C. Koukouvinos and J. Seberry, A search for Hadamard matrices constructed from Williamson matrices, *Bull. Inst. Combin. Appl.*, (to appear).

[42] N. Ito, J. S. Leon and J. Q. Longyear, Classification of $3-(24,12,5)$ designs and 24-dimensional Hadamard matrices, *J. Combin. Theory Ser. A*, 31 (1981), 66–93.

[43] Z. Janko, The existence of a Bush-type Hadamard matrix of order 36 and two new infinite classes of symmetric designs, *J. Combin. Theory Ser. A*, 95 (2001), 360–364.

[44] H. Kharaghani, Arrays for orthogonal designs, *J. Combin. Designs*, 8 (2000), 166-173.

[45] H. Kimura, Hadamard matrices of order 28 with automorphism groups of order two, *J. Combin. Theory Ser. A*, 43 (1986), 98–102.

[46] H. Kimura, On equivalence of Hadamard matrices, *Hokkaido Mathematical Journal*, 17 (1988), 139–146.

[47] H. Kimura, New Hadamard matrices of order 24, *Graphs Combin.*, 5 (1989), 236–242.

[48] H. Kimura, Classification of Hadamard matrices of order 28 with Hall sets, *Discrete Math.*, 128 (1994), 257–268.

[49] H. Kimura, Classification of Hadamard matrices of order 28, *Discrete Math.*, 133 (1994), 171–180.

[50] C. Koukouvinos, Some new orthogonal designs of order 36, *Utilitas Math.*, 51 (1997), 65-71.

[51] C. Koukouvinos, Some new three and four variable orthogonal designs in order 36, *J. Statist. Plann. Inference*, 73 (1998), 21-27.

[52] C. Koukouvinos and S. Kounias, Hadamard matrices of the Williamson type of order 4m, m=pq: An exhaustive search for m=33, *Discrete Math.*, 68 (1988), 45–47.

[53] C. Koukouvinos and S. Kounias, There are no circulant symmetric Williamson matrices of order 39, *J. Combin. Math. Combin. Comput.*, 7 (1990), 161–169.

[54] C. Koukouvinos, M. Mitrouli and J. Seberry, Necessary and sufficient conditions for some two variable orthogonal designs in order 44, *J. Combin. Math. Combin. Comput.*, 28 (1998), 267-286.

[55] C. Koukouvinos, M. Mitrouli, J. Seberry, and P. Karabelas, On sufficient conditions for some orthogonal designs and sequences with zero autocorrelation function, *Australas. J. Combin.*, 13 (1996), 197-216.

[56] C. Koukouvinos and J. Seberry, New weighing matrices and orthogonal designs constructed using two sequences with zero autocorrelation function - a review, *J. Statist. Plann. Inference*, 81 (1999), 153-182.

[57] C. Koukouvinos and J. Seberry, New orthogonal designs and sequences with two and three variables in order 28, *Ars Combin.*, 54 (2000), 97-108.

[58] C. Koukouvinos and J. Seberry, Infinite families of orthogonal designs : I, *Bull. Inst. Combin. Appl.*, 33 (2001), 35-41.

[59] C. Koukouvinos and J. Seberry, Short amicable sets and Kharaghani type orthogonal designs, *Bull. Austral. Math. Soc.*, 64 (2001), 495-504.

[60] C. Koukouvinos and J. Seberry, Orthogonal designs of Kharaghani type: I, *Ars Combin.*, (to appear).

[61] C. Koukouvinos, J. Seberry, A. L. Whiteman, and M. Y. Xia, Optimal designs, supplementary difference sets and multipliers, *J. Statist. Plann. Inference*, 62 (1997), 81-90.

[62] S. Kounias, C. Koukouvinos, N. Nikolaou and A. Kakos, The non-equivalent circulant D-optimal designs for $n \equiv 2 \bmod 4$, $n \leq 54$, $n = 66$, *J. Combin. Theory Ser. A*, 65 (1994), 26-38.

[63] C. Lam, S. Lam and V. D. Tonchev, Bounds on the number of affine, symmetric, and Hadamard designs and matrices, *J. Combin. Theory Ser. A*, 92 (2000), 186–196.

[64] C. Lin, W. D. Wallis, and Zhu Lie, Extended 4-profiles of Hadamard matrices, *Ann. Discrete Math.*, 51 (1992), 175–180.

[65] C. Lin, W. D. Wallis, and Zhu Lie, Equivalence classes of Hadamard matrices of order 32, *Congr. Numerantium*, 95 (1993), 179–182.

[66] C. Lin, W. D. Wallis, and Zhu Lie, Generalized 4-profiles of Hadamard matrices, *J. Comb. Inf. Syst. Sci.*, 18 (1993), 397–400.

[67] C. Lin, W. D. Wallis, and Zhu Lie, Equivalence classes of Hadamard matrices of order 32, *Congr. Numerantium*, 95 (1993), 179–182.

[68] C. Lin, W. D. Wallis, and Zhu Lie, Hadamard Matrices of Order 32, Preprint #92-20, Department of Mathematical Science, University of Nevada, Las Vegas, Nevada.

[69] C. Lin, W. D. Wallis, and Zhu Lie, Hadamard Matrices of Order 32 II, Preprint #93-05, Department of Mathematical Science, University of Nevada, Las Vegas, Nevada.

[70] W.H. Press, B.P. Flannery, S.A. Teukolsky and W.T. Vettering, *Numerical Recipes in Pascal: The Art of Scientific Computing*, Cambridge Univ Press, New York, 1989.

[71] M.R. Schroeder, *Number Theory in Science and Communication*, Springer–Verlag, New York, 1984.

[72] Hadamard matrices of order 100 and 108, *Bull. Nagoya Inst. Technology*, 29 (1977), 147–153.

[73] J. Seberry, and R. Craigen, Orthogonal designs, in *Handbook of Combinatorial Designs*, C.J.Colbourn and J.H.Dinitz (Eds.), CRC Press, (1996), 400-406.

[74] J. Seberry and A.L. Whiteman, New Hadamard matrices and conference matrices obtained via Mathon's construction, *Graphs and Combinatorics*, 4 (1988), 355-377.

[75] J. Seberry and M. Yamada, Hadamard matrices, sequences and block designs, *Contemporary Design Theory:A Collection of Surveys*, eds. J.Dinitz and D.Stinson, J.Wiley, New York, (1992), 431-560.

[76] E. Spence, Regular two-graphs on 36 vertices, *Linear Alg. Appl.*, 226-228 (1995), 459–497.

[77] R.G. Stanton and D.A. Sprott, A family of difference sets, *Canad. J. Math.*, 10 (1958), 73-77.

[78] W. R. Stevens, UNIX Network Programming: Networking APIs: Sockets and XTI, volume 1, Prentice Hall, second edition, 1998.

[79] T. Storer, *Cyclotomy and difference sets*, Lectures in Advanced Mathematics 2, Markham Publishing Company, Chicago, 1967.

[80] V. D. Tonchev, Hadamard matrices of order 36 with automorphism of order 17, *Nagoya Math. J.*, 104 (1986), 163–174.

[81] S. Topalova, Hadamard matrices of order 44 with automorphisms of order 7, *Discrete Math.*, (to appear).

[82] S.A. Tretter, *Introduction to Discrete-time Signal Processing*, John Wiley and Sons, New York, 1976.

[83] R. J. Turyn, An infinite class of Williamson matrices, *J. Combin. Theory Ser. A*, 12 (1972), 319-321.

[84] R. J. Turyn, Hadamard matrices, Baumert-Hall units, four symbol sequences, pulse compression and surface wave encoding *J. Combin. Theory Ser. A*, 16 (1974), 313-333.

[85] W. D. Wallis, A. P. Street, and J. Seberry Wallis, *Combinatorics: Room Squares, Sum-Free Sets Hadamard Matrices*, Lecture Notes in Mathematics, Vol. 292, Springer-Verlag, Berlin, Heidelberg, New York, 1972.

[86] A. L. Whiteman, A family of difference sets, *Illinois J. Math.*, 6 (1962), 107-121.

[87] A. L. Whiteman, An infinite family of Hadamard matrices of Williamson type, *J. Combin. Theory Ser. A*, 14 (1973), 334–340.

[88] J. Williamson, Hadamard's determinant theorem and the sum of four squares, *Duke Math. J.*, 11 (1944), 65-81.

[89] M. Yamada, On the Williamson type j matrices of orders $4 \cdot 29, 4 \cdot 41$ and $4 \cdot 37$, *J. Combin. Theory Ser. A*, 27 (1979), 378–381.

[90] C. H. Yang, On Hadamard matrices constructible by circulant submatrices, *Math. Comput.*, 25 (1971), 181-186.

Chapter 8

CONSTRUCTING A CLASS OF DESIGNS WITH PROPORTIONAL BALANCE *

Ken Gray

Anne Penfold Street

Centre for Discrete Mathematics and Computing
Department of Mathematics
The University of Queensland, 4072, Australia

Abstract   Proportionally balanced designs were introduced by Gray and Matters in response to a need for the allocation of markers of the Queensland Core Skills Test to have a certain property. They have further application in experiments where efficient achievement of the purposes demands that some of the objects of study (varieties) occur more frequently in the experiment than do others. Subsequently, Gray extended the theoretical results relating to such designs and provided further instances, and two general constructions, in the case that the designs comprise blocks of precisely two sizes.

In this paper, we introduce a new construction for proportionally balanced designs when each variety occurs with one of precisely two possible frequencies, and provide some exhaustive lists of possible parameter sets, and sets of blocks for some of the corresponding designs. Two general constructions associated with this method are also discussed.

*Research supported by Australian Research Council Grants A421420658, A421420906, A49937047

1.   Proportionally balanced designs

1.1   Introduction

Proportionally balanced designs, defined below, were introduced by Gray and Matters [4] in 1995. At the time, they were motivated by the properties that were desirable when each marker of the Short-response testpaper of the Queensland Core Skills Test was allocated two units to mark. Put simply, it was desirable that markers were allocated to pairs of units in proportions that reflected the relative numbers of markers allocated in total to each unit.

A subsequent paper by Gray [3] extended the theoretical results relating to proportionally balanced designs, and gave constructions, particular and general, for such designs in some cases where parameter sets of such designs had been determined, of itself a non-trivial question. The following same definitions apply throughout this paper, and are needed for the explanation that then follows of the differences between the methods used in [3] and those used in this paper.

Let V be a finite set of v elements and let $\mathcal{B} = \{B_i\}$ be a family of subsets of V. The subsets are called *blocks* and the pair $(V, \mathcal{B})$ is called a *design* on the set V. The set $\{|B_i| : B_i \in \mathcal{B}\}$ is the set of block *sizes* of the design. A design is said to be *incomplete* if at least one of its blocks is a proper subset of V: otherwise it is called *complete*. We are concerned only with designs based on at least three elements and with all block sizes at least two. In the first instance, we consider the design $(V, \mathcal{B})$ with v elements as being based on $V = \{1, 2, \ldots, v\}$ and let r_i denote the *replication number* of the element i; that is, r_i is the number of blocks containing element i. Similarly, we let λ_{ij} denote the number of blocks containing the pair of distinct elements i and j.

A design in which all the blocks have the same size and all the elements occur in the same number of blocks is called a *block design*. Further, if each pair of elements occurs in precisely λ blocks, for some constant λ, the block design is also said to be *balanced*; see for example Street and Street [6]. A balanced incomplete block design on v elements, with block size k and with each pair of elements occurring in λ blocks, is usually denoted as a $(v, k, \lambda)BIBD$. Sometimes in this paper, the balanced block designs used to construct proportionally balanced designs are complete. Hence, for our purposes here all of the balanced block designs used, most of which are incomplete, will be denoted as $BiBDs$.

Discussions later in this paper regarding some general constructions of proportionally balanced designs require the following additional results relating to $BiBD$s (see, for example, [6], pp 28-9). In a given $BiBD$ with b blocks, let b_i be the number of blocks intersecting a particular block in precisely i elements, for $i = 0, 1, 2, \ldots, k$. Then:

(a) $\sum_{i=0}^{k} b_i = b - 1$;

(b) $\sum_{i=0}^{k} i b_i = k(r - 1)$;

(c) $\sum_{i=0}^{k} \binom{i}{2} b_i = \binom{k}{2}(\lambda - 1)$.

The proportionally balanced designs given in [3] were, in general, constructed by taking a $BiBD$ with $\lambda = 1$ and then incorporating additional elements and supplementing the blocks of this $BiBD$ with additional blocks until a proportionally balanced design with the given parameters was produced. In contrast, the proportionally balanced designs in this paper have been constructed by commencing with a $BiBD$ with $\lambda \geq 2$ that has a block repeated sufficiently often that, on deletion of all but one instance of this block, there is a subset of the elements such that any two elements occur together precisely once. It is this deleted set of blocks that is then supplemented with additional blocks to produce the proportionally balanced design.

1.2 Proportionally balanced designs with blocks of two sizes: basic definitions and results

The definitions and results that follow are those needed for this paper. Other theoretical results pertaining to proportionally balanced designs are given in [3].

Definition 1 A design D is called *proportionally balanced* if

(i) $\lambda_{ij} > 0$, for all $i \neq j$;

(ii) $r_i/r_j = \lambda_{im}/\lambda_{jm}$, for all distinct i, j, m.

Theorem 2 ([4, Theorem 1.2]) *Suppose a design D has blocks all of the same size. Then D is a proportionally balanced design if and only if it is a balanced block design.* $\square$

As in [3], the following notation applicable to proportionally balanced designs is used. Block sizes are written as $k_1, k_2, \ldots$, and the number of blocks of size k_i is written as n_i. The elements of a set X that comprises all the elements with the same replication number are denoted as $x_1, x_2, \ldots$, and we can thus denote the corresponding replication number as r_x. Note also that, since $r_i = r_j (i \neq j)$ implies $\lambda_{im} = \lambda_{jm}$ for all m, we can denote the number of blocks containing both elements $x_i \in X$ and $y_i \in Y$, where X and Y are any two sets not necessarily distinct, simply as λ_{xy}.

The theorem below relates to proportionally balanced designs with blocks of precisely two sizes.

Theorem 3 ([3, Theorem 8]) *Suppose a proportionally balanced design on $V = X \cup Y$, with $x = |X| \geq 2$, $y = |Y| \geq 2$ and $r_x \neq r_y$, has blocks of precisely two sizes, k_1 and k_2. Further suppose that $\lambda_{xx} = 1$, and let $\alpha = \frac{r_y}{r_x}$, where $\alpha \neq 1$ and α is not necessarily an integer. Then:*

(i) $\lambda_{xy} = \alpha$ and $\lambda_{yy} = \alpha^2$;

(ii) $xr_{1,x} + yr_{1,y} = k_1 n_1$ and $xr_{2,x} + yr_{2,y} = k_2 n_2$;

(iii) $n_1 k_1 (k_1 - 1) + n_2 k_2 (k_2 - 1) = x(x - 1) + 2\alpha xy + \alpha^2 y(y - 1)$;

(iv) $r_{1,x}(k_1 - 1) + r_{2,x}(k_2 - 1) = (x - 1) + \alpha y$ and
$r_{1,y}(k_1 - 1) + r_{2,y}(k_2 - 1) = \alpha x + \alpha^2 (y - 1)$;

(v) $r_{1,x} = \frac{1}{\alpha} r_{1,y} + \frac{1-\alpha}{k_2 - k_1}$ and $r_{2,x} = \frac{1}{\alpha} r_{2,y} + \frac{\alpha-1}{k_2 - k_1}$.

Example 4 ([4, Theorem 1.9]) Consider the following case.

X	Y	k_1	k_2	r_x	r_y	λ_{xx}	λ_{xy}	λ_{yy}
$\{x_1, x_2, x_3\}$	$\{y_1, y_2, y_3\}$	3	4	3	6	1	2	4

A proportionally balanced design, on $X \cup Y$, with the above parameters is constructed by taking the following blocks.

$x_1 y_2 y_3$	$x_2 y_1 y_3$	$x_3 y_1 y_2$
$y_1 y_2 y_3$	$y_1 y_2 y_3$	
$x_1 x_2 y_1 y_2$	$x_1 x_3 y_1 y_3$	$x_2 x_3 y_2 y_3$

What now follows is a general approach for finding proportionally balanced designs on $V = X \cup Y$, with blocks of two sizes and the particular parameters $\lambda_{xx} = 1$ and $\lambda_{xy} = \alpha$, for $\alpha \geq 2$, from which we also require $r_y = \alpha r_x$ and $\lambda_{yy} = \alpha^2$. We show that, for $\alpha = 2$, 3 and 4, this method of construction can give rise to only a finite number of feasible parameter sets for the designs; these parameter sets are catalogued. For $\alpha = 2$ and 3, the existence or non-existence of each corresponding proportionally balanced designs is then discussed, with references to the designs that make up a proportionally balanced design with the given parameters provided when possible. Finally, three generalised feasible sets of parameters, dependent on α, are presented in relation to the issue of whether or not the corresponding proportionally balanced designs exist.

2. The method of construction

As foreshadowed in Section 1, the aim is to construct a proportionally balanced design on $V = X \cup Y$ such that $\lambda_{xx} = 1$ and $\lambda_{xy} = \alpha$, and hence $\lambda_{yy} = \alpha^2$ and $r_y = \alpha r_x$, using blocks of only two sizes.

Construction 5 Suppose that, for $k \geq 2$, each of the following exists.

I. a $(v, k, \alpha)BiBD$ with a block, **b** say, that occurs $(\alpha - 1)$ times (at least)

II. a $(v - k, k, \lambda)BiBD$, where $0 < \lambda \leq \alpha^2 - \alpha$

III. a $(v - k, m, \alpha^2 - \alpha - \lambda)BiBD$, where $0 < \lambda \leq \alpha^2 - \alpha$.

Further suppose that the following equation holds.

$$(*) \quad m = k - \frac{(k-1)\alpha(\alpha-1)}{(v-k-1)(\alpha^2-\alpha-\lambda)+\alpha(\alpha-1)}$$

Take V to be a set of v elements which contains the elements in **b**, let X be this subset of elements and let $Y = V \backslash X$.

Then the design that comprises the blocks of I on V, but with $(\alpha - 1)$ occurrences of block **b** deleted, and the blocks of II and III on Y, is a proportionally balanced design on V. Further, this design has block sizes k and m, and the following parameters.

$|X|$	$|Y|$	r_x	r_y	λ_{xx}	λ_{xy}	λ_{yy}
k	$v-k$	$\frac{v\alpha-\alpha k+k-1}{k-1}$	αr_x	1	α	α^2

Proof. Since each element of X appears only in the blocks of I, but $\alpha - 1$ occurrences of block **b** have been deleted, we have $\lambda_{xx} = 1$ and $\lambda_{xy} = \alpha$. Further, $r_x = \frac{(v-1)\alpha}{k-1} - (\alpha - 1) = \frac{v\alpha - \alpha k + k - 1}{k-1}$, and it is clear that $\lambda_{yy} = \alpha + \lambda + \alpha^2 - \alpha - \lambda = \alpha^2$.

It remains to be shown that $r_y = \alpha r_x$.

Now, each element of Y occurs $\frac{(v-1)\alpha}{k-1}$ times in I, $\frac{(v-k-1)\lambda}{k-1}$ times in II, and $\frac{(v-k-1)(\alpha^2-\alpha-\lambda)}{m-1}$ times in III. Therefore,

$$r_y = \frac{(v-1)\alpha}{k-1} + \frac{(v-k-1)\lambda}{k-1} + \frac{(v-k-1)(\alpha^2-\alpha-\lambda)}{m-1}.$$

But by (*),

$$\begin{aligned}
m - 1 &= k - 1 - \frac{(k-1)\alpha(\alpha-1)}{(v-k-1)(\alpha^2-\alpha-\lambda)+\alpha(\alpha-1)} \\
&= \frac{(k-1)(v-k-1)(\alpha^2-\alpha-\lambda)}{(v-k-1)(\alpha^2-\alpha-\lambda)+(\alpha^2-\alpha)},
\end{aligned}$$

yielding

$$\frac{(v-k-1)(\alpha^2-\alpha-\lambda)}{m-1} = \frac{(v-k-1)(\alpha^2-\alpha-\lambda)+(\alpha^2-\alpha)}{k-1}.$$

It follows that

$$\begin{aligned}
r_y &= \frac{(v-1)\alpha}{k-1} + \frac{(v-k-1)\lambda}{k-1} \\
&\quad + \frac{(v-k-1)(\alpha^2-\alpha-\lambda)+(\alpha^2-\alpha)}{k-1} \\
&= \frac{(v-1)\alpha + (v-k-1)(\alpha^2-\alpha)+(\alpha^2-\alpha)}{k-1} \\
&= \frac{(v-1)\alpha + (v-k)(\alpha^2-\alpha)}{k-1} \\
&= \frac{\alpha((v-1)+(v-k)(\alpha-1))}{k-1} \\
&= \frac{\alpha(v\alpha - k\alpha + k - 1)}{k-1} = \alpha r_x,
\end{aligned}$$

as required. $\qquad\square$

The question then remains as to the conditions under which Construction 5 is possible. The lemma that follows provides a basis for addressing this question.

Lemma 6 *Let* $n = \frac{(k-1)\alpha(\alpha-1)}{(v-k-1)(\alpha^2-\alpha-\lambda)+\alpha(\alpha-1)}$, *where the other parameters are the same as those for Construction 5. Then, for Construction 5 to be possible, it is necessary that:*

(i) $\lambda \neq (\alpha^2 - \alpha)$, *and hence* $(\alpha^2 - \alpha - \lambda) \neq 0$, *the significance of this result being that it implies that all of the designs* I, II *and* III *in Construction 5 must be non-empty;*

(ii) n *is a positive integer;*

(iii) $v = k + 1 + \frac{(k-1-n)(\alpha^2-\alpha)}{(\alpha^2-\alpha-\lambda)n}$;

(iv) $v \geq 2k$, *and hence* $\frac{(k-1-n)(\alpha^2-\alpha)}{(\alpha^2-\alpha-\lambda)n} \geq k - 1$;

(v) $0 < n < \frac{(\alpha^2-\alpha)}{(\alpha^2-\alpha-\lambda)}$.

Proof.

(i) $\lambda = (\alpha^2 - \alpha)$ implies $BiBD$ III is empty, which implies that all the blocks of the construction are of the same size k. This is impossible, since in this case $\alpha \geq 2$ implies $r_y \neq r_x$, whereas the definition of proportionally balanced immediately gives $r_x = r_y$ (see [4], Theorem 1.2).

(ii) It is sufficient to observe, from equation $(*)$ in Construction 5, that

$$m = k - \frac{(k - 1)\alpha(\alpha - 1)}{(v - k - 1)(\alpha^2 - \alpha - \lambda) + \alpha(\alpha - 1)},$$

that m and k are both integers and, from the conditions of Construction 5, that each of the terms of the expression must be positive and non-zero.

(iii) This follows immediately from rearranging the equation that defines n.

(iv) For $BiBD$ II to exist, as it must since we require $\lambda > 0$, it is necessary that $v - k \geq k$, and hence $v \geq 2k$.

(v) From (iii) we have $\frac{(k-1-n)(\alpha^2-\alpha)}{(\alpha^2-\alpha-\lambda)n} \geq k-1$. But $(k-1) \geq (k-1-n)$, so we can be sure that $\frac{(k-1)(\alpha^2-\alpha)}{(\alpha^2-\alpha-\lambda)n} \geq k-1$. Since $(\alpha^2-\alpha-\lambda) > 0$ (by (i)), and $n > 0$ and $k \geq 2$, we then have $\alpha(\alpha-1) \geq (\alpha^2 - \alpha - \lambda)n$, and the result follows. $\square$

We are now able to determine all *feasible parameters* for proportionally balanced designs with the blocks of Construction 5, for the cases $\alpha = 2, \alpha = 3$ and $\alpha = 4$. By feasible parameters, we mean (for the given value of α) a set of parameters λ, v, k and m that satisfies all of:

– the necessary conditions for the existence of a $(v, k, \lambda)BiBD$, where $k \leq v$, including $\lambda(v - 1) \equiv 0 \pmod{k - 1}$, and $\lambda v(v - 1) \equiv 0 \pmod{k(k - 1)}$ (see [6], p 291);

– equation $(*)$ of Construction 5, namely

$$m = k - \frac{(k - 1)\alpha(\alpha - 1)}{(v - k - 1)(\alpha^2 - \alpha - \lambda) + \alpha(\alpha - 1)};$$

– the additional necessary conditions of Lemma 6.

These conditions for the feasible parameters will be used in the remainder of this paper without further reference. Note also that the issue of whether or not suitable $BiBDs$ for a given feasible sets of parameters actually exist is also considered throughout this paper.

3. Feasible parameters for $\alpha = 2, 3$ and 4

3.1 The case $\alpha = 2$

Theorem 7 *In the case that $\alpha = 2$, there exists only one set of feasible parameters for Construction 5, given below in terms of the parameters of BiBDs I, II and III.*

λ	v	k	$m = k - n$	I	II	III
1	6	3	2	$(6, 3, 2)$	$(3, 3, 1)$	$(3, 2, 1)$

Proof. Lemma 6 (iv) gives $\frac{(k - 1 - n).2}{(2 - \lambda)n} > k - 1$ and Lemma 6 (v) implies $\lambda = 1$. Therefore, $\frac{(k - 1 - n).2}{n} > k - 1$, which gives $n < \frac{2k - 2}{k + 1} < 2$. Thus we need only consider $n = 1$, in which case Lemma 6 (iii) yields $v = k + 1 + (k - 2).2/1 = 3k - 3$.

Thus design I must be a $(3k - 3, k, 2)BiBD$, which requires $2.(3k - 4) \equiv 0 \pmod{k - 1}$, from which $(-2) \equiv 0 \pmod{(k - 1)}$ and hence $k = 2$ or $k = 3$. Thus I is either a $(3, 2, 2)BiBD$ or else a $(6, 3, 2)BiBD$. In the first case, $v \not\geq 2k$, which contradicts Lemma 6 (iv). In the second case, we require design II to be a $(3, 3, 1)BiBD$ and, by equation $(*)$, design III to be a $(3, 2, 1)BiBD$, each of which satisfies the existence conditions. In this second case the blocks of the design can be taken as follows, where the bold type here denotes that block **123** is the block to be deleted, thus inducing the subsets of elements $X = \{1, 2, 3\}$ and $Y = \{4, 5, \infty\}$. The parameters of this design follow the list of blocks.

$$
\begin{array}{lll}
\textbf{123} & 52\infty & 45\infty & 45 \\
234 & 13\infty & & 5\infty \\
345 & 24\infty & & 4\infty \\
451 & 35\infty & & \\
512 & 41\infty & &
\end{array}
$$

k_1	k_2	r_x	r_y	λ_{xx}	λ_{xy}	λ_{yy}
3	2	4	$8 = 2r_x$	1	2	4

$\square$

3.2   The case $\alpha = 3$

Theorem 8 *In the case that $\alpha = 3$, the only sets of feasible parameters for Construction 5 are those given below in terms of the parameters of BiBDs I, II and III.*

#	λ	v	k	n	$m = k - n$	Design I	Design II	Design III
3.1	2	8	4	1	3	$(8, 4, 3)$	$(4, 4, 2)$	$(4, 3, 4)$
3.2	3	9	4	1	3	$(9, 4, 3)$	$(5, 4, 3)$	$(5, 3, 3)$
3.3	4	7	3	1	2	$(7, 3, 3)$	$(4, 3, 4)$	$(4, 2, 2)$
3.4	5	17	4	1	3	$(17, 4, 3)$	$(13, 4, 5)$	$(13, 3, 1)$
3.5	5	31	6	1	5	$(31, 6, 3)$	$(25, 6, 5)$	$(25, 5, 1)$
3.6	5	8	4	2	2	$(8, 4, 3)$	$(4, 4, 5)$	$(4, 2, 1)$

Proof. Since $0 < \lambda < \alpha^2 - \alpha$ (by the definitions in Construction 5 and Lemma 6 (i)) and $\alpha = 3$, the only values of λ that need to be considered are 1, 2, 3, 4 and 5. By $(*)$, $m = k - n$, where n is defined as in Lemma 6, and $0 < n < \frac{(\alpha^2 - \alpha)}{(\alpha^2 - \alpha - \lambda)}$ by Lemma 6 (v). Hence only values of λ and n defined as above can give rise to feasible sets of parameters.

Bearing in mind that the parameters had also to satisfy the equality of Lemma 6 (iii) and the necessary conditions for the existence of the three $BiBDs$, an exhaustive case by case analysis was carried out. The method of proof in each case is illustrated below, using the case $\lambda = 4$.

When $\lambda = 4$, $n < 6/2 = 3$, so we need only consider $n = 1$ and $n = 2$. Rearranging the equality of Lemma 6 (iii), we have

$$(**) \quad (v - k - 1)(\alpha^2 - \alpha - \lambda)n = (k - 1 - n)(\alpha^2 - \alpha).$$

When $n = 1$, this requires $(v - k - 1).2.1 = (k - 2).6$ and hence $(v - k - 1) = 3(k - 2)$, from which $v = 4k - 5$. But for $BiBD$ I to exist, we require $(4k - 6).3 \equiv 0 \pmod{k - 1}$, from which $6 \equiv 0 \pmod{k - 1}$ and hence k can only be 2, 3, 4 or 7.

When $k = 2$, $v = 3$ and $v \not\equiv 2k$, which is impossible.

When $k = 3, v = 7$, which gives rise to the feasible set of parameters numbered 3.3 in the table, the only set in the case $\lambda = 4$ under discussion.

When $k = 4, v = 11$, but an $(11, 4, 3)BiBD$ cannot exist since $3.11.10 \not\equiv 0 \pmod{12}$.

When $k = 7, v = 23$, but a $(23, 7, 3)BiBD$ cannot exist since $3.23.22 \not\equiv 0 \pmod{42}$.

When $n = 2$, equation $(**)$ requires $(v - k - 1).2.2 = (k - 3).6$ which simplifies to $(v - k - 1).2 = (k - 3).3$. Since $(v - k - 1)$ and $(k - 3)$ are both integers, $2|(k-3)$ and so we write $k = 2j+3$. Then $(v-k-1) = 3j$, giving $v = 5j + 4$.

For the $(5j + 4, 2j + 3, 3)BiBD$ I then to exist and be of the required form, we further require that $3(5j+3) \equiv 0 \pmod{2j+2}$ and $j \geq 0$. Thus $15j+9 \equiv 0 \pmod{2j+2}$ and finally, after writing $15j+9$ as $7(2j+2)-5$, we have $j - 5 \equiv 0 \pmod{2j + 2}$. Since $j > 0$, and hence $j - 5 < 2j + 2$, the only possible values for j are $j = 1$ or $j = 5$. When $j = 1$, $k = 5$ and $v = 9$, which is impossible since $v \not\equiv 2k$. When $j = 5$, $k = 13$ and $v = 29$. But the required $(29, 13, 3)BiBD$ I does not then exist, since $3.29.28 \not\equiv 0 \pmod{13.12}$. $\qquad\square$

The table that follows provides, for each of the numbered parameter sets, details of whether or not all of the corresponding $BiBDs$ I, II and III are known to exist, where $BIBD$ I must contain a repeated block. The non-existence of any one of the three designs establishes that a proportionally balanced design with the parameter set cannot be constructed by this method. The design numbers listed without further comment are those used by Mathon and Rosa in their parameter tables in Chapter 1 of [1]. No attempt is made to reference $(k, k, \lambda), (v, 2, \lambda)$ or $(k+1, k, k-1)BiBD$s, which always exist, or designs which can comprise multiple copies of them.

#	Suitable I exists?	II exists?	III exists?
3.1	No: see Table 1.14 in [1]		
3.2	No: see Table 1.17 in [1]		
3.3	3 copies of # 1	Yes	Yes
3.4	# 148: see Appendix (A)	5 copies of # 3	# 8
3.5	3 copies of # 12	# 370	# 11
3.6	No: as per parameter set 3.1		

3.3   The case $\alpha = 4$

Theorem 9 *In the case that $\alpha = 4$, the only sets of feasible parameters for Construction 5 are those given below in terms of the parameters of BiBDs I, II and III.*

#	λ	v	k	n	$m = k - n$	Design I	Design II	Design III
4.1	3	10	5	1	4	$(10, 5, 4)$	$(5, 5, 3)$	$(5, 4, 9)$
4.2	6	6	3	1	2	$(6, 3, 4)$	$(3, 3, 6)$	$(3, 2, 6)$
4.3	8	7	3	1	2	$(7, 3, 4)$	$(4, 3, 8)$	$(4, 2, 4)$
4.4	8	15	5	1	4	$(15, 5, 4)$	$(10, 5, 8)$	$(10, 4, 4)$
4.5	9	13	4	1	3	$(13, 4, 4)$	$(9, 4, 9)$	$(9, 3, 3)$
4.6	9	28	7	1	6	$(28, 7, 4)$	$(21, 7, 9)$	$(21, 6, 3)$
4.7	9	10	5	2	3	$(10, 5, 4)$	$(5, 5, 9)$	$(5, 3, 3)$
4.8	10	10	3	1	2	$(10, 3, 4)$	$(7, 3, 10)$	$(7, 2, 2)$
4.9	10	31	6	1	5	$(31, 6, 4)$	$(25, 6, 10)$	$(25, 5, 2)$
4.10	10	66	11	1	10	$(66, 11, 4)$	$(55, 11, 10)$	$(55, 10, 2)$
4.11	11	16	3	1	2	$(16, 3, 4)$	$(13, 3, 11)$	$(13, 2, 1)$
4.12	11	133	12	1	11	$(133, 12, 4)$	$(121, 12, 11)$	$(121, 11, 1)$
4.13	11	276	23	1	22	$(276, 23, 4)$	$(253, 23, 11)$	$(253, 22, 1)$
4.14	11	10	5	3	2	$(10, 5, 4)$	$(5, 5, 11)$	$(5, 2, 1)$

Proof. The parameters are found by applying the algorithm used for Theorem 8.   □

To establish whether or not a suitable $BiBD$ I exists we use the following Lemma.

Lemma 10 *Suppose a $(v, k, \alpha)BiBD$ with b blocks and $k \geq 3$ contains a block that occurs (at least) $(\alpha - 1)$ times. Then the following inequality* *is satisfied.*

$$k(r - \alpha + 1) - \binom{k}{2} \leq b - \alpha + 1.$$

Proof. If we have a block occurring precisely $(\alpha - 1)$ times, we can substitute $b_k = \alpha - 2$ in the block intersection equations to obtain the following:

(a) $\sum_{i=0}^{k-1} b_i = b - 1 - (\alpha - 2) = b - \alpha + 1$;

(b) $\sum_{i=0}^{k-1} i b_i = k(r - 1) - k(\alpha - 2) = k(r - \alpha + 1)$;

(c) $\sum_{i=0}^{k-1} \binom{i}{2} b_i = \binom{k}{2}(\alpha - 1) - \binom{k}{2}(\alpha - 2) = \binom{k}{2}$.

Subtracting (b) from (a) we obtain $b_1 + b_2 - (2b_4 + \ldots) = k(r - \alpha + 1) - \binom{k}{2}$, where each term in brackets is positive. Hence $b_1 + b_2 \geq k(r - \alpha + 1) - \binom{k}{2}$. On the other hand, (a) gives $b_1 + b_2 \leq b - \alpha + 1$.

Together, these give the required inequality.

The only other possibility is that a block occurs α times, and a similar argument reveals that the inequality is still satisfied.   □

Of the parameter sets given in Theorem 9, only the sets 4.3, 4.5, 4.8, 4.9, 4.11 and 4.12 satisfy this Lemma. It transpires that proportionally balanced designs exist for all six of these parameter sets.

For set 4.8, we take I to be two copies of a $(10, 3, 2)BiBD$ with a repeated block (see Appendix (B)). We take II as 10 copies of design #1, and III clearly exists.

For set 4.11, we take I to be a $(16, 3, 2)BiBD$ with a block that occurs four times (see Appendix (C)). We take II as 11 copies of design #8, and again III clearly exists.

Proportionally balanced designs for parameter sets 4.3, 4.5, 4.9 and 4.12 are shown to exist later, as examples of Theorem 13.

4.   Constructions motivated by Construction 5

4.1   The case $\lambda = 0$

Construction 5 can also be extended to the case $\lambda = 0$, in which case only the blocks of $BiBD$s I and III are involved. Results (ii) and (iii) of

Lemma 6 still stand, whereas results (iv) and (v) depended on II's being non-empty.

Theorem 11 *Construction 5 extends to the case* $\lambda = 0$*, in which case designs* I *and* III *must be* $(2k - 1, k, \alpha)$ *and* $(k - 1, k - 1, \alpha^2 - \alpha)BiBDs$ *respectively.*

Proof. When $\lambda = 0$, design III is non-empty and, by $(*)$ in Construction 5, we obtain equation (i) $m = k - \frac{k-1}{v-k-1+1} = k - \frac{k-1}{v-k}$. Also, design III requires (ii) $v - k \geq m$.

Together, (i) and (ii) imply $v - k \geq k - \frac{k-1}{v-k}$, from which $(v - k)^2 - k(v - k) + (k - 1) \geq 0$, and hence $(v - 2k + 1)(v - k - 1) \geq 0$. This gives $v \leq k + 1$ or $v \geq 2k + 1$. Since we need $v \geq 2k$, $v \leq k + 1$ has no solutions. Hence $v \geq 2k + 1$ and thus $v - k \geq k - 1$. Now, by (i), $\frac{k-1}{v-k}$ is an integer, which implies $v - k \leq k - 1$. The last two results jointly give $v - k = k - 1$.

Hence designs I and III are $(2k-1, k, \alpha)$ and $(k-1, k-1, \alpha^2-\alpha)BiBDs$ respectively. $\quad\square$

Design III is simply $\alpha^2 - \alpha$ copies of a full block of all $k - 1$ elements, so the only issue left to resolve is the existence of the $(2k - 1, k, \alpha)BiBD$ with the required property for the construction. The following theorem is thus virtually immediate.

Theorem 12 *Suppose a* $(2k - 1, k, \alpha)BiBD$ *exists that contains a block* **b** *which occurs at least* $\alpha - 1$ *times. Then:*

(i) *there exists a proportionally balanced design comprising:*
   - *the blocks of this* $(2k - 1, k, \alpha)BiBD$*, but with* $\alpha - 1$ *occurrences of* **b** *deleted;*
   - $(\alpha^2 - \alpha)$ *copies of the block containing those* $(k - 1)$ *elements not appearing in* **b***;*

(ii) *the only possible set of parameters giving rise to a proportionally balanced design as in* (i) *are those of the design given in the proof of* (i)*.*

Proof.

(i) To show that the parameters of the design are feasible, it suffices to point out that $r_x = \frac{(2k-2)\alpha}{k-1} - (\alpha - 1) = 2\alpha - (\alpha - 1) = \alpha + 1$, and $r_y = 2\alpha + \alpha^2 - \alpha = \alpha^2 + \alpha = \alpha r_x$.

When $k = 4$ and $\alpha = 2$, take the blocks of such a proportionally balanced design as follows.

2456   3560   4601   5012   6123   0234   1345
    013   013

Again, the bold type here denotes the block to be deleted, thus inducing the subsets of elements $X = \{2, 4, 5, 6\}$ and $Y = \{0, 1, 3\}$. Then the design has the following parameters.

k_1	k_2	r_x	r_y	λ_{xx}	λ_{xy}	λ_{yy}
4	3	3	$6 = 2r_x$	1	2	4

(ii) We need only consider $k \geq 3$ and $\alpha \geq 2$. We further require that the $(2k - 1, k, \alpha) BiBD$ have a block $\mathbf{b}$ that occurs $(\alpha - 1)$ times and have seen that this design has replication number $r = 2\alpha$ and hence $b = (2k - 1)2\alpha/k$. From the block intersection equations given in Subsection 1.1, the block intersection sizes satisfy:

(a) $\sum_{i=0}^{k} b_i = b - 1 = \frac{(2k-1)2\alpha}{k} - 1$;

(b) $\sum_{i=0}^{k} ib_i = k(r - 1) = k(2\alpha - 1)$;

(c) $\sum_{i=0}^{k} \binom{k}{2} b_i = \binom{k}{2}(\lambda - 1) = \binom{k}{2}(\alpha - 1)$.

We first note that the condition $\lambda v(v - 1) \equiv 0 \,(\mathrm{mod}\; k(k - 1))$ requires that $\alpha(2k - 1)(2k - 2) \equiv 0 \,(\mathrm{mod}\; k(k-1))$, which implies that $\alpha = k/2$, for even k, or $\alpha \geq k$.

For both cases under discussion, for a block $\mathbf{b}$ to occur $(\alpha - 1)$ times we need, relative to that block, b_k, to be at least $\alpha - 2$. For $b_k = \alpha - 2$, equation (c) gives $\sum_{i=0}^{k-1} \binom{i}{2} b_i = \binom{k}{2}(\alpha - 1) - \binom{k}{2}(\alpha - 2) = \binom{k}{2}$. Similarly, equation (b) can be rewritten as $\sum_{i=0}^{k-1} ib_i = k(2\alpha - 1) - k(\alpha - 2) = \alpha k + k$. Together these give $\sum_{i=0}^{k-1} (i - \binom{i}{2}) b_i = \alpha k + k - \binom{k}{2}$. But, by (a), $\sum_{i=0}^{k} b_i = \frac{(2k-1)2\alpha}{k} - 1$. Therefore $\alpha k + k - \binom{k}{2} \leq \frac{(2k-1)2\alpha}{k} - 1$.

In the case $\alpha \geq k$, this leads to the inequality $\alpha(k - \frac{(2k-1).2}{k}) \leq \binom{k}{2} - 1 - k$, which then leads simply to $k = 3$ as the only possibility for $\alpha > 1$. When $k = 3$, further consideration of the three equations reveals that no solution exists for the block intersection equations.

In the case $\alpha = k/2$ (where $k \geq 4$), (b) $-$ (c) gives

$b_1 + b_2 + [0.b_3 + -2b_4 + \ldots + ((k/2 - 1) - \binom{k/2-1}{2}))b_{(k/2)-1}] = 3k/2$,
where each of the terms in the square brackets is non-positive. On the other hand, (a) gives

$b_1 + b_2 + [b_3 + b_4 + \ldots + (k/2 - 1)] = 3k/2$, where the terms in the brackets are positive.

Together, these imply that $b_1 + b_2 = 3k/2$ and $b_0 = b_3 = b_4 \ldots = b_{(k/2)-1} = 0$. But then $b_2 = \binom{k}{2}$ which, since this requires $\binom{k}{2} \leq 3k/2$, leads to a contradiction for $k > 4$. The case $k = 4$ has already been dealt with in considering $\alpha \geq k$.

The entire proof is completed by observing that, in both cases, no additional parameter sets are obtained from the parallel argument for $b_k = \alpha - 1$.

$\square$

4.2 Other general constructions

Theorem 13 *For any given value of $\alpha > 2$, let q be a positive integer such that $q + 1$ divides $\alpha^2 - \alpha$. Then:*

(i) $v = q^2 + q + 1, k = q + 1, \alpha, \lambda = (\alpha^2 - \alpha)q/(q + 1)$ and $m = q$ are a feasible set of parameters for Construction 5 on $X \cup Y$;

(ii) a proportionally balanced design with the parameters above can be obtained from Construction 5 when q is a prime or prime power.

Proof.

(i) We check easily that the parameters determine designs I, II and III respectively as $(q^2 + q + 1, q + 1, \alpha)$, $(q^2, q + 1, (\alpha^2 - \alpha)q/(q+1))$ and $(q^2, q, (\alpha^2 - \alpha)/(q + 1))BiBDs$, and that the necessary conditions for the existence of the $BiBDs$ are satisfied in all three cases.

For the design to be proportionally balanced we also need $r_y = \alpha r_x$. A standard counting argument for finding the replication number of a balanced design shows that the replication numbers are $(q+1)\alpha$ for design I, $(q-1)(\alpha^2 - \alpha)$ for design II, and $\alpha^2 - \alpha$ for design III. Since the elements of X occur only in design I, and since $\alpha - 1$ blocks, each consisting of these elements, must be deleted, $r_x = \alpha q + 1$. The elements of Y appear in all three designs, so $r_y = (q + 1)\alpha + (q - 1)(\alpha^2 - \alpha) + (\alpha^2 - \alpha) = (\alpha q + 1)\alpha = \alpha r_x$.

(ii) First, when q is a prime or prime power, $BiBD$ III can be taken as $(\alpha^2 - \alpha)/(q+1)$ copies of an *affine plane of order q*. Similarly, a $BiBD$ with the parameters of I can be constructed by taking α *projective planes of order q*, with each such $(q^2+q+1, q+1, 1)BiBD$ also existing when q is a prime or prime power; for further information on affine and projective planes, see, for example, Chapter 8 of [6]. By taking at least $(\alpha - 1)$ of the projective planes with a block in common, we can construct I so that a block occurs at least $(\alpha - 1)$ times as required. Finally, if $(\alpha^2 - \alpha)q/(q+1)$ is a suitable integer, then it must be some multiple of q. Further, $(q^2, q+1, q)BiBDs$ are known to exist for q a prime or prime power (see Lemma 4.1 in [5]). Hence we can always take design II as this $(q^2, q + 1, q)BiBD$, or multiple copies of it.

$\square$

We point out here that, for any given value of $\alpha > 2$, there are at least three values of $q \geq 2$ such that $q + 1$ divides $\alpha^2 - \alpha$ as required for Construction 13, namely $q = \alpha - 2, \alpha - 1$ and $\alpha^2 - \alpha - 1$; these values might not be primes or prime powers.

Example 14 *Let $\alpha = 3$, which implies $q = 5$, so prime. Then designs I, II and III are, respectively, $(31, 6, 3), (25, 6, 5)$ and $(25, 5, 1)BiBDs$, where design I can be taken as three copies of a $(31, 6, 1)BiBD$.*

The design of this example has the set of parameters corresponding to the feasible set #3.5 in Theorem 8. Note that, by this theorem, proportionally balanced designs with the sets of parameters 3.3, 4.3, 4.5, 4.9 and 4.12 must also exist.

Theorem 15 *For any given $\alpha \geq 2$, let $v = 2\alpha + 2, k = \alpha + 1, m = \alpha$ and $\lambda = \alpha - 1$. Then:*

(i) v, k, m and λ are a feasible set of parameters for Construction 5;

(ii) for these parameters, it is impossible for design I of Construction 5 to exist, and hence in this case a proportionally balanced design by means of Construction 5 cannot exist.

Proof.

(i) We easily check that the parameters determine designs I, II and III as $(2\alpha+2, \alpha+1, \alpha), (\alpha+1, \alpha+1, \alpha-1)$ and $(\alpha+1, \alpha, (\alpha - 1)^2)BiBDs$

respectively, and that the necessary conditions for the existence of the $BiBDs$ are satisfied in all three cases. By $(*)$, for the design to be proportionally balanced we need

$$
\begin{aligned}
m &= k - \frac{(k-1)\alpha(\alpha-1)}{(v-k-1)(\alpha^2-\alpha-\lambda)+\alpha(\alpha-1)} \\
&= \alpha+1 - \frac{\alpha^2(\alpha-1)}{\alpha(\alpha-1)^2+\alpha(\alpha-1)} \\
&= \alpha+1 - \frac{\alpha}{(\alpha-1)+1} = \alpha,
\end{aligned}
$$

which is the given value for m.

(ii) We need design I to have a block that occurs $(\alpha-1)$ times. We note that, for this design, $r = 2\alpha+1$ and $b = 2(2\alpha+1) = 2r$, and hence, from the block intersection equations given in Subsection 1.1, the block intersection sizes satisfy:

(a) $\sum_{i=0}^{\alpha+1} b_i = b - 1 = 4\alpha - 1$;

(b) $\sum_{i=0}^{\alpha+1} ib_i = k(r-1) = (\alpha+1).2\alpha$;

(c) $\sum_{i=0}^{\alpha+1} \binom{i}{2}b_i = \binom{k}{2}(\lambda-1) = \binom{\alpha+1}{2}(\alpha-1)$.

For a block **b** to occur $(\alpha - 1)$ times we need, relative to that block, $b_{\alpha+1}$ to be $\alpha - 2$ (at least). For $b_{\alpha+1} = \alpha - 2$, equation (c) gives $\sum_{i=0}^{\alpha} \binom{i}{2}b_i = \binom{\alpha+1}{2}(\alpha-1) - \binom{\alpha+1}{2}(\alpha-2) = \frac{\alpha(\alpha+1)}{2}$. Similarly, equation (b) can be rewritten as $\sum_{i=0}^{\alpha} ib_i = (\alpha+1).2\alpha - (\alpha+1)(\alpha-2) = (\alpha+1)(\alpha+2)$. Rewritten, (b) $-$ (c) then gives $\sum_{i=0}^{\alpha} (ib_i - \binom{i}{2}b_i) = (\alpha+1)(\alpha+2) - \frac{\alpha(\alpha+1)}{2}$, that is, $b_1 + b_2 + 0.b_3 - 2b_4 + \ldots + (\alpha - \binom{\alpha}{2})b_i = \frac{(\alpha+1)(\alpha+4)}{2}$.

Since $(i - \binom{i}{2})$ is non-positive for $i \geq 2$, $b_1 + b_2 \geq \frac{(\alpha+1)(\alpha+4)}{2}$. But $b_1 + b_2 \leq \sum_{i=0}^{\alpha} b_i$, thus requiring $\frac{(\alpha+1)(\alpha+4)}{2} \leq 4\alpha + 1$ (by (a)). Hence $\alpha^2 - 3\alpha + 2 = (\alpha-2)(\alpha-1) \leq 0$, giving $\alpha = 2$ as the only possible value of α.

Completion of the proof simply requires checking that there are no solutions to the block intersection equations when $\alpha = 2$, and observing that a similar argument holds in the only other possible case, that is, $b_{\alpha+1} = \alpha - 1$.

$\square$

4.3 Appendix

A. A $(17, 4, 3)BiBD$ with a block that occurs at least twice is required. A $(17, 4, 3)BiBD$ with a block that occurs three times can be constructed as follows. First take the block occurring three times to be $\infty_1\infty_2\infty_3\infty_4$. Then take the 13 blocks of a $(13, 4, 1)BiBD$ on the 13 elements $\{0, 1, 2, \ldots, 9, A, B, C\}$. Finally, take the 52 blocks obtained by cycling, modulo 13, the starter blocks $\infty_1 014$ $\infty_2 014$ $\infty_3 027$ $\infty_4 027$.

It is easy to see that each element ∞_i, for $i = 1, 2, 3, 4$, occurs with each other element in precisely three blocks. The remaining pairs are accounted for by observing that the final 52 blocks, on deletion of the elements ∞_i for all i, are those of a $(13, 3, 2)BiBD$.

B. A $(10, 3, 2)BiBD$ with a repeated block can be constructed by first taking the block occurring twice to be $\infty_1\infty_2\infty_3$. Then take the seven blocks of a $(7, 3, 1)BiBD$ on the set of elements $\{0, 1, 2, \ldots, 6\}$. Finally, take the 21 blocks obtained by cycling, modulo 7, the starter blocks $\infty_1 01$ $\infty_2 02$ $\infty_3 03$. For a census of $(10, 3, 2)BiBD$s with a repeated block, see [2].

C. A $(16, 3, 4)BiBD$ with a block that occurs at least three times is required. A $(16, 3, 4)BiBD$ with a block that occurs four times can be constructed by first taking the block occurring four times to be $\infty_1\infty_2\infty_3$. Then take three copies of the $(13, 3, 1)BiBD$ on the set of elements $\{0, 1, 2, \ldots, 9, A, B, C\}$ obtained by cycling, modulo 13, the blocks 014 and 027. Finally, take the 78 blocks obtained by cycling, modulo 13, the starter blocks $\infty_1 01$ $\infty_1 02$ $\infty_2 03$ $\infty_2 04$ $\infty_3 05$ $\infty_3 06$.

Acknowledgments

Some helpful advice from Dr Julian Abel is acknowledged.

References

[1] Charles J. Colbourn and Jeffrey H. Dinitz (ed) 1996, *The CRC Handbook of Combinatorial Designs*, CRC Press, Inc, Boca Raton, USA.

[2] B. Ganter, A Gulzow, R. Mathon and A. Rosa, *A complete census of* $(10, 3, 2)$-*block designs and of Mendelsohn triple systems of order ten, IV.* $(10, 3, 2)$ *designs with repeated blocks*, Math. Schriften Kassel **2**

(1978).

[3] Ken Gray, *Proportionally balanced designs: some further results and general constructions*, Utilitas Mathematica **59** (2001), 167–180.

[4] Ken Gray and Gabrielle Matters, *Principles of combinatorial design exploited in the 1993 Queensland Core Skills Test marking operation*, Utilitas Mathematica **48** (1995), 33–64.

[5] Haim Hanani, *Balanced incomplete block designs and related designs*, Discrete Mathematics **11** (1975) 255-369.

[6] A.P. Street and D.J. Street 1987, *Combinatorics of Experimental Design*, Clarendon Press, New York.

Chapter 9

CONSTRUCTIONS USING BALANCED n-ARY DESIGNS

Malcolm Greig

Greig Consulting, #317-130 East 11th St.,
North Vancouver, BC, Canada V7L 4R3

Abstract   We describe how a BIBD construction problem was solved using a Balanced Ternary Design. We explore how this construction can be improved, and how and what balanced n-ary designs can be used in this construction. We settle several open existence cases for BTDs (including both suitable and unsuitable designs for our construction), and provide a number of suitable BnDs.

1.   Introduction

We will assume the reader is familiar with most of our basic terms and definitions; these can be found in standard texts, e.g., [6, 18]. The terms and definitions relating to balanced n-ary designs will be given later.

This article had its origins in another article [2]. There we wanted to construct $(v, 9, 2)$ BIBDs and, at one stage, we were missing some values of the form $v = 36q + 1$ with q a prime power, e.g., $v = 469$. One way to construct such designs would be to first construct a $(9, 2)$ GDD of type 36^q, and then fill in the groups using an extra point, with the fill provided by $(37, 9, 2)$ BIBDs; these latter BIBDs are known, e.g., the 9-th roots in GF(37) give a difference set for one.

One would like some structure to aid in the search for these GDDs, so, considering the point set for the design as $I_{36} \times GF(q)$ and the groups as $I_{36} \times \{y\}$ for $y \in GF(q)$, we decided to look at the structure of the I_{36} part of the points. As a first step, we computed the average density of I_{36} pairs per block - this turned out to be exactly 1. So one apparently feasible structure would place two identical I_{36} elements in every block, and since these doubletons must be spread evenly over the I_{36} elements, we're looking at some multiple of 36 blocks, with an additional restriction that there must be balance between the different I_{36} elements too. Now a structure satisfying these requirements is known: [10, Design 44] — the structure given is of a balanced ternary design having 36 blocks, but the example given does not appear to have a really nice automorphism. This was initially a disappointment to us, as we anticipated using the automorphism in the search of a suitable placement for the $GF(q)$ part of the elements. Eventually we realized that we could place them almost arbitrarily — we just had to use k distinct values in each block, and then multiplying the 36 blocks repeatedly by $(1, x^i)$) for $i = 0, 1, \ldots, q - 2$ with x some primitive generator for $GF(q)$ gave us the base blocks for a difference family over $GF(q)$. We will describe the construction more clearly later.

Some natural questions arose. What was special about the parameters we were using, as once we identify that, we might be able to generalize this construction. Were there any interesting examples in the literature; actually, we already had a partial answer, as two of us had undertaken a similar pair density count approach to another GDD, and decided on a balanced quaternary design for the group part of the element; we could also view some of Hanani's $(v, 7, 3)$ BIBD constructions [21] in this new light, treating his group placement as a ternary design with a base block of $(0, 1, 1, 2, 2, 4, 4)$ (mod 7). But Hanani obtained a BIBD index of three, whereas the index in this ternary design is six!

This was the starting point of this article. Our aims here are to present our construction and some improvements, and to give some (hopefully useful) examples of the key ingredient. A secondary aim is to describe some new results on balanced ternary designs, some of which are related to our main aim.

We should also point out that there is considerable overlap of our basic construction with some work by Buratti; having said that, there is also considerable difference in the details, as well as of presentation. Buratti starts with a difference family construction for GDDs, (a BnD given by a difference family in our terminology) and gives examples of improved indices in the GDDs, whereas we start with a partial difference family (just developing over some algebraic group whose order is the number of groups in our eventual GDD), and show how we may construct and improve upon this. We would prefer to have difference family constructions for our BnDs, and spend some effort in finding these, but we do have constructions that utilize BnDs for which no usable automorphisms are apparent (our initial example discussed above was of this sort). The interested reader would be well advised to study [11].

2.   Balanced n-ary Designs

A classic (binary) block design $(\mathcal{V}, \mathcal{B})$ consists of a set of points, $\mathcal{V}$, and a set of blocks, $\mathcal{B}$, with each block containing some (sub)set of the points. It can be represented by its v by b incidence matrix, N, where $v = |\mathcal{V}|$ and $b = |\mathcal{B}|$. For a binary design, it does not matter if N_{ij} is considered as indicating the presence of the i-th point in the j-th block by 1, and its absence by 0, or else if N_{ij} is considered as indicating the number of times the i-th point is in the j-th block since no repeats are allowed in a block.

An n-ary block design $(\mathcal{V}, \mathcal{B})$ again consists of a set of points, $\mathcal{V}$, and a set of blocks, $\mathcal{B}$, but now each block contains some multi-set of the points, and the entries in the incidence matrix indicate the number of times a point is in the block. In an n-ary block design, the entries of N are restricted to the n numbers $0, 1, \ldots, n - 1$.

The additional restrictions we will place on N are:

$$
\begin{aligned}
\mathbf{1}_v^T N &= k\mathbf{1}_b^T & &\text{(2.1)} \\
N\mathbf{1}_b &= r\mathbf{1}_v & &\text{(2.2)} \\
NN^T &= D + \lambda J & \text{where } D \text{ is diagonal} &\text{(2.3)} \\
D &= dI & &\text{(2.4)} \\
v\lambda &= r(k - 1) & &\text{(2.5)}
\end{aligned}
$$

The introduction of balanced n-ary designs (BnDs) is usually attributed to Tocher [29], although there are earlier instances in the literature. Tocher proposed these designs to ameliorate the integrality constraints imposed by other designs such as BIBDs, PBIBDs, etc., but Tocher's designs only satisfied condition (2.1), that the block size be constant, and condition (2.3), that there be pairwise balance. (To ameliorate the usual integrality constraints Tocher also proposed relaxing the constant block size restriction on binary designs.) However, since then most authors have imposed condition (2.2), that the design be equi-replicate, and indirectly imposed condition (2.4); strictly speaking, the condition they usually impose is that of row-pattern regularity in N (i.e., that each row be a permutation of the first row), but the case usually studied is the ternary case (i.e., $n = 3$, where a BnD is known as a BTD), and, in the presence of condition (2.2), row-pattern regularity is equivalent to condition (2.4) when $n = 3$.

Billington [8, 9, 25, 26] has written some excellent surveys on the history and construction of BnDs. In addition to the main list of small BTDs cited above [10], which is concerned with existence, and gives an example or reference for each design, there are a couple of other papers of interest. Greig and Sarvate [20] give an update of the list, but without the examples in all cases, and also with an error, as [10, Design 106] does exist. Kaski and Östergård [23] have enumerated 143 of the 155 parameter sets listed in [10] with $v \leq 10$. (In the remaining 12 cases, they only give a lower bound on the number of non-isomorphic designs.) We should also mention there is a convention of capitalizing the parameters of BnDs; however, since we're going to carry these parameters into parameters for binary designs, where lower case is the convention, we've decided to use lower case throughout.

3. Our Construction

Let $q \geq k$ be a prime power, and let x be a primitive generator for $GF(q)$. Suppose we have a BnD on the point set I_v satisfying conditions (2.1)–(2.5). We can construct a (k, λ) GDD of type v^q on the point set $I_v \times GF(q)$, as follows.

Construction 3.1 *Given a block in the BnD, say $(b_1, b_2, \ldots, b_k)$, of not-necessarily distinct elements of I_v, we replace the element b_i by (b_i, a_i) where $a_i \in GF(q)$. We pick the a's to be distinct. We repeat this for every block of the BnD; the a's used for any one block must be distinct, but they are allowed to vary from block to block of the BnD. We next multiply each new block by $(1, x^i)$ for $i = 0, 1, \ldots (q - 2)$ to generate $(q - 1)$ blocks from each block of the BnD, and then develop these blocks additively over $GF(q)$ for our GDD.*

Now our claim is that Construction 3.1 produces a (k, λ) GDD of type v^q. We now justify that claim, dealing en passant with the restrictions (2.1)–(2.5).

It is useful to note some facts about a (k, λ) GDD of type v^q; it has $\lambda v^2 q(q-1)/(k(k-1))$ blocks, each point occurs $\lambda v(q-1)/(k-1)$ times, and if $i \neq x$, $j \neq y$, then there are $v\binom{q}{2}$ unordered pairs of the pure form (i, x) with (i, y), each occurring λ times, $2\binom{v}{2}\binom{q}{2}$ unordered pairs of the mixed form (i, x) with (j, y), each occurring λ times, and $q\binom{v}{2}$ unordered pairs of the form (i, x) with (j, x), each occurring zero times (they are groupmates), for a total of $\binom{vq}{2}$ unordered pairs of distinct points.

Each time a point i occurs in a block of the BnD, it generates $q(q-1)$ points of the form (i, x) for all the $x \in GF(q)$, and $(q-1)$ for any particular x. Since the replication number in the GDD is constant over all points, we need the total number of times the point i occurs to be a constant independent of i, thus justifying restriction (2.2). Restriction (2.1) is obviously needed. Now suppose a point i occurs m times in n_{mi} blocks. We now have $\sum_m mn_{mi} = r$ for every i. Part of restriction (2.3) says $\sum_m m^2 n_{mi} = D_i + \lambda$. which merely amounts to a definition of D_i. The point i will also generate $\sum_m \binom{m}{2} n_{mi} q(q-1)$ pure pairs in the GDD, i.e., $(D_i + \lambda - r)\binom{q}{2}$ pure pairs. Since we don't want this to vary with i we impose restriction (2.4). Each time the point pair i with j appears in a block it generates $q(q-1)$ pairs of the form (i, x) with (j, y) — again we don't want this to vary with the pair i, j so we need restriction (2.3). We may now note that $\lambda(v-1) = \sum_m n_{mi} m(k-m)$ and so $\lambda(v-1) = rk - d - \lambda$, i.e., $d = rk - \lambda v$.

So now each pure pair happens equally often, as does each mixed pair. The latter happens λ times per pair, and we have now justified restrictions (2.1)–(2.4). Using these restrictions, and pre- and post-multiplying by the unit vector we find that $vr = bk$ and $v(d + \lambda v) = bk^2$ for the BnD, from which we get $b\binom{k}{2} = (d + \lambda v - r)v/2$.

Our construction generates $b\binom{k}{2}q(q-1)$ pairs in total, and we have an index of λ for the $\binom{v}{2}q(q-1)$ mixed pairs, so there remain $(d-r+\lambda)\binom{q}{2}v/2$ pure pairs in total, for an average index of $(d-r+\lambda)$ per pair, i.e, exactly this for any arbitary pure pair. We noted above that $\lambda(v-1) = rk-d-\lambda$, from which we deduce that the pure index is $r(k-1) - \lambda(v-1)$, and since our mixed index is λ, restriction (2.5) ensures we do have the required GDD.

It is clear from the above discussion that if we only have restrictions (2.1)–(2.4), (which are satisfied by all the BnDs usually studied) and lack restriction (2.5), then Construction 3.1 produces a PBIBD(3) with a rectangular association scheme and index triple

$$(\lambda_R, \lambda_C, \lambda_3) = (r(k-1) - \lambda(v-1), 0, \lambda).$$

In the three class rectangular association scheme, the points are laid out in an R by C rectangular grid, and pairs of points in the same row are row associates, pairs in the same column are column associates, and other pairs are 3rd associates. We might note that a GDD is a PBIBD(2) with a group divisible association scheme (like the rectangular, but the rows are ignored - there's not even a requirement that every column have the same number of rows); the term GDD is reserved for the subclass whose index double is of the form $(0, \lambda)$.

There is one final point. If we look at the pure pairs generated by the point i, then we have $\lambda_R \binom{q}{2} = \sum_m n_{mi} \binom{m}{2} q(q-1)$ and so $\lambda_R = \sum_m n_{mi} m(m-1)$ is even, and under restriction (2.5) $\lambda_R = r(k-1) - \lambda(v-1) = \lambda$ is even.

4.   Some Examples

One important catalogue of balanced ternary designs (or BTDs) is the list given by Billington and Robinson [10], and in this section we look at that list more closely.

We examined Billington and Robinson's list to determine if the design satisfied restriction (2.5). In Table 4.1, we only list Billington and Robinson's design number, and its existence status. We will later give further details of the BTDs we consider more useful. Designs 113, 205 (given in Theorem 7.1.a) and 344 (given in Example 7.1) are the only known designs with $k > 9$ in Table 4.1.

Billington and Robinson list all parameter sets that satisfy restrictions (2.1)–(2.4). They also impose the (necessary) condition that $b \geq v$ (Fisher's rank condition, which follows from restrictions (2.3) and (2.4) if $d > 0$), and impose $k \leq v$. This latter condition excludes the combinatorially isomorphic complementary design. In a BnD the complementary incidence matrix is given by $(n-1)J - N$, and the complementary design has the same v and b, but $r \mapsto (n-1)b - r$, $k \mapsto (n-1)v - k$, and $\lambda \mapsto (n-1)^2 b - 2(n-1)r + \lambda$. Billington and Robinson also require that $k > n$ to avoid trivialities, and that any row of the incidence matrix contains at least one of each of 0, 1 and 2. (This merely excludes mildly disguised BIBDs.) Restriction (2.5) does not necessarily hold for the complementary design if restriction (2.5) holds for the original design, and vice versa, but there were no examples of designs with $v > k$ where restriction (2.5) holds found by complementing designs in the table. There are such examples excluded from the table as not

Table 4.1. Billington and Robinson's designs satisfying restriction (2.5)

#	Exists	#	Exists	#	Exists	#	Exists
1	Yes	3	Yes	5	No	7	Yes
8	Yes	10	No	13	Yes	17	No
25	Yes	27	Yes	28	No	33	Yes
36	No	37	Yes	40	Yes	44	Yes
50	Yes	54	Yes	57	Yes	59	Yes
62	Yes	65	??	73	Yes	82	No
85	Yes	92	??	113	Yes	114	Yes
116	Yes	117	??	118	??	119	??
120	No	124	Yes	128	Yes	132	Yes
135	??	136	??	140	Yes	145	Yes
147	Yes	150	??	151	No	156	Yes
162	??	166	Yes	174	No	183	??
191	No	205	Yes	210	??	211	No
225	Yes	234	??	261	Yes	267	Yes
271	??	272	??	274	No	281	Yes
285	??	290	Yes	296	Yes	301	??
307	??	312	Yes	315	Yes	319	??
326	??	331	Yes	344	Yes		

truly ternary (the difference set $(0, 0, 1, 1, 2, 2, 4, 4)$ in Z_7 is an example). Billington and Robinson [10, Result 2.2] impose two further necessary conditions, given in Lemma 4.1.

Lemma 4.1 *Given the incidence matrix of a* $(v, b; \rho_1, \rho_2, r; k, \lambda)$ *BTD where every row contains* ρ_i *instances of* i. *Then it is necessary that:*

a *if* $\lambda = 2$ *or* 3, *then* $b \geq v\rho_2$;

b $\lfloor \frac{k}{2} \rfloor b \geq v\rho_2$.

Finally, Billington and Robinson's list is restricted to $r \leq 15$. They also give the row pattern as $0^{b-\rho_1-\rho_2} 1^{\rho_1} 2^{\rho_2}$, and parameterize the design as $(v, b; \rho_1, \rho_2, r; k, \lambda)$.

We now take the opportunity to update the existence status of Billington and Robinson's full list.

Table 4.2. Billington and Robinson's list

#	v	b	ρ_1	ρ_2	r	k	λ	Exists	Status
62	20	40	8	1	10	5	2	Yes	Example 4.1
64	27	30	8	1	10	9	3	Yes	[17]
94	28	28	9	1	11	11	4	Yes	[17]
106	10	22	3	4	11	5	4	Yes	[23], Example 4.4
121	27	27	10	1	12	12	5	Yes	[20]
153	15	15	6	3	12	12	9	Yes	[15]
164	32	32	4	4	12	12	4	Yes	[17]
197	12	26	3	5	13	6	5	No	Theorem 5.2
203	19	19	1	6	13	13	8	No	[16]
209	18	42	12	1	14	6	2	Yes	[28]
212	61	61	12	1	14	14	3	Yes	[15]
223	12	28	10	2	14	6	6	Yes	Example 4.2
245	16	28	6	4	14	8	6	Yes	[28]
262	20	28	2	6	14	10	6	No	Theorem 5.1
266	18	18	2	6	14	14	10	Yes	[16]
296	10	30	9	3	15	5	6	Yes	[20]
302	18	18	9	3	15	15	12	Yes	[15]
332	10	25	3	6	15	6	7	No	[23], Theorem 5.3
334	14	30	3	6	15	15	11	Yes	Example 4.3
342	50	50	1	7	15	15	4	Yes	[20]

Example 4.1 The following example was used by Hanani [21] to construct a $(5,1)$ GDD of type 20^7, over $I_4 \times Z_5$. In this, and many subsequent, examples we write the point (a, b) in the compressed form ab.

$$\{00, 22, 32, 03, 23\} \quad \{00, 32, 22, 23, 03\}$$
$$\{10, 02, 12, 03, 03\} \quad \{10, 32, 02, 13, 33\}$$
$$\{20, 12, 22, 13, 13\} \quad \{20, 22, 12, 33, 33\}$$
$$\{30, 02, 32, 33, 13\} \quad \{30, 12, 02, 23, 23\}$$

This is a $(20; 40; 8, 1, 10; 5; 2)$ BTD and is listed as Design 62 and unknown in [10]. If we ignore the second element over every point, we get an equi-replicate 4-ary design which is not "regular", since the pattern for point 2 is $0^3 2^5$, and for the other three points it is $0^1 1^3 2^2 3^1$. This B4D has $v = 4$, $b = 8$, $r = 10$, $k = 5$, and $\lambda = 10$, and also satisfies restriction (2.5). Its incidence matrix follows.

$$\begin{pmatrix} 2 & 2 & 3 & 1 & 0 & 0 & 1 & 1 \\ 0 & 0 & 2 & 2 & 3 & 1 & 1 & 1 \\ 2 & 2 & 0 & 0 & 2 & 2 & 0 & 2 \\ 1 & 1 & 0 & 2 & 0 & 2 & 3 & 1 \end{pmatrix}$$

Example 4.2 A $(12, 28; 10, 2, 14; 6, 6)$ BTD with point set $Z_3 \times Z_2 \times Z_2$.

$$(000, 000, 100, 111, 210, 211) \quad (000, 011, 101, 101, 200, 210)$$
$$(000, 001, 100, 110, 200, 200) \quad (000, 010, 100, 101, 200, 211)$$
$$(000, 000, 001, 010, 101, 110)$$

The first four blocks are developed only over $Z_2 \times Z_2$; the last base block is full.

Example 4.3 A $(14, 30; 3, 6, 15; 7, 6)$ BTD with point set $I_2 \times Z_7$.

$$(00, 11, 11, 12, 12, 14, 14) \quad (00, 00, 01, 01, 14, 14, 10)$$
$$(00, 00, 02, 02, 11, 11, 10) \quad (00, 00, 04, 04, 12, 12, 10)$$
$$(00, 01, 02, 03, 04, 05, 06) \quad (00, 01, 02, 03, 04, 05, 06)$$

The last two (identical) blocks are fixed.

Greig and Sarvate previously claimed, in [20], that Design 106 did not exist. Kaski and Östergård [23] showed that Design 106 did exist, and there were nine non-isomorphic solutions [23]. We take this opportunity of presenting one of their solutions.

Example 4.4 A $(10, 22; 3, 4, 11; 5, 4)$ BTD with point set $(I_2 \times Z_3) \cup \{\infty\}$.

$$(00, 00, 10, 10, 11) \quad (00, 00, 20, 20, 21)$$
$$(00, 11, 11, 22, 22) \quad (00, 12, 12, 22, 22)$$
$$(00, 12, 12, 21, 21) \quad (\infty, 00, 00, 01, 01)$$
$$(\infty, \infty, 10, 11, 12) \quad (\infty, \infty, 20, 21, 22)$$

The last two blocks are each taken twice and are fixed.

5. Some Non-existence Results

One source of non-existence results is via the Bruck-Ryser-Chowla conditions for square (i.e., $v = b$) designs. The BRC conditions are necessary (and sufficient) conditions for the existence of a *rational* matrix N satisfying restrictions (2.1)–(2.4), so they apply to the n-ary case as well as the binary, as was noted in [8, Theorem 3.4]; practically the only difference is that $d = rk - \lambda v$ no longer reduces to $r - \lambda$ as it does in the binary case.

However, there is a scarcity of non-existence results for combinatorial designs, so it is worth commenting on Lemma 4.1. In essence, there one considers the designs formed by points of each multiplicity seperately. Lemma 4.1.a considers the doubletons, and notes that the total number of pairs they generate must be a multiple of 4, and is minimized if the doubletons are spread as evenly as possible over the blocks. Lemma 4.1.a does exclude some BTDs, e.g., an $(8, 18; 3, 3, 9; 4, 3)$ BTD. Presumably some generalization along these lines would exclude some parameter sets whenever $\lambda \not\equiv 0 \pmod 4$ — we do have a result along these lines.

Theorem 5.1 *No* $(20, 28; 2, 6, 14; 10, 6)$ *BTD exists.*

Proof: Even if the doubletons are spread as evenly as possible (i.e., 4 in 20 blocks and 5 in 8 blocks), there are 200 doubleton by doubleton pairs, and since this exceeds $\binom{v}{2}$ some pair of points must get more than one doubleton pair, and so this pair generates at least $8 > \lambda$ pair instances. ∎

Lemma 4.1.b merely says (in a disguised form) that if the block size is odd, then there must be enough singletons in the design to put one in every block, i.e., that $b \le v\rho_1$. The inequality in Lemma 4.1.b is implied by restrictions (2.1)–(2.4) when k is even. Lemma 4.1.b does exclude some BTDs, e.g., a $(3, 7; 1, 3, 7; 3, 4)$ BTD.

If λ is odd, then we must have at least $\binom{v}{2}$ pairs of singletons in the design. We note that spreading the singletons as evenly as possible over the blocks would minimize the number of pairs, and this can drive that number below the $\binom{v}{2}$ bound. There are two factors which force some spread of singletons: firstly, an odd block size; secondly, and much stronger, closeness of v to b. If $v = b$, and we have a square design, then $NN^T = N^T N$ by a result of Ryser's. In particular, the diagonal of $N^T N$ is constant, and so every block has the same pattern of multiplicities, and so the singletons are equally distributed over the blocks. In the square case, this argument excludes, e.g., a $(11, 11, 2, 3, 8; 8, 5)$ BTD. If we did not have exact squareness, then the integralized Rayleigh quotient argument of Greig and van Rees can be used to derive bounds on the diagonal elements of $N^T N$, showing they cannot vary overmuch from block to block. Actually, their upper bound is the stronger, and this would say a block cannot have too many doubletons.

Briefly, the integralized Rayleigh quotient argument mentioned above is simply that, if a matrix Q has eigenvalues $\theta_1 \geq \theta_2 \geq \cdots \theta_m$, and θ_1 has an eigenvector y then $\theta_m x^t x \leq x^T Q x \leq \theta_2 x^T x$ for any vector x such that $x^T y = 0$. Also, for any vector z of order m, we have $z^T(z-1) \geq -m/4$, but if z has integral elements, then $z^T(z-1) \geq 0$. We will take $Q = N^T N$, and we know $y = 1$ is an eigenvector, and that the eigenvalues of Q are the same as those of $N N^T$ (i.e., kr once, and $(kr - v\lambda)$ with multiplicity $v - 1$), plus $b - v$ zeros. See [30] for a detailed example of this argument being applied.

Theorem 5.2 *No* $(12, 26; 3, 5, 13; 6, 5)$ *BTD exists.*

Proof: Since λ is odd, we need every pair of points to have an odd number of singleton versus singleton pairs. Since a point only has three singletons, this number must be one or three, but if it were three then all remaining pairs for those points would be doubleton versus doubleton pairs and $\lambda - 3 \not\equiv 0 \pmod 4$, so there must be exactly one singleton versus singleton pair for every pair of points, and a total of $\binom{12}{2}$ pairs altogether. Moreover, every block must contain an even number of singletons, so the distribution of singletons over the blocks must be $6^x 4^{12-3x} 2^{-6+3x} 0^{20-x}$ for some $2 \leq x \leq 4$. Now let N_s be the incidence matrix of the singletons; we have shown that $N_s N_s^T = (\rho_1 - 1)I + J$, so N_s has rank 12, but there are only $6 + x \leq 10$ non-zero columns in N_s. ∎

Kaski and Östergård [23] showed the non-existence of Design 332 by exhaustive search. There is a theoretical argument which also shows this non-existence.

Theorem 5.3 *No* $(10, 25; 3, 6, 15; 6, 7)$ *BTD exists.*

Proof: Since k is even, every block contains an even number of singletons, and there are 30 of these in total, so the distribution of singletons over the blocks must be $6^x 4^y 2^{15-2y-3x} 0^{10+y+2x}$. Now we cannot have more than one doubleton versus doubleton pair on any pair of points, so $1 \cdot (15 - 2y - 3x) + 3 \cdot (10 + y + 2x) \leq \binom{10}{2}$ and so $x = y = 0$, and the singleton distribution is $2^{15} 0^{10}$. But we also need an odd number (at least one) of singleton versus singleton pairs for every pair of points, and there are only 15 such pairs. ∎

See Billington [8] for further arguments based on singleton/doubleton placement.

6.   An Improved Construction

Naturally, for use in constructing (v, k, λ) BIBDs, (k, λ) GDDs with all groups of size $k - 1$ or k are particularly attractive. One example in the literature that was familiar was given by Hanani in the Remark following [21, Lemma 4.19].

Theorem 6.1 *If $k \equiv 3 \pmod 4$ is a prime power, and $q \geq k$ is an odd prime power, then a $(k, (k - 1)/2)$ GDD of type k^q exists.*

Remark 8 The restriction that $k \equiv 3 \pmod 4$ is too strong, and k odd suffices. (However, Hanani did use Design 7 in this fashion to give $(5, 2)$ GDDs of type 5^q.)

Example 6.1 We construct a $(7, 3)$ GDD of type 7^q for $q \geq 7$, an odd prime power. Let x be a primitive generator for $GF(q)$, and let:

$$B_0 = ((0, 0)(1, x)(1, -x)(2, x^2)(2, -x^2)(4, x^3)(4, -x^3))$$

Now let $B_i = (1, x^{2i}) \cdot B_0$. Then B_i for $i = 0, 1, \ldots (q - 3)/2$ is our difference family over $Z_7 \times GF(q)$.

In terms of our original construction using Design 25, we have improved the index from $\lambda = 6$ to 3, but at a cost of restricting the q's we may use (although here excluding just even q is a pretty mild restriction). The integrality constraints on the GDD impose the same restriction on the q's, so this does not matter here, but could be a nontrivial cost in some generalizations. Again in terms of our original construction, we have not used an arbitary selection of a's, but have taken $a_1 = 0$, $a_{2i} = x^i$ and $a_{2i+1} = -a_{2i}$ i.e., we have plus/minus pairs for all the doubletons and zero for the singleton in a quadratic residue set, a $(k, k; 1, (k - 1)/2, k; k, k - 1)$ BTD. As noted in Remark 8 we could use any quadratic residue set to achieve this. Combining this with the corrected version of [10, Corollary 2.7] gives us another way of generalizing Theorem 6.1.

Theorem 6.2 *If $k + 1$ is the order of a skew Hadamard matrix, and $q \geq k > 2$ is an odd prime power, then a $(k, (k - 1)/2)$ GDD of type k^q exists.*

Now in the choice of a's, we took $a_1 = 0$, $a_{2i} = x^i$ and $a_{2i+1} = -a_{2i}$. Hanani gives some examples of carefully chosen a_{2i}'s, but retaining $a_1 = 0$ and $a_{2i+1} = -a_{2i}$, which enabled him to construct $(7, 1)$ GDDs of type 7^q for some q's. In generated the difference family, he was able to just retain every third base block and still achieve balance. Here again we need an extra restriction on q, effectively that the number of base blocks be divisible by three. (For $k = 5$ Hanani adopted another structure for the a's which enabled him to retain just the first half of the difference family, and so obtain $(5, 1)$ GDDs of type 5^q). One naturally wonders whether one can always select these a's nicely, and can one similarly take every fourth block for $k = 9$, and fifth block for $k = 11$, etc.. This was explored (for $k \leq 11$) in [19] and the brief answer is one can, except for a number of smaller values of q, which number generally grows with k (some explicit results are given in Remark 9). For the small failures, one could patch together n failed (k, λ) families, and with nicely chosen failures get a $(k, n\lambda)$ family. Alternatively, one could take every nth block in n failed families, and try patching these together for a (k, λ) family. For more details and examples, see [19]; also [11] uses the same approach in a different, but related, context. A further note here: these structured choices of the a's impose fairly large automorphism groups on the GDDs and, consequently, the searches for the a's can be very efficient.

So Theorem 6.1 gives us an instance where we can guarantee we can reduce the index, and one would naturally like to generalize this.

Theorem 6.3 *Suppose a BnD satisfying restrictions (2.1)–(2.5) exists, and suppose that $q \geq k$ is a prime power, and $q \equiv 1 \bmod f$. Suppose also that, with at most one exception per block, the multiplicity of every point in the BnD is a multiple of f, and that, in the possible exceptional case, the multiplicity of that point is one more than a multiple of f (including the zero multiple), then a $(k, \lambda/f)$ GDD of type v^q exists.*

Proof: In the basic construction, in the block $\{b_1, \ldots, b_k\}$ we replace a point b_i by the point (b_i, a_i), where we will pick $a_i \in GF(q)$ such that the a's in any block are distinct. Here, we can refine this by taking the subset of identical points $\{b_i, b_{i+1}, \ldots, b_{i+f-1}\}$ and replacing b_{i+j} by the point $(b_i, a_i w^j)$ where w is an f-th root in $GF(q)$, and if there is a point left over, say b_k, we replace it by $(b_k, 0)$. Again we pick the a_i's to ensure that the second elements of all the points in an amended block are distinct.

We now generate the base blocks from these amended blocks by multiplying by $(1, x^i)$ for $i = 0, 1, \ldots, (q-1)/f - 1$. Since the generating block is invariant under multiplication by $(1, w)$, our result follows. ∎

7.   Quadratic Residues

In Section 6, we used the quadratic residue set to give a useful BTD. A more complete description for the quadratic residue cases is given in Theorem 7.1.

Theorem 7.1 *Suppose $p = 2t + 1$ is an odd prime power. Let S denote the non-zero squares in $GF(p)$, and let N denote the non-squares. Then, using & as the concatenation operator, we have the following $(v, b; \rho_1, \rho_2, r; k, \lambda)$ BTD difference families over $GF(p)$, all of which satisfy restriction (2.5):*

a   *a $(p, p; 1, t, p; p, p - 1)$ BTD given by $B_s = (\{0\}$ & S & $S)$;*

b   *a $(p, p; 1, t, p; p, p - 1)$ BTD given by $B_n = (\{0\}$ & N & $N)$;*

c   *a $(p, 2p; 2, 2t, 2p; p, 2p - 2)$ BTD given by $B = B_s,\ B_n$;*

d   *if $p \equiv 3$ (mod 4), a $(p, p; 0, t + 1, p + 1; p + 1, p + 1)$ BTD:*

$$B_{s+} = (\{0\}\ \&\ \{0\}\ \&\ S\ \&\ S) ;$$

e   *a $(p, 2p; 0, 2t + 2, 2p + 2; p + 1, 2p + 2)$ BTD:*

$$B = B_{s+},\ (\{0\}\ \&\ \{0\}\ \&\ N\ \&\ N) .$$

Theorem 7.2 *Suppose p is an odd prime power, and $q \geq p$ is an odd prime power. Then the following GDDs exist:*

a   *a $(p, (p - 1)/2)$ GDD of type p^q; if $p \equiv 3$ (mod 4) then this design has $p\lfloor (q-1)/4 \rfloor$ 2-resolution classes, and so is 2-resolvable if $q \equiv 1$ (mod 4);*

b   *a $(p, p - 1)$ GDD of type p^q.*

Proof: Only the resolvability needs comment. In part (b) the underlying BTD is formed from complementary pairs (in the ternary sense), and each pair covers GF(p) twice. Developing each pair over GF(q) gives a 2-resolution set, which is then developed over GF(p).

For part (a), we already have noted that each base block is invariant under multiplication by $(1, -1)$. Multiplication of the underlying BTD by -1 gives the ternary complement, and so, if we were to use this complement in place of the corresponding block, it would generate the same differences in $GF(p) \times GF(q)$. Consequently, replacing every other block by its complement, i.e., multiplication by $(-1, 1)$, (allows us to pair up base blocks as in part (b); the residue of q modulo 4 determines whether there is an odd number of base blocks (and so an unpaired one) or not. ■

Theorem 7.3 *Suppose p is an odd prime power, and $q \geq p$ is an odd prime power. Then the following GDDs exist:*

  a *if $p \equiv 3$ (mod 4) and $q > p$, a $(p + 1, (p + 1)/2)$ GDD of type p^q;*

  b *if $q > p$, a $(p + 1, p + 1)$ GDD of type p^q.*

Moreover, if $\beta = \gcd(k/2, (q - 1)/2)$ and $\alpha = k/(2\beta)$, then this GDD is an α-frame.

Proof: Only the resolvability needs comment. Let x be a primitive element for GF(q). For our a's we take $a_{2i-1} = x^i$ for $i = 1, 2, \ldots, k/2$ and $a_{2i} = -a_{2i-1}$ otherwise (although it really is only the distribution over the cosets of order 2β and the plus/minus placement within the doubletons that matters). Then we generate the family by multiplying by $(1, x^i)$ for $i = 0, 1, \ldots, (q - 3)/2$. If we take every β-th base block, then we get a set of base blocks that covers every non-zero element of GF(q) α times, and the development over GF(p) gives us a base holey resolution class missing $GF(p) \times \{0\}$. In case (b), this is done seperately for each of the two generating blocks. ■

Remark 9 The cases where constructing a frame seems natural are also the cases where $\lambda = 1$ is possible. By carefully choosing the a's, it is possible to construct a frame, and thence the following RBIBDs:

  a a $(5q + 1, 6, 1)$ RBIBD for $q \equiv 1$ (mod 12) with q a prime power for $q < 5000$, with the possible exception of $q = 13$ and 37; (a $(186, 6, 1)$ RBIBD constructed by other means is known; only in two cases, $q = 49$ and $q = 73$ were four sets of a's used, otherwise two sufficed);

b a $(7q+1, 8, 1)$ RBIBD for $q \equiv 1 \pmod{8}$ with q a prime power for $q < 4096$, with the possible exception of $q = 9$, 17, 25 and 89; (a $(v, 8, 1)$ RBIBD constructed by other means is known in the first two cases; in all cases a single set of a's sufficed);

c a $(9q + 1, 10, 1)$ RBIBD for $q \equiv 1 \pmod{20}$ with q a prime power for $q \leq 6561$, with the possible exception of $q = 41$, 61, 101, 121, 181 and 241; (in all cases two sets of a's sufficed).

Details are given in [19]; even when using two sets of a's, they were regularly patterned, so the first set determined the second.

The corrected version of [10, Corollary 2.7] (the skew condition was omitted from their statement, but is obvious in their proof); gives us a way of generalizing Theorem 6.1. (There is an altered version of this result in [16, Result 6], but the alteration is incorrect also.) The value of this theorem is that skew Hadamard matrices are known for order 2 and all $v + 1 \leq 184$ with $v + 1 \equiv 0 \pmod{4}$ (see [13, Remark IV.24.15]).

Theorem 7.4 *If* $v \equiv 3 \pmod{4}$, *then a skew Hadamard matrix of order* $v + 1$ *exists if and only if a a* $(v, v; 1, (v - 1)/2, v; v, v - 1)$ *BTD exists.*

Remark 10 Using this skew Hadamard difference set in place of the quadratic residue set of Theorem 7.1 allows us to generate the corresponding results in Theorems 7.2 and 7.3, except for the 2-resolution sets in Theorem 7.2.a.

Example 7.1 We can use the Kronecker product for skew Hadamard matrices to produce a skew Hadamard matrix of order 16, and so a $(15, 15; 1, 7, 15; 15, 14)$ BTD (i.e., Design 344). We present this on the point set $(I_2 \times Z_7) \cup \{\infty\}$.

$$(\infty, \infty, 10, 11, 11, 12, 12, 14, 14, 21, 21, 22, 22, 24, 24)$$
$$(10, 10, 11, 11, 12, 12, 14, 14, 20, 21, 21, 22, 22, 24, 24)$$
$$(\infty, 20, 20, 21, 21, 22, 22, 23, 23, 24, 24, 25, 25, 26, 26)$$

The skewness in the 3 (mod 4) case is forced by the fact there is only one singleton, yet $\lambda \equiv 2 \pmod{4}$. In the comparable $\lambda \equiv 0 \pmod{4}$

case this forces symmetry on the incidence matrix (assuming the singletons are placed on the diagonal). We have seen, in Theorem 7.4, that we do not need a k to be prime power in the $k \equiv 3 \pmod 4$ case. There is a comparable, but much weaker, result in the $k \equiv 1 \pmod 4$ case [15, Theorem 3.5].

Theorem 7.5 *If* $v \equiv 1 \pmod 4$, *then a conference matrix of order* $v+1$ *exists if and only if a a* $(v, v; 1, (v-1)/2, v; v, v-1)$ *BTD exists.*

The value of this theorem is that conference matrices are known for some non-prime power orders, e.g., 46 and 226. A necessary (and conjecturally sufficient) condition for existence in this case is that v be the sum of two squares (see [14, Theorem IV.52.3.2, 52.6, 52.15–16]). (This condition is always satisfied here if v is a prime power.) However, this condition excludes many composite values; the only values less than 256 not excluded are 45, 65, 85, 117, 145, 153, 185, 205, 221, 225 and 245.

8.   Another Improvement

We begin this section with an example using [10, Design 37] to produce a $(4, 3)$ GDD of type 4^5. We will initially produce the GDD with index 6 using our basic construction.

Example 8.1 Let Design 37, a $(4, 8; 2, 3, 8; 4, 6)$ be given over $Z_3 \cup \{\infty\}$ by:

$$(\infty, 0, 1, 2) \quad (\infty, 0, 1, 2) \quad (\infty, \infty, 0, 0) \quad (1, 1, 2, 2)$$

Note the first two block are fixed.

We construct the following base blocks over $(Z_3 \cup \{\infty\}) \times Z_5$:

$((\infty, 1)(0, 2)(1, 3)(2, 4))$	$((\infty, 1)(0, 2)(1, 3)(2, 4))$
$((\infty, 2)(0, 4)(1, 1)(2, 3))$	$((\infty, 2)(0, 4)(1, 1)(2, 3))$
$((\infty, 4)(0, 3)(1, 2)(2, 1))$	$((\infty, 4)(0, 3)(1, 2)(2, 1))$
$((\infty, 3)(0, 1)(1, 4)(2, 2))$	$((\infty, 3)(0, 1)(1, 4)(2, 2))$
$((\infty, 1)(\infty, 4)(0, 2)(0, 3))$	$((\infty, 2)(\infty, 3)(0, 4)(0, 1))$
$((\infty, 4)(\infty, 1)(0, 3)(0, 2))$	$((\infty, 3)(\infty, 2)(0, 1)(0, 4))$
$((1, 1)(1, 4)(2, 2)(2, 3))$	$((1, 2)(1, 3)(2, 4)(2, 1))$
$((1, 4)(1, 1)(2, 3)(2, 2))$	$((1, 3)(1, 2)(2, 1)(2, 4))$

These base blocks generate a $(4, 6)$ GDD of type 4^5, where the first 8 blocks are fixed over Z_3.

Example 8.2 Noting that every base block in Example 8.1 is repeated, we can immediately write down the base blocks for a $(4, 3)$ GDD of type 4^5 over $(Z_3 \cup \{\infty\}) \times Z_5$ where the first 4 blocks are fixed over Z_3.

$$((\infty, 1)(0, 2)(1, 3)(2, 4)) \qquad ((\infty, 2)(0, 4)(1, 1)(2, 3))$$
$$((\infty, 4)(0, 3)(1, 2)(2, 1)) \qquad ((\infty, 3)(0, 1)(1, 4)(2, 2))$$
$$((\infty, 1)(\infty, 4)(0, 2)(0, 3)) \qquad ((\infty, 2)(\infty, 3)(0, 4)(0, 1))$$
$$((1, 1)(1, 4)(2, 2)(2, 3)) \qquad ((1, 2)(1, 3)(2, 4)(2, 1))$$

These base blocks generate a $(4, 6)$ GDD of type 4^5.

Example 8.2 has some extra properties due to the underlying BnD, which was repeated elements of $AG(2, 2)$ plus some complete singleton blocks. Now $AG(2, 2)$ has a partition into parallel classes, and a complete block is a trivial parallel class, so this resolvability carries over into our construction. Specifically, the first two rows of base blocks only miss the group $(Z_3 \cup \{\infty\}) \times \{0\}$, as do the last two rows. A GDD with such partial parallel classes is known as a frame. This design is something more: each base block in the first two rows develops into a complete parallel class, so this design is also a semi-frame. (A semi-frame is a GDD that has a partition into some complete parallel clases and some holey parallel classes, see [27] for more examples and details.)

Now, although we can use a $(4, 4, 3)$ BIBD to fill in the groups to give a $(20, 4, 3)$ BIBD, we cannot use a $(4, 4, 3)$ RBIBD to complete all the partial parallel classes of the frame, as there are 4 partial parallel classes per hole, but a $(4, 4, 3)$ RBIBD only has three parallel classes. Now is where we use the fact that Example 8.2 is actually a semi-frame, and we can use some parallel classes of the RBIBD to complete the partial parallel classes of the semi-frame, use the complete parallel classes of the semi-frame for more complete parallel classes, and finally build more parallel classes from the unused parallel classes of the $(4, 4, 3)$ RBIBD (there are actually no unused parallel classes here).

So what we have shown is that if $q > 4$ is an odd prime power, then a $(4q, 4, 3)$ RBIBD exists. This is not a new result: Baker [5] has already shown that $(4n, 4, 3)$ RBIBDs exist for all $n \geq 1$, and in fact the above forms part of Baker's proof.

Construction 8.1 *Partition our BnD's block set into two parts, and suppose in the left part compute f_L as we computed f in Theorem 6.3, and in the right part we suppose f_R that every block appears with a multiplicity that is some multiple of f_R. If $f = \gcd(f_L, f_R)$ then the result of Theorem 6.3 holds with the f in Theorem 6.3 replaced by the f we have just computed.*

Proof: In the left part, invariance under multiplication by $(1, w)$ shows our basic construction (with the choice of a's as in Theorem 6.3) will generate f copies and clearly so will the right part, so our result follows. ∎

We note that any BnD with $v = k$ and $d = rk - \lambda v > r$ can be made to satisfy restriction (2.5) by the addition of $d - r$ complete singleton blocks.

Baker's constructions only work when k is even. We will generalize a part of his work to the case when k is a prime power. Note the Design 37 we started with consisted of replacing each element of $AG(2, p)$ by p copies, then adding $p^3 - p^2 - p$ singleton blocks, with $p = 2$.

Theorem 8.2 *Let p and q be prime powers with $q \equiv 1 \pmod{p}$ and $q > p^2$. Then a $(p^2, p^2 - 1)$ GDD of type $(p^2)^q$ exists; this GDD is also a frame, and a semi-frame with up to $(p^2 - p - 1 - i)(q - 1)$ complete parallel classes and $p + 1 + i$ cycles of partial classes (a cycle misses each group in turn), for any i such that $0 \leq i \leq p^2 - p - 1$. Moreover, a $(qp^2, p^2, p^2 - 1)$ RBIBD exists.*

Proof: The general construction follows the outline given earlier. It is mostly the counting that needs verifying.

We note that the left part of the BnD i.e., the $AG(2, p)$, has a replication count of $p + 1$ which is inflated by p, and its $p^2 + p$ blocks each generate $(q - 1)/p$ base blocks, which form $p + 1$ cycles of partial classes. The right part of the BnD i.e., the $p^3 - p^2 - p$ singleton blocks which we reduce to $p^2 - p - 1$, then let each generate $q - 1$ base blocks, for a total of $(p^2 - p - 1)(q - 1)$ on the right part, and a grand total of $(p^2 - 1)(q - 1)$ each to be developed over $GF(q)$. Now a partial parallel class contains $(q - 1)$ blocks and a cycle contains $q(q - 1)$. A complete parallel class contains q blocks.

Finally, we note that the trivial $(p^2, p^2, 1)$ RBIBD is used to fill the groups and complete the partial parallel classes to give the RBIBD. ∎

9.   An Application

In this section we describe how one of the suitable BnDs was used to assist in constructing all $(v, 7, 14)$ BIBDs. This application also illustrates how useful α-resolvability can be.

Let us begin by noting that

$$(0, 0, 0, 0, 1, 1, 2)$$

gives a base block over Z_3 of a $(3, 7, 14)$ BnD which satisfies restrictions (2.1)–(2.5), and, using the improvement in Theorem 6.3 we have a $(7, 7)$ GDD of type 3^q if $q \geq 7$ is an odd prime power. Moreover, this is given by a difference family over $Z_3 \times GF(q)$, namely $B_0, B_1, \ldots, B_{(q-3)/2}$ where

$$B_0 = \big((0, 1)(0, -1)(0, x)(0, -x)(1, x^2)(1, -x^2)(2, 0)\big)$$

and $B_i = (1, x^i) \cdot B_0$ for x a primitive generator of $GF(q)$, and since each base block generates a k-resolution set, this design is k-resolvable, where $k = 7$ here.

In [3], the above construction was the main tool in showing that the neccessary conditions for the existence of a $(v, 7, 14)$ BIBD, namely $v \geq 7$ and $v \equiv 1 \pmod 3$, were sufficient. Most of the remaining designs we needed were in the literature. Hanani [21] had already shown that $v \equiv 1 \pmod 6$ was sufficient for the existence of a $(v, 7, 7)$ BIBD, and also constructed designs for $v \leq 43$ plus $v = 52$. In addition, he had shown a $(7, 2)$ GDD of type 3^7 (a TD) existed, and there is also a $(7, 2)$ GDD of type 3^8 [22, Theorem 15.7.4] (an alternative construction is to reduce modulo 24 Bose's affine relative difference set over Z_{48}, an application of Theorem 10.4 to [7]).

The basic outline is we first use the truncated transversal construction to show that any we can construct a $(\{7, 8, 9\}, 1)$ GDD on v points with group sizes taken from $M = 2$–47, 50, 51, 58 for any $v \neq 1$. Next, we already have $(7, 14)$ GDDs of type 3^n for $n = 7$ and 8; we can get a $(7, 14)$ GDD of type 3^9 by using our $(7, 2)$ GDD of type 3^8 as ingredient in an application of Wilson's fundamental construction (WFC) on the trivial $(9, 8, 7)$ BIBD. Another application of WFC shows we can construct a $(7, 14)$ GDD with group sizes taken from $3M$, so if we could construct a $(3m + 1, 7, 14)$ BIBD for every $m \in M$ with m odd we would be done.

At this point we use our $(7, 7)$ GDD of type 3^q, and add new fixed points to the first r of the 7-resolution sets for $r \in \{0, 2, 4\}$ (except $r = 4$

is too many if $q = 7$) to get a $(\{7,8\}, 7)$ GDD of type $3^q r^1$. Using this in WFC, with our $(7,2)$ GDDs as ingredients, gives a $(7,14)$ GDD of type $9^q (3r)^1$ and again we may fill the groups using an extra point with $(n, 7, 14)$ BIBDs with $n \in \{10, 1, 7, 13\}$. This covers all the designs we need except that we lacked $q = 15$ and also missed $v = 46, 58, 76$. Now Baker [4] had constructed a $(7,1)$ GDD of type 3^{15}, and the whole design formed a 7-resolution set; one can also put 13 of Baker's designs together, and thence obtain a $(46, 7, 14)$ BIBD without much bother, filling different pairs of the groups, each time using an extra point. At this point we could see what damage not being able to use 19 or 25 in M caused us — $q = 19$ was used for 154–160 and 162, but all these odd values could be obtained by taking $q = 49$ and $r \in \{8, 10, 12\}$, so this would turn out to be no problem here. Alternatively, one could construct $(v, 7, 14)$ BIBDs for $v = 58$ and 76, and leave no open cases. This turned out not to be too onerous; $v = 58$ took a short computer run, and $v = 76$ was solved by hand; both solutions had a fair amount of structure.

10. More Difference Families

In this section we give some examples of BnDs with v a divisor of $k(k-1)$. In recursive constructions of BIBDs, GDDs with group sizes that are a factor of $k(k-1)$ are especially useful, and we briefly justify this view.

Let $m = k(k-1)/\gcd(k(k-1), \lambda)$. Then it is a fact that the integrality constraints for a (v, k, λ) BIBD are satisfied, or not, depending on the residue class of $v \pmod{m}$, and that 1 $\pmod{m}$ and $k \pmod{m}$ are two (not necessarily distinct) satisfactory residue classes, and these are the only satisfactory residue classes if k is a prime power. This is more fully documented in [1]. (We might remark for clarification that we are saying, e.g., that for a $(v, 6, 1)$ BIBD the satisfactory residue classes are only 1, 6, 16, 21 modulo 30; we are not precluding that they might collapse, as they do here into 1, 6 modulo 15.)

We refrain from linking the size to λ in our examples, as we have already indicated some ways of getting our GDDs at indices different from that given by the underlying BnD.

We start with some generic examples.

Lemma 10.1 *Let* $k = pt$ *with* $k > p > 1$. *If a* (k, t, λ) *BIBD exists with* $v \geq p(t-1)/(p-1)$, *then there exists a* (k, k, λ') *BnD satisfying* *restrictions (2.1)–(2.5) with* $\lambda' = p^2\lambda + \alpha$, *where* $\alpha = p\lambda(pk - 2k + 1)/(t-1)$.

Proof: If N is the incidence matrix of the BIBD, then the (partitioned) incidence matrix for the BnD is given by:

$$pN : J$$

where J_α is the k by α all ones matrix.

The value of $r(t-1) - \lambda k$ is $\lambda > 0$, so restriction (2.5) is not satisfied for the initial BIBD (that difference between the right- and left-hand sides is always λ for a BIBD). Replacing the singletons of the BIBD by p-tons alters this difference to α. Each block of singletons added reduces this difference by 1 (which is true whenever $v = k$). ■

We have already seen how designs constructed in this way can be used in Baker's approach to RBIBD construction. It is probably true (as a consequence of the Hadamard existence conjecture) that a that a $(4n - 1, 2n - 1, n - 1)$ BIBD exists for all $n > 0$; by a result of Kokay and van Rees, this design exists if and only if a $(4n, 2n, 2n - 1)$ RBIBD exists [24]. Consequently the case $p = 2$ of our construction yields RBIBDs here, although this result is entirely subsumed in Baker's [5].

Lemma 10.2 *There exists a* $(k - 1, k, k^2 + k)$ *BnD satisfying restrictions (2.1)–(2.5).*

Proof: The (partitioned) incidence matrix for the BnD is given by:

$$N_1 : N_2 : \cdots : N_k : kI$$

where $N_i = I + J$ for all i. ■

The design of Lemma 10.2 has a high index. This is compensated somewhat by the simplicity of the structure. In fact, there are currently 91 unknown $(v, 9, 1)$ BIBDs and the last three direct solutions used this BnD [12].

In many of our examples in the remainder of this section we will use p^n to denote n copies of the point p.

Lemma 10.3 *If k is a square, then there exists:*

a a $(2, k, (k^2 - k)/2)$ BnD satisfying restrictions (2.1)–(2.5) (such a BnD only exists if k is a square);

b a $(3, k, (k^2 - k)/3)$ BnD satisfying restrictions (2.1)–(2.5).

Proof: The difference set over Z_2 is given by $(0^x, 1^{k-x})$ where $x = (k - \sqrt{k})/2$.

The difference set over Z_3 is given by $(0^x, 1^x, 2^{k-2x})$ where $x = (k \pm \sqrt{k})/3$; the value of x is chosen to be integral, (such a choice is always possible), and if $x \equiv 0 \pmod 3$ there are two valid choices for x.

Solutions for part (b) can exist even when k is not a square. ∎

Theorem 10.4 *Let a $(|G|, k, \lambda)$ balanced n-ary design satisfying restrictions (2.1)–(2.5), be given by a difference family, $\mathcal{D}$, over a group G, and let H be a normal subgroup of G. Then there exists a difference family over group G/H for a $(|G/H|, k, \lambda|H|)$ balanced $(n|H|)$-ary design satisfying restrictions (2.1)–(2.5).*

Proof: Replace all points in $\mathcal{D}$ by their coset. ∎

The following simple lemma turns out to be quite helpful in constructions.

Lemma 10.5 *The total number of pairs amongst the identical elements within the $|\mathcal{D}|$ base blocks in a difference family of a (v, k, λ) BnD satisfying restrictions (2.1)–(2.5) is $\binom{k}{2}|\mathcal{D}|/v$, i.e., $\lambda/2$.*

Example 10.1 Suppose one desired a difference family for a $(4, 5, \lambda)$ BnD satisfying restrictions (2.1)–(2.5). One needs $10|\mathcal{D}|/4$ pairs, so one needs at least two base blocks. By trial and error, one can find a solution of $(0, 0, 0, 1, 3)$ $(0, 1, 1, 3, 3)$ over Z_4. There are 3 and $1 + 1$ pairs from these two blocks which generate a $(4, 5, 10)$ BnD.

Application of Theorem 10.4 also yields a $(2, 5, 20)$ BnD satisfying restrictions (2.1)–(2.5), given by $(0, 0, 0, 1, 1)$ $(0, 1, 1, 1, 1)$ over Z_2.

Table 10.3. More Examples

v	λ	f	Design
$k = 4$			
Z_2	6	3	$(0^3, 1^1)$
Z_3	4	2	$(0^2, 1^2)$
I_4	6	2	Double $(4, 2, 1)$ single $(4, 4, 2)$ BIBDs
Z_4	12	1	$(0^2, 1^2)$ $(0^2, 2^2)$ $(0^2, 1^1, 3^1)$ $(0^2, 1^1, 3^1)$
$GF(4)$	12	1	$(00^2, 01^2)$ $(00^2, 10^2)$ $(00^2, 11^2)$ $(00^1, 01^1, 10^1, 11^1)$
Z_6	2	1	$(0^2, 1^1, 4^1)$
$k = 5$			
Z_2	10	2	$(0^4, 1^1)$ $(0^3, 1^2)$
$GF(4)$	5	1	$(00^3, 01^1, 10^1)$ $(00^1, 01^2, 10^2)$
Z_5	2	2	$(0^1, 1^2, 4^2)$
I_{10}	2	1	Does not exist [10, Design 5]
Z_{10}	4	1	$(0^2, 1^1, 3^1, 6^1)$ $(0^2, 2^1, 4^1, 9^1)$
$I_2 \times Z_5$	6	1	$(00^2, 03^1, 04^1, 13^1)$ $(00^2, 01^1, 11^1, 12^1)$ $(00^2, 02^1, 10^1, 14^1)$
			$(10^2, 01^1, 12^1, 14^1)$ $(10^2, 04^1, 11^1, 13^1)$ $(10^2, 00^1, 02^1, 03^1)$
$GF(4) \times Z_5$	2	1	See Example 4.1
I_{30}	2	1	Unknown [10, Design 271]
$k = 6$			
Z_5	6	2	$(0^2, 1^2, 4^2)$ $(0^2, 2^2, 3^2)$
I_6	10	2	Double $(6, 3, 2)$ single $(6, 6, 2)$ BIBDs
I_{15}	2	1	Does not exist [10, Design 10]
Z_{15}	4	1	$(0^2, 1^1, 2^1, 4^1, 7^1)$ $(0^2, 4^1, 7^1, 9^1, 10^1)$
I_{30}	2	1	Unknown [10, Design 117]
$k = 7$			
Z_2	42	2	$(0^5, 1^2)$ $(0^5, 1^2)$ $(0^5, 1^2)$ $(0^4, 1^3)$
Z_3	14	2	$(0^4, 1^2, 2^1)$
Z_7	6	2	$(0^1, 1^2, 2^2, 4^2)$
I_{14}	6	1	Unknown [10, Design 234]
I_{21}	2	1	Does not exist [10, Design 17]
Z_{21}	4	1	$(0^2, 1^1, 2^1, 5^1, 12^1, 15^1)$ $(0^2, 2^1, 5^1, 6^1, 9^1, 11^1)$
I_{42}	2	1	Unknown [10, Design 211]
$k = 8$			
Z_2	56	2	$(0^2, 1^6)$ $(0^4, 1^4)$
Z_7	8	2	$(0^2, 1^2, 2^2, 4^2)$
I_8	14	2	Double $(8, 4, 3)$ single $(8, 8, 2)$ BIBDs
I_{14}	4	1	Does not exist [10, Design 36]
I_{14}	6	1	Unknown [10, Design 150]
I_{28}	2	1	Does not exist [10, Design 28]

Table 10.4. More Examples (continued)

v	λ	f	Design
$k = 9$			
Z_2	18	2,3	$(0^6, 1^3)$
Z_3	24	4	$(0^1, 1^4, 2^4)$
Z_3	24	2	$(0^5, 1^2, 2^2)$
$GF(4)$	5	3	$(00^3, 01^3, 10^3)$
$GF(9)$	8	2	$(00^1, 01^2, 21^2, 02^2, 12^2)$
I_{12}	6	1	See [10, Design 57]
I_{12}	8	1	Unknown [10, Design 162]
I_{12}	10	1	Unknown [10, Design 319]
Z_{12}	12	1	$(0^1, 1^1, 3^2, 6^2, 7^1, 11^2)$ $(0^1, 5^1, 6^2, 7^2, 9^2, 11^1)$
$I_3 \times Z_6$	4	1	$(11^2, 13^1, 10^1, 21^1, 32^1, 34^2, 35^1)$
			$(12^1, 14^2, 15^1, 21^2, 23^1, 20^1, 34^1)$
			$(11^1, 21^2, 22^1, 25^1, 31^2, 33^1, 30^1)$
I_{24}	4	1	Unknown [10, Design 135]
I_{36}	2	1	See [10, Design 44]
$k = 10$			
Z_2	90	2,3	$(0^3, 1^7)$ $(0^4, 1^6)$
Z_3	60	2	$(0^6, 1^2, 2^2)$ $(0^2, 1^4, 2^4)$
$GF(9)$	20	2	$(00^2, 01^2, 21^2, 02^2, 12^2)$ $(00^2, 10^2, 20^2, 11^2, 22^2)$
I_{10}	18	2	Double $(10, 5, 4)$ single $(10, 10, 2)$ BIBDs
I_{15}	6	1	Does not exist [10, Design 82]
I_{45}	2	1	Unknown [10, Design 65]
$k = 11$			
Z_2	220	2	$(0^3, 1^8)$ $(0^4, 1^7)$ $(0^4, 1^7)$ $(0^5, 1^6)$
Z_{11}	10	2	$(0^1, 1^2, 3^2, 4^2, 5^2, 9^2)$
I_{55}	2	1	Unknown [10, Design 92]
$k = 12$			
Z_2	198	2	$(0^4, 1^8)$ $(0^4, 1^8)$ $(0^5, 1^7)$
Z_2	264	2	$(0^4, 1^8)$ $(0^4, 1^8)$ $(0^4, 1^8)$ $(0^6, 1^6)$
Z_3	44	2	$(0^6, 1^4, 2^2)$
Z_{11}	20	2	$(0^1, 1^2, 3^2, 4^2, 5^2, 9^2)$ $(0^1, 2^2, 6^2, 7^2, 8^2, 10^2)$
I_{12}	22	2	Double $(12, 6, 5)$ single $(12, 12, 2)$ BIBDs
I_{22}	6	1	Does not exist [10, Design 151]
I_{33}	4	1	Unknown [10, Design 136]
I_{66}	2	1	Does not exist [10, Design 120]

Table 10.5. More Examples (continued)

v	λ	f	Design
$k = 13$			
Z_2	156	2,3	$(0^4, 1^9)$ $(0^6, 1^7)$
Z_3	52	2	$(0^6, 1^5, 2^2)$
Z_{13}	12	2	$(0^1, 1^2, 3^2, 4^2, 9^2, 10^2, 12^2)$
I_{26}	6	1	Does not exist [10, Design 191]
I_{39}	4	1	Unknown [10, Design 183]
I_{78}	2	1	Does not exist [10, Design 174]
$k = 14$			
Z_2	182	1	$(0^4, 1^{10})$ $(0^5, 1^9)$ $(0^6, 1^8)$ $(0^7, 1^7)$
Z_{13}	28	2	$(0^2, 1^2, 3^2, 4^2, 9^2, 10^2, 12^2)$
			$(0^2, 2^2, 5^2, 6^2, 7^2, 8^2, 11^2)$
I_{14}	26	2	Double $(14, 7, 6)$ single $(14, 14, 2)$ BIBDs
I_{91}	2	1	Does not exist [10, Design 211]
$k = 15$			
Z_2	420	2	$(0^5, 1^{10})$ $(0^5, 1^{10})$ $(0^6, 1^9)$ $(0^7, 1^8)$
Z_2	420	2	$(0^4, 1^{11})$ $(0^6, 1^9)$ $(0^7, 1^8)$ $(0^7, 1^8)$
Z_3	140	2	$(0^6, 1^5, 2^4)$ $(0^8, 1^5, 2^2)$
$(I_2 \times Z_7) \cup \{\infty\}$	14	1	See Example 7.1
I_{21}	10	1	Unknown [10, Design 326]
I_{35}	6	1	Unknown [10, Design 301]
I_{105}	2	1	Does not exist [10, Design 274]
$k = 16$			
Z_2	120	2,3,5	$(0^{10}, 1^6)$
Z_3	80	4	$(0^8, 1^4, 2^4)$
Z_5	48	4	$(1^4, 2^4, 3^4, 4^4)$
Z_{15}	16	2	$(3^2, 6^2, 7^2, 9^2, 11^2, 12^2, 13^2, 14^2)$
I_{16}	30	2	Double $(16, 8, 7)$ single $(16, 16, 2)$ BIBDs

Acknowledgments

The first draft of this article was prepared whilst I held an Ethel Raybould Visiting Fellowship at the University of Queensland and I thank Liz Billington and the Department of Mathematics for their support and hospitality.

References

[1] R.J.R. Abel, I. Bluskov and M. Greig, Balanced incomplete block designs with block size 8, *J. Combin. Des.* **9** (2001), 233–268.

[2] R.J.R. Abel, I. Bluskov and M. Greig, Balanced incomplete block designs with block size 9, preprint.

[3] R.J.R. Abel and M. Greig, Balanced incomplete block designs with block size 7, *Des. Codes Cryptogr.* **13** (1998), 5–30.

[4] R.D. Baker, Elliptic semi-planes. I. Existence and classification, *Congr. Numer.* **19** (1977) 61–73.

[5] R.D. Baker, Resolvable BIBD and SOLS, *Discrete Math.* **44** (1983), 13–29.

[6] T. Beth, D. Jungnickel and H. Lenz, *Design Theory*, (2nd ed.), Cambridge University Press, Cambridge, England, (1999).

[7] R.C. Bose, An affine analogue of Singer's theorem, *J. Indian Math. Soc.* **6** (1942) 1–15.

[8] E.J. Billington, Balanced n-ary designs: a combinatorial survey and some new results, *Ars Combin.* **17A** (1984), 37–72.

[9] E.J. Billington, Designs with repeated elements in blocks: a survey and some recent results, *Congr. Numer.* **68** (1989), 123–146.

[10] E.J. Billington and P.J. Robinson, A list of balanced ternary designs with $R \leq 15$, and some necessary existence conditions, *Ars Combin.* **16** (1983), 235–258.

[11] M. Buratti, Old and new designs via difference multisets and strong difference families, *J. Combin. Des.* **7** (1999), 406–425.

[12] M. Buratti and F. Zuanni, G-invariantly resolvable Steiner 2-designs which are 1-rotational over G, *Bull. Belg. Math. Soc. Simon Stevin* **5** (1998), 221–235; Addendum, *ibid.* **7** (2000), 311.

[13] R. Craigen, Hadamard matrices and designs, in: *The CRC Handbook of Combinatorial Designs*, (eds. C.J. Colbourn and J.H. Dinitz), CRC Press, Boca Raton, FL, 1996, 370–377.

[14] R. Craigen, Weighing matrices and conference matrices in: *The CRC Handbook of Combinatorial Designs*, (eds. C.J. Colbourn and J.H. Dinitz), CRC Press, Boca Raton, FL, 1996, 496–504.

[15] J.F. Dillon and M.A. Wertheimer, Balanced ternary designs derived from other combinatorial designs, *Congr. Numer.* **47** (1985), 285–298.

[16] J.D. Fanning, Symmetric balanced ternary designs with $\rho_1 = 1$ or 2, *Aequationes Math.* **47** (1994), 143–149.

[17] J.D. Fanning, The coexistence of some binary and N-ary designs, *Ars Combin.* **45** (1997), 217–227.

[18] S.C. Furino, Y. Miao and J. Yin, *Frames and Resolvable Designs*, CRC Press, Boca Raton, FL, (1996).

[19] M. Greig, Some group divisible design constructions, *J. Combin. Math. Combin. Comput.* **27** (1998), 33–52.

[20] M. Greig and D.G. Sarvate, Some constructions of block designs, *J. Combin. Math. Combin. Comput.* **28** (1998), 149–161.

[21] H. Hanani, Balanced incomplete block designs, *Discrete Math.* **11** (1975), 255–369.

[22] M. Hall, Jr., *Combinatorial Theory*, (2nd ed.), John Wiley and Sons, New York, (1986).

[23] P. Kaski and P.R.J. Östergård, Enumeration of balanced ternary designs, preprint.

[24] W. Kocay and G.H.J. van Rees, Some non-isomorphic $(4t + 4, 8t + 6, 4t + 3, 2t + 2, 2t + 1)$-BIBDs, *Discrete Math.* **92** (1991), 159–172.

[25] E.J. Morgan, Construction of balanced n-ary designs, *Utilitas Math.* **11** (1977), 3–31.

[26] E.J. Morgan, Construction of balanced designs and related identities, *Lecture Notes in Math.* **748** (1979), 79–91.

[27] R. Rees, Semiframes and nearframes, *Combinatorics 88: Proc. of the International Conference on Incidence Geometries and Combinatorial Structures*, Universita degli Studi di Napoli (1989), vol. 2, 359–367.

[28] K. Sinha, A construction of balanced ternary designs, *Ars Combin.* **33** (1992), 276–278.

[29] K.D. Tocher, The design and analysis of block experiments, *J. Roy. Statist. Soc. Ser. B* **14** (1952), 45–100.

[30] G.H.J. van Rees, $(22, 33, 12, 8, 4)$-BIBD, an update, in *Computational and Constructive Design Theory*, (W.D. Wallis, ed.), Kluwer Academic Publ., 1996, 337–357.

Chapter 10

SETS OF STEINER TRIPLE SYSTEMS OF ORDER 9 REVISITED

T. S. Griggs

Department of Pure Mathematics,
The Open University,
Walton Hall, Milton Keynes, United Kingdom MK7 6AA

A. Rosa

Department of Mathematics and Statistics,
McMaster University,
Hamilton, Ontario, Canada L8S 4K1

Abstract   We determine all minimal large sets of 8 Steiner triple systems of order 9 (STS(9)); there are precisely four pairwise nonisomorphic solutions. We also classify all maximal sets of STS(9) which mutually intersect in the same number of triples (uniformly intersecting sets).

1.   Introduction

In this paper we are concerned with minimal large sets and maximal uniformly intersecting sets of Steiner triple systems of order 9. We start with some basic definitions. Recall that a Steiner triple system of order v (briefly STS(v)) is a pair $(V, \mathcal{B})$ where V is a v-set, and $\mathcal{B}$ is a collection of 3-subsets of V called *triples* such that each 2-subset of V is contained in exactly one triple. A family $(V, \mathcal{B}_1), \ldots, (V, \mathcal{B}_q)$ of q Steiner triple systems of order v, all on the same set V, is a *large set* of STS(v) if every 3-subset of V is contained in *at least one* STS(v) of the family.

This is a wider definition than that used by some authors who require that every 3-subset occurs in precisely one system, i.e. $B_i \cap B_j = \emptyset$, $1 \leq i < j \leq q$. We call such families large sets of *mutually disjoint* (MD) Steiner triple systems. It is well known that large sets of MD STS(v) contain precisely $v - 2$ systems and exist for all admissible $v \equiv 1$ or $3 \ (mod\ 6), v \neq 7$ [7], [8], [9], [10].

In [6], Lindner and Rosa began the study of large sets of *mutually almost disjoint* (MAD) Steiner triple systems. These are large sets in which $|B_i \cap B_j| = 1$, $1 \leq i < j \leq q$. For $v \neq 7$, the number of systems q in a large set of MAD STS(v) equals v or $v + 1$ (with one extra possibility $q = 15$ for $v = 13$) [4]. Large sets of v MAD STS(v) may be constructed from Steiner quadruple systems, SQS($v + 1$) [6] and exist for all admissible $v \equiv 1$ or $3 \ (mod\ 6)$. Large sets of $v + 1$ MAD STS(v) are known to exist for $v = 13$ and $v = 15$ [4] but not for $v = 9$ [6].

The above naturally raises the question of whether it is possible to obtain large sets of $v - 1$ STS(v) and motivates the following definition which can also be found in [4]. A large set of STS(v) is said to be *nearly disjoint* (ND) if $|B_i \cap B_j| \leq 1$, $1 \leq i < j \leq q$. It was further shown in [4] that large sets of $v - 1$ ND STS(v) exist for all $v = 2^n - 1, n \geq 3$. But there is no large set of 8 ND STS(9)! However, this does not mean that there is no large set of 8 STS(9) in which some of the intersections between systems have cardinality different from 0 or 1. Indeed, if one begins with either of the two nonisomorphic large sets of 7 MD STS(9) [1], and then chooses the eighth system appropriately, such a large set will be constructed. But this large set is not *minimal* in that if any of the Steiner triple systems is removed then the systems that remain no longer form a large set. It is one of the aims of this paper to enumerate minimal large sets of 8 STS(9). We find that there are precisely 4 pairwise nonisomorphic solutions.

Another concept that is of interest is that of sets of *uniformly intersecting* Steiner triple systems. These are families of systems $(V, \mathcal{B}_1), \ldots,$ $(V, \mathcal{B}_q)$ which have the property that $|B_i \cap B_j| = d$, $1 \leq i < j \leq q$, for some given integer d, $d < v(v - 1)/6$. They are not necessarily large sets; indeed the question of whether there exist large sets of uniformly intersecting STS(v) for values of d other than 0 or 1 is open. We are interested in *maximal* sets of uniformly intersecting STS(v), i.e. sets which cannot be extended. For $v = 7$ the classification is complete. Cayley [2] determined that the maximum number of disjoint STS(7) is two; all such pairs are unique to within isomorphism. There is a unique maximal set of 15 MAD STS(7) [6] and it is easily seen that the maximum number of STS(7) intersecting in 3 triples is also two. For $v = 9$ the maximal sets of disjoint systems have been classified by Cooper [3]. We extend this classification to maximal large sets of STS(9) intersecting in 1, 2, 3, 4, or 6 triples, the only other possibilities.

2.   Large sets

There are 840 distinct STS(9) on the same base set. A brute force attack on the problem of finding all minimal large sets of 8 such systems is not sensible even if possible. But by using a number of easy lemmas and the classification of pairs of STS(9) given in [5], the search space can be reduced considerably. We start with the easiest of these.

Lemma 1 *In a minimal large set of 8 STS(9),*
a) 72 triples occur precisely once and 12 triples occur precisely twice,
b) the 12 triples occurring twice themselves form an STS(9).

Proof. Consider any pair $\{a, b\}$. There are precisely 7 distinct triples containing $\{a, b\}$ and hence 6 of these must occur in one system of the large set and one must occur in two systems. □

We will call the STS(9) formed from the 12 triples which occur twice the *cross* STS(9) of the large set. We will also use the *system intersection graph* (SIG), actually a multigraph, of the large set. The vertices of the graph are the systems of the large set, and two vertices are joined by n edges if the systems they represent have n triples in common. Trivially the SIG has 8 vertices and 12 edges.

The next lemma summarizes the results of Kramer and Mesner [5] concerning pairs of STS(9).

Lemma 2 *There are precisely 8 nonisomorphic pairs of distinct STS(9) on the same base set. The number of triples in the intersection of these pairs are respectively 0, 0, 1, 2, 3, 3, 4, and 6.*

Proof. See [5]. □

We are now in a position to describe our strategy. Denote the eight systems of the large set by $A, B, C_1, \ldots, C_6$. We then consider in turn the largest intersection of any two systems of the large set and assume that this occurs between systems A and B. The next result is easy to prove and illustrates the strategy well.

Lemma 3 *No two systems of a minimal large set of 8 STS(9) can intersect in 6 triples.*

Proof. Without loss of generality, system A can be assumed to contain the triples $123, 456, 789, 147, 258, 369, 159, 267, 348, 168, 249, 357$ (here and throughout the paper set brackets and intermediate commas will be omitted for clarity). Also without loss of generality, system B can be assumed to contain the triples $123, 456, 789, 147, 269, 358, 159, 248, 367, 168, 257, 349$. Moreover, these are the only two systems which contain the 6 common blocks $123, 456, 789, 147, 159, 168$. Hence the cross system is either system A or system B, say, the former. But then there are 6 edges of the SIG joining vertices A and B and a further 6 edges joining vertex A to vertices $C_i, 1 \leq i \leq 6$. Hence the systems $B, C_1, \ldots, C_6$ form a large set of 7 MD STS(9), contradicting the minimality requirement of the large set of 8 STS(9).   □

The argument used in the previous lemma, that if the cross system is also one of the systems of the large set of 8 STS(9), then the other 7 systems form an MD large set, is an elementary observation but will be used in later lemmas to eliminate various structures.

Lemma 4 *No two systems of a minimal large set of 8 STS(9) can intersect in 4 triples.*

Proof. Without loss of generality, let system A contain the triples $123, 456, 789, 147, 258, 369, 159, 267, 348, 168, 249, 357$ and system B contain the triples $123, 489, 567, 147, 269, 358, 159, 278, 346, 168, 245, 379$. The triples containing the element 1 are those which are common to both systems. The proof is in three parts, the last two of which rely on computer searches.

i) System A cannot intersect any system $C_i, 1 \leq i \leq 6$ in 4 triples. If it intersects it in 3 triples then from the analysis of pairs of STS(9) in [5] these can be either a parallel class (impossible) or a triangle (possible). But in the latter case 7 triples of the cross system are specified and it must be system A. Similarly if it intersects it in 2 triples then these two triples intersect and are therefore in different parallel classes of system A. Again the cross system must be system A. Similarly the above argument can be repeated for system B.

ii) Now assume that all of the systems $C_i, 1 \leq i \leq 6$, are disjoint from both systems A and B. There are 48 such systems but in any collection of six such systems, there are more than 8 edges in the SIG (computer search).

iii) Hence without loss of generality system A has an intersection of one triple with system C_1, say 456. It then follows that the triple 789 is also in the cross system. There are then two possibilities; the cross system is either system A or is 123, 456, 789, 147, 269, 358, 159, 248, 367, 168, 257, 349 which has two triples in common with system B. Thus the SIG has 4 edges between vertices A and B, 2 edges between vertex A and vertices $C_i, 1 \leq i \leq 6$ and 2 edges between vertex B and vertices $C_i, 1 \leq i \leq 6$. There are 208 systems with an intersection of zero or one triple with both systems A and B but in any collection of six such systems there are more than 4 edges in the SIG (computer search). □

We next consider the cases where the intersection between systems A and B is of cardinality 3; either a parallel class or a triangle. The argument follows the same line as the previous lemma but is slightly more involved. We sketch the outline of the proof; it is easy to fill in the details.

Lemma 5 *There are precisely 4 pairwise nonisomorphic minimal large sets of 8 STS(9) in which two of the systems intersect in 3 triples.*

Proof. i) Suppose that system A and system B intersect in 3 triples, either a parallel class or a triangle. Then it is easy to show that neither system A nor system B can intersect any of the systems $C_i, 1 \leq i \leq 6$ in 3 triples.

ii) Now assume that systems A and B intersect in a parallel class, and that all of the systems $C_i, 1 \leq i \leq 6$ are disjoint from them. There are 42 such systems. A computer search reveals 7 solutions but further analysis shows that these fall into just 2 isomorphism classes of 6 large sets and 1 large set, respectively. Details of the large sets are given in Section 3. Repeating this in the case where systems A and B intersect in a triangle, there are 50 systems disjoint from them both. But there are no large sets in this case (computer search).

iii) Hence without loss of generality system A has an intersection of at least one triple with system C_1. There are a number of different possibilities to consider here in both cases where system A intersects system B in either a parallel class or in a triangle. But in all cases either the cross system is system A or the SIG has 3 edges between vertices A and B, 3 edges between vertex A and vertices $C_i, 1 \leq i \leq 6$, and 3 edges between vertex B and vertices $C_i, 1 \leq i \leq 6$. There are 420 systems with an intersection of less than 3 triples with both systems A and B in the case where the latter intersect in a parallel class, and 404 systems where they intersect in a triangle. A computer search reveals 54 solutions in the former case but none in the latter. The 54 solutions again fall into just two isomorphism classes of 36 and 18 large sets, respectively. Details of these systems are again given in Section 3. □

If there exist further minimal large sets of 8 STS(9), the maximum intersection between any pair of systems is at most 2 triples. Therefore choose systems A and B to intersect in 2 triples and determine all systems which intersect them both in 2 or less triples. There are 424 such systems. An exhaustive computer search reveals 8 solutions. However, in all cases the 12 edges of the SIG are all incident with either system A or system B thus showing that this is the cross system and that the large sets are not minimal. Finally, consider the case where systems A and B intersect in just one triple. There are 188 systems which intersect them both in either zero or one triple. Again an exhaustive computer search reveals 8 solutions for a SIG containing 12 edges. However, upon further analysis in all cases there is a triple intersection between two of the systems $C_i, 1 \leq i \leq 6$, i.e. the large sets found are isomorphic to those already discovered. We state the above two results formally.

Lemma 6 *There is no minimal large set of 8 STS(9) in which the maximum intersection between any two systems is 2 triples.*

Proof. Computer search.   □

Lemma 7 *There is no minimal large set of 8 nearly disjoint (ND) STS(9).*

Proof. Computer search, also reported in [4].   □

We collect together the results of the above lemmas into a single theorem.

Theorem 8 *There exist precisely 4 pairwise nonisomorphic minimal large sets of 8 STS(9).*

3.   Results

In this section we give details of the 4 minimal large sets of 8 STS(9) identified in Section 2 including the cross system, automorphism group, and system intersection graph. Each individual system will be represented in compact notation

$$
\begin{array}{ccc}
a & b & c \\
d & e & f \\
g & h & i
\end{array}
$$

to represent the triples $abc, def, ghi, adg, beh, cfi, aei, bfg, cdh, afh, bdi, ceg$.

Large set No.1

A	B	C_1	C_2	C_3	C_4	C_5	C_6
1 2 3	1 2 3	1 2 4	1 2 5	1 2 6	1 2 7	1 2 8	1 2 9
4 5 6	4 5 6	3 5 9	3 8 9	3 5 9	3 8 6	3 5 6	3 8 6
7 8 9	8 9 7	8 7 6	6 4 7	7 4 8	5 4 9	9 7 4	4 7 5

The cross system is

$$
\begin{array}{ccc}
1 & 2 & 3 \\
4 & 5 & 6 \\
9 & 7 & 8
\end{array}
$$

The automorphism group is the group $C_3 \times S_3$ of order 18 generated by the permutations (1 5 3 4 2 6)(7 8 9) and (1 4 7)(2 5 8)(3 6 9).

The system intersection graph has 2 components; one component $(3K_2)$ consists of 3 edges joining systems A and B, while in the other $(K_{3,3})$ there is one edge joining each of systems C_1, C_2, C_6 to each of systems C_3, C_4, C_5.

Large set No.2

A	B	C_1	C_2	C_3	C_4	C_5	C_6
1 2 3	1 2 3	1 2 4	1 2 5	1 2 6	1 2 7	1 2 8	1 2 9
4 5 6	4 5 6	3 8 9	3 7 8	3 9 7	3 5 6	3 4 5	3 6 4
7 8 9	8 9 7	6 7 5	4 9 6	5 8 4	9 4 8	7 6 9	8 5 7

The cross system is

$$
\begin{array}{ccc}
1 & 2 & 3 \\
4 & 5 & 6 \\
9 & 7 & 8
\end{array}
$$

This is probably the most interesting of the 4 large sets. The automorphism group is of order 432 and is identical to the automorphism group of the cross system.

The system intersection graph has 4 components with 3 edges each $(4 \times (3K_2))$, joining respectively A and B, C_1 and C_4, C_2 and C_6, and C_3 and C_5, and representing in each case a parallel class of the cross system.

Large set No.3

A	B	C_1	C_2	C_3	C_4	C_5	C_6
1 2 3	1 2 3	1 2 4	1 2 5	1 2 6	1 2 7	1 2 8	1 2 9
4 5 6	4 5 6	3 5 7	3 8 6	3 9 8	3 4 9	3 7 4	3 5 4
7 8 9	8 9 7	9 6 8	4 9 7	7 5 4	6 8 5	5 6 9	8 7 6

The cross system is

$$
\begin{array}{ccc}
1 & 2 & 3 \\
4 & 5 & 6 \\
7 & 9 & 8
\end{array}
$$

The automorphism group is the group C_3 of order 3 generated by the permutation (1 2 3)(4 6 5).

The system intersection graph is connected, has 3 edges joining systems A and B, and further edges joining the following pairs of systems: A and C_1, C_1 and C_3, C_3 and B, A and C_5, C_5 and C_2, C_2 and B, A and C_6, C_6 and C_4, and C_4 and B.

Large set No.4

A	B	C_1	C_2	C_3	C_4	C_5	C_6
1 2 3	1 2 3	1 2 4	1 2 5	1 2 6	1 2 7	1 2 8	1 2 9
4 5 6	4 5 6	3 7 8	3 8 9	3 9 7	3 5 6	3 4 5	3 6 4
7 8 9	8 9 7	5 9 6	6 7 4	4 8 5	9 4 8	7 6 9	8 5 7

The cross system is

$$
\begin{array}{ccc}
1 & 2 & 3 \\
4 & 6 & 5 \\
8 & 9 & 7
\end{array}
$$

The automorphism group is the group S_3 of order 6 generated by the permutations (1 2 3)(7 9 8) and (2 3)(4 5)(8 9).

The system intersection graph is connected, has 3 edges joining systems A and B, and further edges joining the following pairs of systems: C_1 and A, C_1 and B, C_1 and C_6, C_2 and A, C_2 and B, C_2 and C_4, C_3 and A, C_3 and B, and C_3 and C_5.

Finally in this section we recall that there exist precisely 2 pairwise nonisomorphic large sets of 7 MD STS(9), first constructed by Bays [1]. Details of these in the same format as given above as well as of their automorphism groups are given in [5].

For $v = 7$, we have already remarked that there is no large set of 5 MD STS(7). There is a unique minimal large set of 6 STS(7), first constructed in [4]. For completeness we also list the systems of this large set (each row is an STS(7)) and give its automorphism group which previously has not been determined.

Systems:

B_1:	1 2 3	1 4 5	1 6 7	2 4 6	2 5 7	3 4 7	3 5 6
B_2:	1 2 3	1 4 6	1 5 7	2 4 7	2 5 6	3 4 5	3 6 7
C_1:	1 2 4	1 3 7	1 5 6	2 3 5	2 6 7	3 4 6	4 5 7
C_2:	1 2 5	1 3 6	1 4 7	2 3 4	2 6 7	3 5 7	4 5 6
C_3:	1 2 6	1 3 5	1 4 7	2 3 7	2 4 5	3 4 6	5 6 7
C_4:	1 2 7	1 3 4	1 5 6	2 3 6	2 4 5	3 5 7	4 6 7

The large set is nearly disjoint (ND); $|B_1 \cap B_2| = 1$, $|C_i \cap C_j| = 1, 1 \le i < j \le 4$; $B_i \cap C_j = \emptyset, 1 \le i \le 2, 1 \le j \le 4$.

The automorphism group is the group S_4 of order 24 generated by the permutations (1 2 3)(5 6 7) and (2 3)(4 5 7 6). It acts naturally on the systems C_1, C_2, C_3, C_4. Even permutations stabilize the systems B_1, B_2 and odd permutations interchange them.

4. Uniformly intersecting sets

In this section we classify all maximal sets of uniformly intersecting STS(9). The first two cases, i.e. when every pair of distinct systems intersect in 6 and 4 triples respectively, can be done by hand. In the other cases we use a computer search. Starting in turn with one of the remaining 6 classes of nonisomorphic pairs of distinct STS(9) intersecting in 3 or less triples (see Lemma 2), all maximal sets of uniformly intersecting

STS(9) are determined. Isomorph rejection was done by hand. We now present the results with the systems expressed in the same format as the large sets given in the previous section. Orders of automorphism groups are also determined.

Intersection size = 6

A representative pair of STS(9) intersecting in 6 triples is given in the proof of Lemma 3; $123, 456, 789, 147, 258, 369, 159, 267, 348, 168, 249, 357$: system A, and $123, 456, 789, 147, 269, 358, 159, 248, 367, 168, 257, 349$: system B. These intersect in the parallel class $123, 456, 789$ and the 3 further triples containing the element 1. Also, as observed in the proof of Lemma 3, these are the only two systems which intersect in these 6 triples. Thus, for the set to be extended, any further system must intersect each of the above systems either in the same parallel class and 3 further triples all containing a different element, or, inter alia, in a different parallel class. It is easily verified that this latter possibility cannot occur. The former possibility also cannot occur if the further triples all contain either of the elements 2 or 3 which occur in a triple with the element 1 in the parallel class. However, if, without loss of generality, we choose the element 4 then the STS(9) $123, 456, 789, 147, 268, 359, 169, 257, 348, 158, 249, 367$ (system C) intersects system A in the given parallel class and the 3 further triples containing the element 4. Moreover, it intersects system B in the given parallel class and the 3 further triples containing the element 7. The set of STS(9) can then be further extended by $123, 456, 789, 147, 259, 368, 169, 248, 357, 158, 267, 349$ (system D) which also contains the given parallel class and intersects systems A, B, C in 3 further triples containing the elements $7, 4, 1$, respectively. It is now clear that the set of STS(9) cannot be extended further with any systems intersecting systems A, B, C, D in 6 triples. Hence to within isomorphism there exists a unique maximal set of STS(9) mutually intersecting in 6 triples and consisting of 4 systems. These are given by

A	B	C	D
1 2 3	1 2 3	1 2 3	1 2 3
4 5 6	4 6 5	4 6 5	4 5 6
7 8 9	7 9 8	7 8 9	7 9 8

The automorphism group is of order 48 and is generated by the permutations (2 3), (5 6), (8 9), (1 4)(2 5 3 6) and (1 4 7)(2 5 8)(3 6 9).

Intersection size = 4

A representative pair of STS(9) intersecting in 4 triples is given in the proof of Lemma 4; $123, 456, 789, 147, 258, 369, 159, 267, 348, 168, 249,$ 357: system A, and $123, 489, 567, 147, 269, 358, 159, 278, 346, 168, 245,$ 379: system B. These intersect in the 4 triples containing the element 1. Now consider a further system which intersects system A in 4 triples containing a point other than the element 1. It is easily verified that this cannot intersect system B in 4 triples. Therefore any further systems intersecting both system A and system B in 4 triples do so in the 4 triples containing the element 1. It is easily verified that there are two further systems; $123, 469, 578, 147, 256, 389, 159, 248, 367, 168, 279, 345$: system C, and $123, 458, 679, 147, 289, 356, 159, 246, 378, 168, 257, 349$: system D. Moreover, systems C and D intersect in the 4 triples containing the element 1. Hence to within isomorphism there exists a unique maximal set of STS(9) mutually intersecting in 4 triples and consisting of 4 systems. These are given by

A	B	C	D
1 2 3	1 2 3	1 2 3	1 2 3
4 5 6	4 9 8	4 6 9	4 8 5
7 8 9	7 6 5	7 5 8	7 9 6

The automorphism group is of order 192 and is generated by the permutations $(2\ 3)(5\ 6)(8\ 9)$, $(4\ 7)(5\ 8)(6\ 9)$, $(2\ 7)(3\ 4)(5\ 9)$, $(4\ 6\ 5)(7\ 8\ 9)$ and $(6\ 9\ 8\ 5)$.

Intersection size = 3

There are 5 nonisomorphic maximal sets of STS(9) intersecting in 3 triples. Of these, 3 consist of 3 systems and the other 2 consist of 5 systems and 8 systems, respectively. We list each of these in turn.

1. 3 systems

$$
\begin{array}{ccc}
A & B & C \\
1\,2\,3 & 1\,2\,3 & 1\,2\,3 \\
4\,5\,6 & 4\,5\,6 & 4\,5\,6 \\
7\,8\,9 & 8\,9\,7 & 9\,7\,8
\end{array}
$$

The parallel class $123, 456, 789$ is common to all 3 systems. The automorphism group is of order 324 and is generated by the permutations $(1\ 2)(4\ 5)(7\ 8)$, $(4\ 7\ 5\ 8\ 6\ 9)$, $(4\ 5\ 6)$ and $(1\ 4\ 7\ 2\ 5\ 8\ 3\ 6\ 9)$.

2. 3 systems

$$
\begin{array}{ccc}
A & B & C \\
1\,2\,3 & 1\,2\,3 & 1\,2\,3 \\
4\,5\,6 & 7\,5\,4 & 6\,5\,7 \\
7\,8\,9 & 6\,8\,9 & 4\,8\,9
\end{array}
$$

The triangle $123, 258, 159$ is common to all 3 systems. The automorphism group is a nonabelian group of order 18 generated by the permutations $(4\ 6\ 7)$, $(1\ 2\ 5)(3\ 8\ 9)$ and $(1\ 2)(4\ 6)(8\ 9)$.

3. 3 systems

$$
\begin{array}{ccc}
A & B & C \\
1\,2\,3 & 1\,2\,3 & 1\,2\,8 \\
4\,5\,6 & 7\,5\,4 & 6\,9\,3 \\
7\,8\,9 & 6\,8\,9 & 7\,5\,4
\end{array}
$$

The intersection between systems A and B is the triangle $123, 258, 159$, between systems B and C is the triangle $167, 246, 457$, and between systems A and C is the triangle $348, 369, 789$.

The automorphism group is the group C_9 of order 9 generated by the permutation $(1\ 4\ 9\ 2\ 7\ 3\ 5\ 6\ 8)$.

4. 5 systems

A	B	C	D	E
1 2 3	1 2 3	1 2 3	1 2 3	1 2 3
4 5 6	4 5 6	4 5 8	4 6 8	4 7 9
7 8 9	8 9 7	7 9 6	7 5 9	5 6 8

Systems A and B intersect in the parallel class $123, 456, 789$. The intersections between all other pairs of systems are triangles.

The automorphism group is the group C_2 of order 2 generated by the permutation $(2\ 3)(4\ 6)(7\ 9)$.

5. 8 systems

A	B	C	D	E	F	G	H
1 2 3	1 2 3	1 2 3	1 2 3	1 2 3	1 2 6	1 2 7	1 2 9
4 5 6	4 7 5	4 5 6	4 8 5	4 9 5	3 5 4	3 4 9	3 7 5
7 8 9	8 9 6	8 9 7	6 9 7	8 6 7	7 8 9	6 5 8	8 4 6

In this maximal set, there are 8 triples $123, 148, 159, 247, 258, 349, 357,$ 789 which occur in 5 of the systems. These form a maximum partial triple system on the set $\{1, 2, 3, 4, 5, 7, 8, 9\}$. The automorphism group of this structure is the stabilizer of the point in the automorphism group of the STS(9) from which it is obtained and is of order 48. This extends to be the automorphism group of the maximal set. The 4 triples $167, 269, 368, 456$, which with the above triples form an STS(9), occur in two of the systems; there are 48 triples occurring in one system, and 24 triples which do not occur at all. Systems A and C, B and H, D and F, E and G intersect in a parallel class. The intersections between all other pairs of systems are triangles.

Intersection size $= 2$

There are 17 nonisomorphic maximal sets of STS(9) intersecting in 2 triples. Of these 2 consist of 4 systems, 5 consist of 5 systems, 9 consist of 6 systems and there is one consisting of 7 systems. Again we list these.

1. 4 systems

A	B	C	D
1 2 3	1 2 3	1 2 3	1 2 3
4 5 6	4 8 9	4 7 5	4 6 7
7 8 9	7 6 5	9 6 8	8 9 5

The automorphism group is the group S_4 of order 24 generated by the permutations (2 3)(4 5 7 6)(8 9) and (1 3)(4 5)(8 9) and acts naturally on the systems A, B, C, D.

2. 4 systems

A	B	C	D
1 2 3	1 2 3	1 2 3	1 2 3
4 5 6	4 8 9	4 8 5	4 6 9
7 8 9	7 6 5	6 7 9	5 8 7

The automorphism group is the group S_3 of order 6 generated by the permutations (1 2 3)(4 8 9)(5 7 6) and (1 3)(4 9)(5 7). The group stabilizes system B and acts naturally on the systems A, C, D.

3. 5 systems

A	B	C	D	E
1 2 3	1 2 3	1 2 3	1 2 4	1 2 4
4 5 6	4 8 9	4 7 5	3 5 9	3 6 7
7 8 9	7 6 5	6 9 8	7 8 6	5 8 9

The only automorphism is the identity.

4. 5 systems

A	B	C	D	E
1 2 3	1 2 3	1 2 3	1 2 5	1 2 9
4 5 6	4 8 9	4 7 5	3 6 4	3 4 6
7 8 9	7 6 5	6 9 8	9 8 7	5 8 7

The only automorphism is the identity.

5. 5 systems

A	B	C	D	E
1 2 3	1 2 3	1 2 3	1 2 7	1 2 9
4 5 6	4 8 9	4 7 5	3 6 8	3 6 4
7 8 9	7 6 5	6 9 8	4 5 9	8 7 5

The only automorphism is the identity.

6. 5 systems

A	B	C	D	E
1 2 3	1 2 3	1 2 3	1 2 4	1 2 6
4 5 6	4 8 9	4 7 5	3 5 6	3 5 9
7 8 9	7 6 5	6 9 8	7 9 8	4 7 8

The only automorphism is the identity.

7. 5 systems

A	B	C	D	E
1 2 3	1 2 3	1 2 3	1 2 4	1 2 7
4 5 6	4 8 9	4 7 5	3 5 9	3 8 5
7 8 9	7 6 5	9 6 8	7 8 6	4 9 6

The only automorphism is the identity.

8. 6 systems

A	B	C	D	E	F
1 2 3	1 2 3	1 2 3	1 2 3	1 2 4	1 2 7
4 5 6	4 8 9	4 7 5	4 6 7	3 6 7	3 5 4
7 8 9	7 6 5	6 9 8	8 5 9	5 8 9	6 8 9

The automorphism group is the group C_2 of order 2 generated by the permutation (4 7)(5 6).

9. 6 systems

A	B	C	D	E	F
1 2 3	1 2 3	1 2 3	1 2 3	1 2 4	1 2 6
4 5 6	4 8 9	4 7 5	4 6 8	3 5 9	3 5 7
7 8 9	7 6 5	6 9 8	5 9 7	7 8 6	9 8 4

The automorphism group is the group C_6 of order 6 generated by the permutation (1 8 7 3 5 9).

10. 6 systems

A	B	C	D	E	F
1 2 3	1 2 3	1 2 3	1 2 3	1 2 5	1 2 7
4 5 6	4 8 9	4 7 5	4 6 8	3 9 4	3 5 4
7 8 9	7 6 5	6 9 8	5 9 7	6 7 8	6 8 9

The only automorphism is the identity.

11. 6 systems

A	B	C	D	E	F
1 2 3	1 2 3	1 2 3	1 2 3	1 2 5	1 2 9
4 5 6	4 8 9	4 7 5	4 6 8	3 9 4	3 8 4
7 8 9	7 6 5	6 9 8	5 9 7	6 7 8	6 5 7

The automorphism group is the group C_2 of order 2 generated by the permutation (1 3)(5 8)(7 9).

12. 6 systems

A	B	C	D	E	F
1 2 3	1 2 3	1 2 3	1 2 4	1 2 4	1 2 9
4 5 6	4 8 9	4 7 5	3 5 6	3 6 7	3 7 4
7 8 9	7 6 5	6 9 8	7 9 8	5 8 9	5 6 8

The automorphism group is the group C_3 of order 3 generated by the permutation (2 6 7)(3 9 8).

13. 6 systems

A	B	C	D	E	F
1 2 3	1 2 3	1 2 3	1 2 4	1 2 7	1 2 7
4 5 6	4 8 9	4 7 5	3 6 7	3 5 4	3 6 8
7 8 9	7 6 5	6 9 8	5 8 9	6 8 9	4 5 9

The automorphism group is the group C_2 of order 2 generated by the permutation (1 9)(5 6).

14. 6 systems

A	B	C	D	E	F
1 2 3	1 2 3	1 2 3	1 2 6	1 2 7	1 2 7
4 5 6	4 8 9	4 7 5	3 5 9	3 8 5	3 6 8
7 8 9	7 6 5	6 9 8	4 7 8	4 9 6	4 5 9

The automorphism group is the group C_2 of order 2 generated by the permutation (2 4)(8 9).

15. 6 systems

A	B	C	D	E	F
1 2 3	1 2 3	1 2 3	1 2 4	1 2 5	1 2 5
4 5 6	4 8 9	4 7 5	3 5 6	3 6 4	3 9 4
7 8 9	7 6 5	6 9 8	7 9 8	9 8 7	6 7 8

The only automorphism is the identity.

16. 6 systems

A	B	C	D	E	F
1 2 3	1 2 3	1 2 4	1 2 4	1 2 7	1 2 7
4 5 6	4 8 9	3 5 6	3 6 9	3 6 5	3 5 9
7 8 9	7 6 5	7 9 8	7 8 5	4 9 8	4 8 6

The automorphism group is the group S_3 of order 6 generated by the permutations (2 4 7)(5 6 9) and (4 7)(5 6).

17. 7 systems

A	B	C	D	E	F	G
1 2 3	1 2 3	1 2 3	1 2 3	1 2 4	1 2 5	1 2 9
4 5 6	4 8 9	4 7 5	4 6 7	3 7 8	3 8 7	3 4 6
7 8 9	7 6 5	9 6 8	5 9 8	9 6 5	6 4 9	5 8 7

In this maximal set, there are 7 triples $123, 147, 168, 248, 267, 346, 378$ which occur in 4 of these systems. These form an STS(7) on the set $\{1, 2, 3, 4, 6, 7, 8\}$. There are 56 triples occurring in one system and 21 triples which do not occur at all. The automorphism group is the group C_7 generated by the permutation (1 2 6 3 7 4 8).

Intersection size = 1

There are 7 nonisomorphic maximal sets of STS(9) intersecting in 1 triple. Of these 4 consist of 4 systems, 2 consist of 5 systems and there is one consisting of 9 systems which is the unique large set of MAD STS(9).

1. 4 systems

A	B	C	D
1 2 3	1 2 3	1 2 3	1 2 4
4 5 6	4 5 7	4 9 6	3 5 8
7 8 9	8 9 6	5 7 8	9 6 7

The automorphism group is the group C_3 of order 3 generated by the permutation (1 2 3)(4 7 9)(5 8 6).

2. 4 systems

A	B	C	D
1 2 3	1 2 3	1 2 4	1 2 5
4 5 6	4 5 7	3 5 7	3 9 8
7 8 9	8 9 6	9 6 8	4 6 7

The automorphism group is the group C_2 of order 2 generated by the permutation (1 9)(2 4)(3 5)(6 8).

3. 4 systems

A	B	C	D
1 2 3	1 2 3	1 2 4	1 2 8
4 5 6	4 5 7	3 5 7	3 9 4
7 8 9	8 9 6	9 6 8	5 7 6

The only automorphism is the identity.

4. 4 systems

A	B	C	D
1 2 3	1 2 3	1 2 4	1 2 6
4 5 6	4 5 7	3 5 7	3 4 8
7 8 9	8 9 6	9 6 8	9 7 5

The automorphism group is the group C_2 of order 2 generated by the permutation (1 7)(2 5)(6 9).

5. 5 systems

A	B	C	D	E
1 2 3	1 2 3	1 2 3	1 2 4	1 2 5
4 5 6	4 5 7	4 9 7	3 5 8	3 9 8
7 8 9	8 9 6	5 6 8	9 6 7	7 6 4

The automorphism group is the group C_4 of order 4 generated by the permutation (2 6 3 8)(4 5 7 9).

6. 5 systems

A	B	C	D	E
1 2 3	1 2 3	1 2 3	1 2 4	1 2 6
4 5 6	4 5 7	4 9 7	3 5 8	3 7 9
7 8 9	8 9 6	5 6 8	9 6 7	4 8 5

The automorphism group is the group C_2 of order 2 generated by the permutation (2 3)(5 7)(6 9).

7. 9 systems

A	B	C	D	E	F	G	H	I
1 2 3	1 2 3	1 2 3	1 2 4	1 2 5	1 2 6	1 2 7	1 2 8	1 2 9
4 5 6	4 5 7	4 9 6	3 7 9	3 6 9	3 7 5	3 6 9	3 6 5	3 7 5
7 8 9	8 9 6	5 7 8	6 8 5	8 4 7	9 4 8	5 8 4	7 4 9	4 8 6

In this large set of 9 MAD STS(9), there are 12 triples $123, 479, 568,$ $145, 278, 369, 167, 259, 348, 189, 246, 357$ which occur in 3 of the systems and form an STS(9). We will extend our terminology and refer to this as the cross system. The remaining 72 triples occur in one system. The automorphism group of the large set is a subgroup of index 3 of the automorphism group of the cross system and is of order 144. It is generated by the permutations (2 4 7 8 3 5 4 9), (1 7 8 6 2 5 4 9) and (1 6 7)(2 8 9 3 5 4).

Intersection size = 0

To conclude, and in order to make this paper complete so that readers do not have to consult further papers for information, we give the results of Bays [1] and Cooper [3] on maximal sets of nonintersecting STS(9). There are 6 nonisomorphic such sets. Of these, one consists of 4 systems, 3 consist of 5 systems and 2 consist of 7 systems; the latter are the large sets of MD STS(9).

1. 4 systems

A	B	C	D
1 2 3	1 2 7	1 2 8	1 2 9
8 4 7	6 3 5	5 3 4	8 3 5
9 6 5	4 8 9	9 7 6	7 6 4

The only automorphism is the identity.

2. 5 systems

A	B	C	D	E
1 2 3	1 2 4	1 2 5	1 2 6	1 2 8
8 4 7	9 3 8	9 3 7	5 3 8	9 3 5
9 6 5	7 6 5	6 8 4	9 4 7	4 7 6

The automorphism group is the group C_2 of order 2 generated by the permutation (1 2)(3 4)(5 6)(7 9).

3. 5 systems

A	B	C	D	E
1 2 3	1 2 4	1 2 5	1 2 6	1 2 7
8 4 7	9 3 8	9 3 7	4 3 8	4 3 5
9 6 5	7 6 5	6 8 4	9 5 7	6 9 8

The automorphism group is the group C_2 of order 2 generated by the permutation (1 6)(3 5)(4 7)(8 9).

4. 5 systems

A	B	C	D	E
1 2 3	1 2 4	1 2 5	1 2 7	1 2 8
8 4 7	9 3 8	6 3 4	5 3 4	6 3 5
9 6 5	7 6 5	9 8 7	8 9 6	4 7 9

The automorphism group is a nonabelian group of order 20 generated by the permutations (1 4 2 5 8) and (1 4 8 5)(3 9 7 6).

5. 7 systems

A	B	C	D	E	F	G
1 2 3	1 2 4	1 2 5	1 2 6	1 2 7	1 2 8	1 2 9
4 5 6	3 8 5	3 6 9	3 4 8	3 6 5	3 5 9	3 6 4
7 9 8	6 7 9	7 4 8	9 7 5	8 9 4	4 7 6	5 8 7

The automorphism group is of order 42 and is generated by the permutations (5 9)(1 6 4 3 7 8) and (1 6 7 3 8 2 4).

6. 7 systems

A	B	C	D	E	F	G
1 2 3	1 2 4	1 2 5	1 2 6	1 2 7	1 2 8	1 2 9
4 5 6	3 8 7	3 8 4	3 5 8	3 9 6	3 6 7	3 5 6
7 9 8	9 5 6	6 9 7	7 4 9	5 4 8	4 9 5	8 7 4

The automorphism group is of order 54 and is generated by the permutations (5 8)(2 4 6 3 7 9), (1 5 8)(2 6 7)(3 4 9) and (1 2 3)(4 5 6)(7 9 8).

Acknowledgments

Most of this work was done while the first author was visiting the Department of Mathematics and Statistics, McMaster University, Hamilton, Ontario. He would like to thank the University for its hospitality. Research of the second author is supported by NSERC of Canada Grant No.OGP007268. Both authors would like to thank Rudi Mathon of University of Toronto for verifying computationally some of the automorphism groups given in this paper.

References

[1] S. Bays, Une question de Cayley relative au problème des triades de Steiner. *Enseignement Math.* **19** (1917), 57–67.

[2] A. Cayley, On the triadic arrangements of seven and fifteen things. *London, Edinburgh and Dublin Philos. Mag. and J. Sci.* **37** (1850), 50–53 (*Collected Mathematical Papers I*, 481–484).

[3] D. S. Cooper, Maximal disjoint Steiner triple systems of order 9 and 13. *Unpublished manuscript.*

[4] T. S. Griggs and A. Rosa, Large sets of Steiner triple systems in *Surveys in Combinatorics* (Ed. P. Rowlinson), London Math. Soc. Lecture Note Ser. **218** (1995), 25–39.

[5] E. S. Kramer and D. M. Mesner, Intersections among Steiner systems. *J. Combinat. Theory (A)* **16** (1974), 273–285.

[6] C. C. Lindner and A. Rosa, Construction of large sets of almost disjoint Steiner triple systems. *Canad. J. Math.* **27** (1975), 256–260.

[7] J. X. Lu, On large sets of disjoint Steiner triple systems I-III. *J. Combinat. Theory (A)* **34** (1983), 140-182.

[8] J. X. Lu, On large sets of disjoint Steiner triple systems IV-VI. *J. Combinat. Theory (A)* **37** (1984), 136-192.

[9] J. X. Lu, On large sets of disjoint Steiner triple systems VII. *Unpublished manuscript.*

[10] L. Teirlinck, A completion of Lu's determination of the spectrum for large sets of disjoint Steiner triple systems. *J. Combinat. Theory (A)* **57** (1991), 302–305.

Chapter 11

SOLVING ISOMORPHISM PROBLEMS FOR t-DESIGNS

Reinhard Laue

*Universität Bayreuth, Germany**

Abstract   Designs usually provide examples for very hard isomorphism problems. So, instead of solving the general isomorphism problem here the problem is solved for many cases where the designs are constructed with a common prescribed group of automorphisms by Kramer and Mesner's approach [30]. The method is based on an analysis of group actions and improves on Burnside's table of marks approach [15]. In particular, no knowledge of the full subgroup lattice of the symmetric group on the point set is needed.

1.   Introduction

In 1976 Kramer and Mesner [30] in a seminal paper formalized a method to construct t-designs out of orbits of a given group of automorphisms. They demonstrated the power of their approach by substantial numbers of t-designs found and stated:

With appropriate hypergraphs and groups we would not be surprised if one could eventually construct t-designs with "large" t.

With our system DISCRETA, which is based on this strategy, we actually could construct new Steiner 5-designs [13], [14] and many simple 6-, 7-, and 8-designs with small parameters, see [5] [6] [7] [8] [10] [11] [12] [35] [39] [34] [33], and for a database of parameter sets see [16].

*supported by the Deutsche Forschungsgemeinschaft, Ke 201/17-2.

In particular, in [33] some simple 8-designs obtained by this method were extended employing Alltop's construction [2] to simple 9-$(28, 14, \lambda)$ designs for $\lambda = 3204, 3240, 4608, 5076$, and 5148. So, this confirms the expectation of Kramer and Mesner.

But the bigger the success in solving the construction problem the bigger is the next problem. After reporting of having found 324 solutions to a particular problem until some print quota was exceeded Kramer and Mesner continued:

How many of these are nonisomorphic is a question we might be afraid to consider.

In fact, we could determine many designs with the same prescribed group of automorphisms. In particular, all 8-$(27, 13, \lambda)$ designs for $\lambda = 3204, 3240$, and 4608 with prescribed group of automorphisms $ASL(3, 3)$ could be obtained. There are 1076436, 1236842, and 471356178 respectively of such 8-designs. Obviously, testing them algorithmically for isomorphism would be an enormous task. In fact, the number of isomorphism types in the three cases is just half the number of solutions. The purpose of this article is to explain the techniques that can be used to solve such isomorphism problems.

There are different approaches to solve isomorphism problems. One can search for invariants that distinguish the designs. Several kinds of intersection numbers have been used by A. Betten [4]. But often these invariants do not separate all different isomorphism types. B. D. McKay first transforms the problem into an equivalent graph theoretic isomorphism problem and then uses his program NAUTY to solve it. The program is made available by B. D. McKay and was used for example by Khosrovshahi, Mohammad-Noori, and Tayfeh-Rezaie in their recent classification of all smallest 6-designs with non-trivial automorphism group [29].

However such a purely algorithmic approach does not give much insight and of course is limited in its reach, depending on the state of the technology.

We here focus on algebraic ways of classifying objects up to isomorphism that in some cases can handle infinite series of isomorphism problems. In other cases, integrating theoretical insights into an algorithmic approach, makes problems accessible that concern enormous numbers of objects. In particular, the problem stated in Kramer and Mesner's paper can easily be solved.

In this paper we restrict our attention to t-designs though most of the techniques make no use of the special nature of these objects, for a more general setting see [33]. The presentation here extends [33] by new results.

2. Basics

A t-(v, k, λ) design $\mathcal{D} = (\mathcal{V}, \mathcal{B})$ consists of a point set V of v elements and of a collection $\mathcal{B}$ of k-element subsets of V, called blocks, such that each t-element subset of V is contained in exactly λ blocks. $\mathcal{D}$ is simple if no block appears more than once in $\mathcal{B}$. We throughout this paper only consider simple designs and we usually identify $\mathcal{D}$ with its collection of blocks.

The high regularity of these objects makes it difficult to distinguish different isomorphism types of t-designs with the same parameters. We therefore present some algebraic tools that in our search for t-designs with large t have been proven useful in solving this problem.

An isomorphism $\phi : \mathcal{D}_1 \longrightarrow \mathcal{D}_2$ of a t-design $\mathcal{D}_1$ onto a t-design $\mathcal{D}_2$ can be understood as a renaming of the points. The blocks of $\mathcal{D}_1$ by such a renaming of their elements become the blocks of $\mathcal{D}_2$. Thinking of the structure of a t-design as the only important issue means to abstract from the actual given naming of the points. So, classifying t-designs up to isomorphism means to concentrate on the structural properties of these designs. The *isomorphism problem* is to decide whether any two given designs are isomorphic.

If the two designs $\mathcal{D}_1$ and $\mathcal{D}_2$ are defined on the same point set V then such a renaming is a permutation of V. Applying all permutations to $\mathcal{D}_1$ yields all designs defined on V and isomorphic to $\mathcal{D}_1$. The isomorphism class of $\mathcal{D}_1$ on V thus is just an orbit of the full symmetric group S_V on the set of all designs. Usually, such a set of objects is so large that just forming the orbits is unfeasible. Remind that the size of the acting group S_V is $v!$ in the case of our t-designs.

Not each renaming of the points of $\mathcal{D}_1$ produces a design different from $\mathcal{D}_1$. These renamings are called automorphisms of $\mathcal{D}_1$. They form a group called the automorphism group $Aut(\mathcal{D}_1)$ of $\mathcal{D}_1$. Any two renamings $\mathcal{D}_1 \longrightarrow \mathcal{D}_2$ that only differ by an automorphism will give the same design $\mathcal{D}_2$. So, the number of different t-designs defined on V and isomorphic to $\mathcal{D}_1$ is the index of $Aut(\mathcal{D}_1)$ in S_V.

The automorphism group as an abstract group is a poor invariant for designs [20] because usually many designs will have the same automorphism group. But it turns out that taking into account the action will turn the automorphism group into the clue to the solution of the isomorphism problem in many cases.

The basic observation is that if two designs $\mathcal{D}_1$ and $\mathcal{D}_2$ have the same automorphism group A then any permutation ϕ mapping $\mathcal{D}_1$ onto $\mathcal{D}_2$ has to normalize A. This group theoretic term means that such a ϕ has the property that $\phi^{-1}A\phi = A$. All such elements ϕ form a group called the normalizer $N_{S_V}(A)$ of A in S_V.

So, instead of searching all permutations of V for an isomorphism of $\mathcal{D}_1$ onto $\mathcal{D}_2$ only the usually much smaller group $N_{S_V}(A)$ suffices. Indeed, as we mentioned above only such permutations that lie in different cosets of A in $N_{S_V}(A)$ need be considered. On the other hand no permutation outside of A will map $\mathcal{D}_1$ onto itself by the definition of A. So, each orbit of $N_{S_V}(A)$ on the set of designs with automorphism group A has the same length $l = |N_{S_V}(A)|/|A|$.

Obviously, dividing the number of designs with automorphism group A by this length l will give the number of orbits, that is the number of isomorphism types. In order to get a set of representatives for these types one still has to form the orbits.

The method formulated by Kramer and Mesner assumes a group of automorphisms A of the desired t-designs and the designs are assembled from orbits of this group on the set of all orbits of A on the k-element subsets. The algebraic approach to a solution of the isomorphism relies on the following observation.

Lemma 2.1: *Let A be the group of all automorphisms of each of the designs in a set Δ. Then any permutation mapping a design $\mathcal{D}_1 \in \Delta$ onto a design $\mathcal{D}_2 \in \Delta$ has to lie in the normalizer $N_{S_V}(A)$ of A in the symmetric group on the underlying point set V.*

Proof. Assume $\mathcal{D}_1, \mathcal{D}_2 \in \Delta$ and $\mathcal{D}_1^g = \mathcal{D}_2$ for some $g \in G$. For each $a \in Aut(\mathcal{D}_1)$ we have $g^{-1}ag \in Aut(\mathcal{D}_2)$, since

$$\mathcal{D}_2^{g^{-1}ag} = \mathcal{D}_1^{ag} = \mathcal{D}_1^g = \mathcal{D}_2.$$

But $Aut(\mathcal{D}_1) = A = Aut(\mathcal{D}_2)$ such that $g^{-1}ag$ lies in A. This shows that $g \in N_{S_V}(A)$.   □

There are two extremal cases to consider. The easy extreme case is the situation where A is a maximal subgroup different from the alternating group of S_V. In this case A must be equal to its normalizer such that all designs with automorphism group A must be pairwise non-isomorphic. Group theorists have determined all maximal subgroups of the symmetric groups such that all these easy cases can be easily identified. So, in many cases the author and his research group could classify large numbers of t-designs up to isomorphism without any additional work. One such case was the first 7-designs with small parameters. There are exactly 4996426 isomorphism types of $7-(33,8,10)$ designs with automorphism group $P\Gamma L(2,32)$.

The difficult extreme case appears if A consists only of the identity. No satisfying solution has been found in general for this case.

A general theory will be developed to cope with the ordinary cases. A first attempt in the spirit of Burnside's table of marks from the beginning of the 20th century gives at least a theoretically satisfying answer. But for the problems in question this solution is still infeasible. It requires for a given group of automorphisms the knowledge of all overgroups in the symmetric group. If we look at up to 40 points this knowledge is rarely available.

We therefore go back to a technique already known to Jordan in the end of the 19th century. Combined with a recursive splitting of the problem into cases with growing known automorphism groups it in each case allows to reduce the search for automorphisms to the normalizer of a prime power group. The overgroups of the given group of automorphisms A can be directly constructed. No catalogue of previously computed overgroups from group theoretic sources is needed. Of course, since this is an algorithmic approach, limits will be reached soon. But it turned out that finding the designs with large t that we wanted to classify yielded more severe limitations.

Some special cases can be exploited to avoid extensive computations. But in general the computer package GAP is used to solve several of the group theoretic problems involved.

An automatic use of this theory as part of our software package DISCRETA is under development. The author thanks his collaborators A. Betten, E. Haberberger and A. Wassermann for their enthusiasm in contributing to this system and also his colleague A. Kerber for fruitful discussions. It should be noticed that much of the theory was stimulated

by the practical problems that came out of the project of constructing
t-designs with large *t* up to isomorphism.

3. Moebius Inversions

Usually we cannot directly apply the Lemma 2.1 to designs obtained
by prescribing a group of automorphisms A. Some of these designs may
have an automorphism group that is strictly bigger than A. So, in a first
step one should take each proper overgroup B of A in S_V and construct
the designs that admit B. Each of these designs has to be removed from
those obtained by prescribing A. The remaining set Δ then can be dealt
with using the Lemma. The algorithmic version is obvious by Lemma
2.1.

If one is only interested in the number of isomorphism types one can
use the principle of inclusion and exclusion to the mere numbers of solutions obtained for each subgroup between A and S_V. This is the back-
ground of Burnside's table of marks [15]. The Lemma has been used by
the author to give a general constructive proof in [32]. The inclusion and
exclusion computation can be done by a matrix inversion. First form
the inclusion matrix ζ of the lattice of overgroups of A in S_V , where
$\zeta(U_1, U_2) = 1$ if and only if $U_1 \leq U_2$ and $\zeta(U_1, U_2) = 0$ else. If the U_i are
sorted so that $U_i \leq U_j$ implies $i \leq j$ then the matrix is upper triangular.
So, it is invertible with inverse matrix μ.

If $a(B)$ denotes the number of designs with full automorphism group
B then summing over all these numbers for all overgroups including B
yield the number of designs $f(B)$ invariant under B. So, multiplying the
row of B of the inclusion matrix with the vector a of the numbers $a(B_i)$
of all B_i results in $f(B)$.

As a matrix equation this reads
$$\zeta \times a = f$$

such that multiplying with the inverse matrix μ from the left results
in the requested vector a. The entries $a(B_i)$ still have to be divided
by the the length of each orbit of $N_{S_V}(B_i)$ which is the index of B_i in
$N_{S_V}(B_i)$. This is achieved by multiplying a with a matrix n^{-1} that has
entry $\frac{1}{|N_{S_V}(B_i)/B_i|}$ on the i-th row of the diagonal and zero elsewhere.
So, the equation reads
$$n^{-1} \times \mu \times f = I$$

where I is the vector of numbers of isomorphism types.

Figure 3.1. Schmalz's example

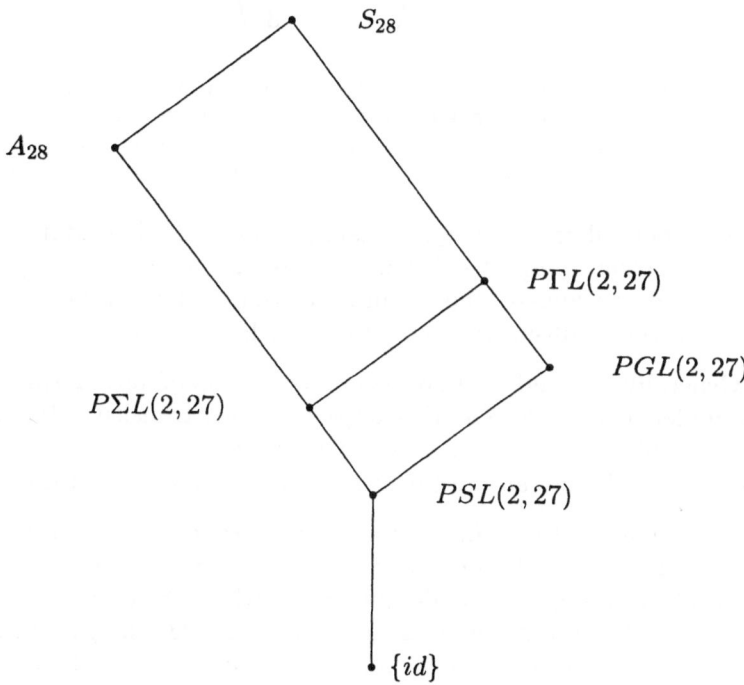

We demonstrate this computation by an example from Schmalz[39]. Among other results he found the numbers of 5-$(28, 6, 9)$ designs where A is one of the groups $PSL(2, 27), PGL(2, 27), P\Sigma L(2, 27)$, $P\Gamma L(2, 27)$ appearing in the following equation. In all these cases $N_{S_{28}}(A) = P\Gamma L(2, 27)$.

$$
\begin{pmatrix}
\frac{1}{6} & 0 & 0 & 0 \\
0 & \frac{1}{3} & 0 & 0 \\
0 & 0 & \frac{1}{2} & 0 \\
0 & 0 & 0 & 1
\end{pmatrix}
\cdot
\begin{pmatrix}
1 & -1 & -1 & 1 \\
0 & 1 & 0 & -1 \\
0 & 0 & 1 & -1 \\
0 & 0 & 0 & 1
\end{pmatrix}
\cdot
\begin{pmatrix}
68976931 \\
369 \\
31 \\
3
\end{pmatrix}
$$

$$
=
\begin{pmatrix}
11496089 \\
122 \\
14 \\
3
\end{pmatrix}
$$

68976931   is the number of designs invariant under $PSL(2, 27)$,

369   is the number of designs invariant under $PGL(2, 27)$,

31   is the number of designs invariant under $P\Sigma L(2, 27)$,

3   is the number of designs invariant under $P\Gamma L(2, 27)$.

No other subgroup of S_{28} containing any such A is admitted as the automorphism group of a 5-(28, 6, 9) design. In fact, only S_{28} and A_{28} had to be considered. The number of isomorphism types is therefore given by the resulting vector on the right hand side.

Generally, A_V and S_V never occur as a group of automorphisms of an incomplete design, since both groups are highly transitive. If only one k-element subset is contained in a design with a k-transitive automorphism group then already all k-element subsets are contained in the design.

So, starting with $PSL(2, p^f)$ as a given group of automorphisms only the overgroups strictly smaller than A_V and S_V may occur as full automorphism groups of the designs found with $PSL(2, p^f)$. It is known that $P\Gamma L(2, p^f)$ is a maximal subgroup of S_V [36] and its intersection $P\Sigma L(2, p^f)$ with A_V is a maximal subgroup in A_V. Thus, only the interval $[PSL(2, p^f), P\Gamma L(2, p^f)]$ of the subgroup lattice has to be considered. This is isomorphic to the subgroup lattice of the factor group $P\Gamma L(2, p^f)/PSL(2, p^f)$ which itself is isomorphic to $C_f \times C_2$. So, the Moebius inversion can be done easily in infinitely many cases. In fact for odd f it coincides with the number theoretic inversion on the lattice of divisors of $2f$. It is remarkable that we can describe the solution to the problem of counting isomorphism types in this way though we do not yet know the number of designs that have to be classified. We want to keep this feature also in our more advanced methods as far as possible. With respect to a general theory we remark that this approach does not even make use of the fact that we want to classify designs.

There has been some effort to reduce the knowledge needed on the subgroup lattice in the Moebius inversion approach. So, Rota and Smith

[38] noticed that only stabilizers of the permuted objects have to be considered. But it is not clear how the stabilizers can be found. So, this nice formalism can only be applied in the case of small point sets, because the subgroup lattices required are not completely known for larger sets.

We develop a tool that can determine the required overgroups without falling back upon lists of subgroups. In addition, this tool does not always require the full automorphism group of the constructed t-designs. The technique dates back to Jordan [26]. It only had to be looked at from a different point of view for our purpose. With respect to our goal of deciding upon isomorphisms this is very helpful. It allows to decide upon isomorphism without knowing the full automorphism group of the designs. But often a knowledge of the full automorphism group of a design is of interest. So, some attempt is made to also construct large parts of the full automorphism group from the given smaller one. In many important cases the resulting group then is the full automorphism group.

The approach we follow is of an algorithmic nature, while the table of marks approach above has the flavour of a complete mathematical solution.

Theorem 3.1: *Let $\mathcal{D}_1, \mathcal{D}_2$ be two isomorphic designs admitting some group of automorphisms A and $g \in S_V$ such that $\mathcal{D}_1^g = \mathcal{D}_2$. Let a Sylow subgroup P of A be already a Sylow subgroup of $Aut(\mathcal{D}_2)$. Then $\mathcal{D}_1^n = \mathcal{D}_2$ for some $n \in N_{S_V}(P)$.*

Proof. Since $P^g \leq Aut(\mathcal{D}_1)^g = Aut(\mathcal{D}_1^g) = Aut(\mathcal{D}_2)$ and also $P \leq Aut(\mathcal{D}_2)$, there is some $h \in Aut(\mathcal{D}_2)$ such that $P^g = P^h$ by the Sylow Theorem. Then $gh^{-1} = n \in N_{S_V}(P)$ and $g = nh$. Therefore $\mathcal{D}_2 = \mathcal{D}_1^g = \mathcal{D}_1^{nh}$ and $\mathcal{D}_1^n = \mathcal{D}_2^{h^{-1}} = \mathcal{D}_2$. $\qquad\square$

Assume our prescribed automorphism group A contains a Sylow subgroup P of S_V. Then by the theorem two designs $\mathcal{D}_1, \mathcal{D}_2$ having A as an automorphism group may be mapped one onto the other by a permutation g only if already some $n \in N_{S_V}(P)$ maps $\mathcal{D}_1$ onto $\mathcal{D}_2$. If even $N_{S_V}(P)$ is contained in A then any designs fixed by A are pairwise not isomorphic. That means that in this case the solutions of the system of linear equations given by the Kramer-Mesner matrix form a full set of representatives from all isomorphism types of designs admitting automorphism group A.

As an example we consider the original problem from the Kramer-Mesner paper, see also [24], of 2-(13, 5, 45) designs. There the group A was a semidirect product of the cyclic group $P = C_{13}$ by its group of automorphisms of order 6, acting on 13 points. So, A contains a Sylow 13-subgroup of the full symmetric group S_{13}. By Theorem 3.1, two designs invariant under A may only be mapped one onto the other by some permutation g if g lies in $N_{S_V}(P)$. This normalizer is the semidirect product of C_{13} by its full automorphism group isomorphic to C_{12}. This semidirect product usually is called the holomorph $Hol(C_{13})$ of C_{13}. The index of A in $Hol(C_{13})$ is 2 such that designs invariant under A but not invariant under $Hol(C_{13})$ are grouped into isomorphic pairs under the action of this normalizer. Those invariant under $Hol(C_{13})$ are pairwise non-isomorphic. More generally, one can reduce the Burnside matrix of S_V to that of $N_{S_V}(P)$ in these cases.

Theorem 3.2: *Let P be a prime power subgroup of S_V and let t-(v, k, λ) be an admissible set of parameters of t-designs. Let Δ be the set of t-(v, k, λ) designs $\mathcal{D}$ of which the automorphism group contains P as a normal Sylow subgroup. Then the Burnside matrix of $N_{S_V}(P)$ suffices to determine the numbers of different isomorphism types in Δ weighted by their automorphism groups.*

Instead of the obvious proof we demonstrate the use of the Theorem by the problem mentioned by Kramer and Mesner.

The number of isomorphism types of block-designs with these automorphism groups can be obtained by Moebius-inversion from the numbers of invariant designs as we showed above.

$$
\begin{pmatrix} \frac{1}{6} & 0 & 0 & 0 \\ 0 & \frac{1}{3} & 0 & 0 \\ 0 & 0 & \frac{1}{2} & 0 \\ 0 & 0 & 0 & 1 \end{pmatrix} \cdot \begin{pmatrix} 1 & -1 & -1 & 1 \\ 0 & 1 & 0 & -1 \\ 0 & 0 & 1 & -1 \\ 0 & 0 & 0 & 1 \end{pmatrix} \cdot \begin{pmatrix} 136876801 \\ 24643 \\ 890 \\ 28 \end{pmatrix}
$$

$$
= \begin{pmatrix} 22825216 \\ 8205 \\ 431 \\ 28 \end{pmatrix}
$$

Figure 3.2.   Kramer and Mesner's example

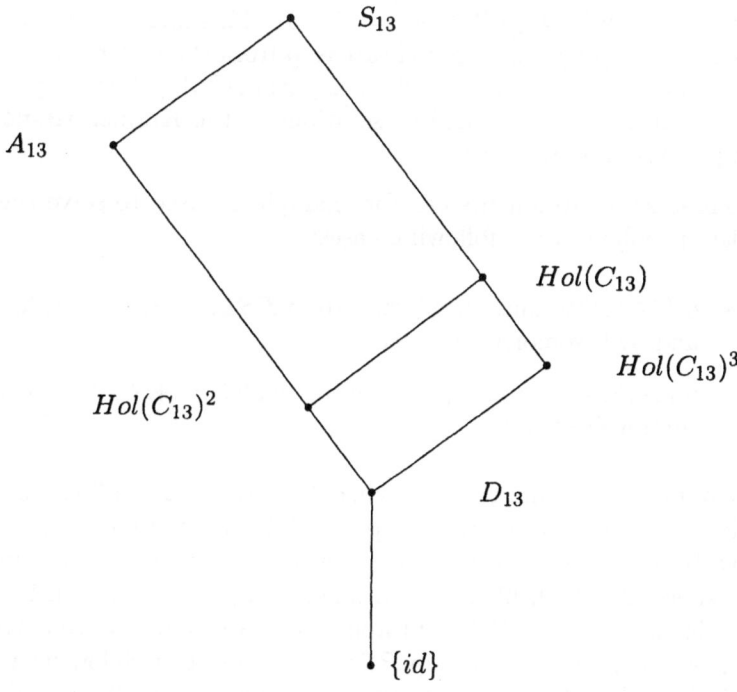

136876801   is the number of designs invariant under D_{13},
    24643   is the number of designs invariant under the
            subgroup $Hol(C_{13})^3$ of index 3 in $Hol(C_{13})$,
      890   is the number of designs invariant under the
            subgroup $Hol(C_{13})^2$ of index 2 in $Hol(C_{13})$,
       28   is the number of designs invariant under $Hol(C_{13})$.

Taking $A = PGL(3,5)$, a group of order $(5^2 + 5 + 1)(5^2 - 1)(5 - 1)5^3$ acting on the set of 31 one-dimensional subspaces of $V(3,5)$ as a prescribed automorphism group we found 8-$(31,10,93)$ and 8-$(31,10,100)$ designs. To solve the isomorphism problem for these designs we used that A contains a Sylow 31-subgroup P of S_{31}. The normalizer of P in the full symmetric group is the holomorph of P, i. e. the semidirect product of P with its automorphism group. This normalizer is not contained in A but $|N_{S_{31}}(P) \cap A| = 3 \times 31$. This intersection has 10 right cosets in $N_{S_{31}}(P)$. For representatives g from the nontrivial right cosets we formed $\langle A, A^g \rangle$ and in each case obtained A_{31}. Thus, by the above theory, all designs obtained as solutions of the Kramer-Mesner system are pairwise non-isomorphic.

These same arguments can for example be used to solve the isomorphism problem in the following cases:

- 6-$(22,8,60)$, automorphism group $PSL(2,19) + +$, 2296 solutions and 574 isomorphism types,

- 6-$(19,7,4)$, automorphism group $Hol(C_{17}) + +$, 2 solutions and 1 isomorphism type.

An important infinite series where Theorem 3.2 can be applied is provided by the the projective groups $PGL(2,p)$ for primes p. These groups have been used to construct infinite series of 5-designs and in a series of papers [21], [22], [23], [37] Grannel, Griggs and Mathon have shown that in many cases $PSL(2,p)$ appears as a group of automorphisms of Steiner 5-systems. The group $PSL(2,p)$ contains a Sylow p-subgroup P of S_{p+1} and $PGL(2,p)$ contains the required normalizer of P in S_{p+1}. So, any designs admitting $PGL(2,p)$ are pairwise non-isomorphic. For designs not admitting $PGL(2,p)$ but $PSL(2,p)$ only the action of

$$PGL(2,p)/PSL(2,p) \cong C_2$$

on the set of solutions of the Kramer-Mesner system has to be taken into account. This explains why in [21], [22], [23], and [39] a representative from the non trivial coset of $PSL(2,p)$ in $PGL(2,p)$ already suffices to distinguish between the isomorphism types of Steiner systems constructed with automorphism group $PSL(2, p)$.

In some cases one can show that P must be a Sylow subgroup of the full automorphism groups of the designs invariant under A even if P is not a Sylow subgroup of the full symmetric group. A computational approach would be to just prescribe every possible p-group in S_V that contains P as a maximal subgroup. But similar to a result of Aschbacher [3] on symmetric designs there is a result for general t-designs which in minimal situations can be used to prove that P must be a Sylow subgroup of the full automorphism groups of the designs.

Theorem 3.4: *Let $\mathcal{D}$ be a t-(v, k, λ) design and $g \in Aut(\mathcal{D})$ with v_g fixed points. Let $o(g) = p$ be a prime such that $\lambda < p$ and $k < p + t$. Then if $t \le v_g$ all blocks of $\mathcal{D}$ consisting of fixed points of g form a t-(v_g, k, λ) design.*

Proof Let $C_V(g)$ be the set of fixed points of g. Let $T \subseteq C_V(g)$ and $\mathcal{D}_T = \{B \mid |b \in \mathcal{D}, \ T \subseteq B\}$. Then g acts on $\mathcal{D}_T$ which is of size λ. By the assumptions $\lambda < p$ such that g leaves each $B \in \mathcal{D}_T$ invariant. Since at least the t points of T are fixed, g has to act on the remaining point sets $B \backslash T$ of size $k - t$. Again the assumption $k - t < p$ implies that g fixes also all these point sets. Thus, all blocks from $\mathcal{D}_T$ are subsets of $C_V(g)$. $\square$

Usually, one is interested in designs with small λ such that $\lambda < p$ is often satisfied. Also, $k - t < p$ is a restriction that will be fulfilled in many cases. Only the assumption that g has at least t fixed points seems to be very special. So, we use the result to produce contradictory situations in a proof. If it is known that no t-(v_g, k, λ) design can exist for all $v - p > v_g \ge t$ then it follows that $v_g < t$. So, in this case g consists of as many p-cycles as possible, say c, and a few additional fixed points.

One such situation appears if $t < p$ and p does not divide $\binom{v}{t}$. Then g in its action on all t subsets must have some invariant t-subset T which because of $t < p$ must consist of fixed points of g.

We say that a prime p is *large* for t-(v, k, λ) if $t < p$, $k - t < p$, $\lambda < p$, $v < p^2$.

Theorem 3.5: *Let p be large for t-(v, k, λ) and v chosen minimal among these parameter sets such that $p \mid |Aut(\mathcal{D})|$ for some t-(v, k, λ)*

design $\mathcal{D}$. *Then for this* $\mathcal{D}$ *all Sylow* p*-subgroups of* $Aut(\mathcal{D})$ *have order* p.

Proof By Theorem 3.4 an automorphism g of order p of $\mathcal{D}$ has only $c < t$ fixed points. All other points lie in cycles of length p. Suppose, $Aut(\mathcal{D})$ has a subgroup P of order p^2. We show that then no element other than the identity of P leaves any block of $\mathcal{D}$ invariant. Otherwise, some element g of order p leaves such a block B invariant. Then g acts on B, a set of k elements. Since the number of fixed points of g in B is smaller than t, and $t < p$ then g must not fix any t-subset of B such that p divides $\binom{k}{t}$. But then p does not divide $b = \binom{v}{t}/\binom{k}{t}\lambda$, since p divides the enumerator at most to the first power. Then there must be orbits of P on the set of blocks whose length is not a multiple of p. But each orbit length of P must be a power of p, such that there exists a block B which is invariant under P. Since $2p > k$, the symmetric group on B contains no subgroup of order p^2. So there is some element g of order p in P acting trivially on B. This contradicts Theorem 3.4. Now each orbit of P on $\mathcal{D}$ has length p^2 such that the number of blocks b is divisible by p^2. But since $t < p$ the number $\binom{v}{t}$ is a product of t subsequent factors of which only one may be a multiple of p. Thus, we have a final contradiction.

Let P be a group of automorphisms of order p^2 of $\mathcal{D}$. Then if p^2 does not divide v there exists an orbit of size smaller than p^2. We assume that P has no fixed points. Then there exists an orbit of size p such that there exists a point stabilizer of order p. Any non-identity element of this fixes p elements which is impossible as before.   □

An example for this situation occurs with the $6-(14,7,4)$ designs admitting automorphism group C_7, see [18]. Here 7 is large such that C_7 has to be a Sylow subgroup of the full automorphism group. In fact, C_7 already is the full automorphism group in these cases. So, to decide the isomorphism problem for these designs it suffices to consider the action of the normalizer of C_7 in S_{14} on the set of designs. The Kramer Mesner matrix has size 429×492 and there are exactly 168 solutions. Since this number is two times the index of C_7 in its normalizer which is an extension of its centralizer $C_7 \times C_2$ by its automorphism group C_6, they form just two orbits under that normalizer. So, there are just two isomorphism types of designs of this kind, confirming the result of Tayfeh and Khosrovshahi.

If we let M_{11} act diagonally on two copies of the underlying set of 11 points, formally we have the action of $M_{11} \times < Id_2 >$, then we get four solutions for a 5-$(22, 7, 40)$ design with this prescribed automorphism group. The prime $p = 11$ is a large prime for this parameter set. Also this v is minimal in the sense of the Theorem 3.5. Therefore any two of the four designs may only be isomorphic if some element from the normalizer of a Sylow 11-subgroup of the automorphism group maps one design onto the other. Adding the permutation of order two which interchanges the two actions of M_{11} we get a group which is larger by a factor of 2. This group is still an automorphism group of two of the designs. The new element obviously lies in the normalizer of a Sylow 11-subgroup. It interchanges two of the designs found. The remaining two designs admit the interchange of the two actions as an automorphism. These designs are not isomorphic because the normalizer of the Sylow subgroup contains no element mapping one design onto the other. It is also possible to distinguish them by their intersection numbers.

The construction of overgroups which may leave some of the designs with prescribed group A invariant relies on the following simple observation.

Lemma 3.6: *Let $\mathcal{D}_1, \mathcal{D}_2$ be two isomorphic designs admitting some group of automorphisms A and let $g \in S_V$ such that $\mathcal{D}_1^g = \mathcal{D}_2$. Then $B = < A, A^g >$ is a group of automorphisms of $\mathcal{D}_2$.*

Proof Since $A^g \leq Aut(\mathcal{D}_1)^g = Aut(\mathcal{D}_1^g) = Aut(\mathcal{D}_2)$, both A and A^g lie in $Aut(\mathcal{D}_2)$. $\qquad \qquad \square$

An algorithm now is easily obtained from the following consequence of these results.

Theorem 3.7: *Let Δ be the set of t-(v, k, λ)-designs invariant under a prescribed group A. Let P be a Sylow p-subgroup of A. Then after removing every design from Δ that is invariant under any subgroup B such that either*

$$A < B < N_{S_v}(A) \ or$$

$$B = < A, A^n > \ for \ some \ n \in N_{S_v}(P) \ with \ A^n \neq A$$

two designs in Δ are isomorphic if and only if some element from $N_{S_v}(A)$ maps one onto the other.

Proof If a Sylow subgroup P of A is not yet a Sylow subgroup of $Aut(\mathcal{D})$ for some $\mathcal{D} \in \Delta$ then P is a maximal subgroup of some bigger

p-group Q which lies in $Aut(\mathcal{D})$. By some well known results on finite p-groups, see [25] then P is a normal subgroup of Q. So, $Q = <P, n>$ for some $n \in N_{S_v}(P)$. If $n \in N_{S_v}(A)$ then $\mathcal{D}$ is removed from Δ. If n is not in $N_{S_v}(A)$ then

$$A \; < \; <A, A^n> \; \leq \; <A, n> \; \leq \; <A, Q> \; \leq \; Aut(\mathcal{D}).$$

So, also in this case $\mathcal{D}$ is removed from Δ. Thus, after removing each such $\mathcal{D}$ for the remaining designs P is a Sylow subgroup of their full automorphism group and the previous Theorem applies.   □

The algorithm first constructs the groups B that appear in the theorem and then finds the designs invariant under such a group B. Since each such B is bigger than A, the problem of finding the designs with such a group of automorphisms is easier to solve than it had been for A initially. In many cases, such groups B will not be admitted at all as a group of automorphisms because they are rather big groups. Compared to the table of marks description above we here do not need to know all overgroups of the given group A. The subgroup $N_{S_v}(P)$ from which we have to take the conjugating elements n contains only a small fraction of all initially needed permutations. We usually take for P a large Sylow subgroup of A, often of a large prime order. Then it is easy to describe the normalizer. Also, the conjugating elements n may be chosen from different cosets of $N_{S_v}(P) \cap N_{S_v}(A)$ in $N_{S_v}(P)$. The last remark then concerns the designs removed. Of course, also for them the isomorphism problem has to be solved in just the same way. But one has to restrict to representatives B from the conjugacy classes of subgroups. This simplifies the computation but requires some further group theoretic checking.

In [24] this approach has been used to classify the 3-$(20, 4, 1)$ designs which admit the automorphism group A of the Dodecahedron as a group of automorphisms up to isomorphism. In that case a weaker version of the last Theorem was sufficient because a Sylow 5-subgroup of A already is a Sylow 5-subgroup of the full automorphism groups of all these Steiner quadruple systems. This could be shown by checking that no group of order 25 is allowed as a group of automorphisms of any 3-$(20, 4, 1)$ design.

The new Theorem will be used for an algorithmic determination of isomorphism types of t-designs in a new version of DISCRETA which also employs GAP [19] for some group theoretic computations.

In many cases, the recursive application of this Theorem to the prescribed groups B and their designs allowed to construct the full automorphism group of the designs. In particular, prescribing the automorphism group of the icosahedron which is isomorphic to A_5 as a group of automorphisms of a 5-$(12, 6, 1)$ Witt design and P a Sylow 5-subgroup ends up with the Mathieu group M_{12}. As well prescribing $PSL(2, 23)$ as a group of automorphisms of a 5-$(24, 8, 1)$ Witt design and P a Sylow 11-subgroup ends up with the Mathieu group M_{24}.

4. Extensions of designs

Some constructions of t-designs from smaller t-designs admit to replace the smaller ones by isomorphic copies. So, one can produce a large number of new t-designs for which the isomorphism problem has to be solved. We have examined the extension method of van Leijenhorst and Tran van Trung [41], [40] in this respect. Hopefully, other constructions as for example the large set recursions by Ajoodani-Namini and Khosrovshahi [1], [28] could be handled in a similar way.

We shortly review the idea and results of this approach. Let $V = \{1, \cdots, v\}$ and $V^+ = V \cup \{v + 1\}$. Then a

$$t\text{-}(v, k - 1, \lambda) \text{ design } \mathcal{D}_1 = (V, \mathcal{B}_1)$$

and a

$$t\text{-}(v, k, \lambda \tfrac{(v-k+1)}{(k-t)}) \text{ design } \mathcal{D}_2 = (V, \mathcal{B}_2)$$

extend to a

$$t\text{-}(v + 1, k, \lambda \tfrac{(v-t+1)}{(k-t)}) \text{ design } \mathcal{D} = (V^+, \mathcal{B})$$

where $\mathcal{B}$ consists of $\mathcal{B}_2$ and all $B \cup \{v + 1\}$ for $B \in \mathcal{B}_1$, denoted as $\mathcal{B} = \mathcal{B}_2 \cup \mathcal{B}_1 * \{v + 1\}$.

Obviously, the designs $\mathcal{D}_1, \mathcal{D}_2$ may be replaced by others with the same parameters. We may start with only the two designs $\mathcal{D}_1, \mathcal{D}_2$ and replace them by isomorphic copies. If we want to solve the isomorphism problem for two designs constructed in this way we may first rename the points so that the first operand is the given design $\mathcal{D}_1$. So, the isomorphism problem may be restricted to extensions $\mathcal{D}(\pi) = (V^+, \mathcal{B}(\pi))$ with $\mathcal{B}(\pi) = \mathcal{B}_1 * \{v + 1\} \cup \mathcal{B}_2^\pi$ where π is a permutation on V applied to the blocks in $\mathcal{B}_2$.

Now, there are $v!$ designs to classify corresponding to the different permutations π on V.

Let on the other hand some permutation α_1 map $\mathcal{D}(\pi_1)$ onto $\mathcal{D}(\pi_2)$ and fix $v+1$. Since $\mathcal{D}_1$ is obtained from $\mathcal{D}(\pi_1)$ and from $\mathcal{D}(\pi_2)$ by forming the derived design at $v+1$, α_1 is an automorphism of $\mathcal{D}_1$. Also the residual designs at $v+1$, are mapped one onto the other such that π as a permutation on V is an automorphism of $\mathcal{D}_2$.

Theorem 4.1: *Let $\mathcal{D}_1$ be a t-$(v, k-1, \lambda)$ design with automorphism group A_1 and $\mathcal{D}_2$ be a t-$(v, k, \lambda\frac{(v-k+1)}{(k-t)})$ design with automorphism group A_2 both defined on the same point set V. Then there exists an isomorphism*

$$\phi : \mathcal{D}(\pi_1) \longrightarrow \mathcal{D}(\pi_2)$$

fixing $v+1$ if and only if

$$A_1 \pi_1 A_2 = A_1 \pi_2 A_2$$

in S_V.

Proof If α_1 is an automorphism of $\mathcal{D}_1$ and α_2 is an automorphism of $\mathcal{D}_2$ then π and $\alpha_1 \pi \alpha_2$ define isomorphic extensions. Each such isomorphism as a permutation on V^+ fixes the new point $v+1$.

Let on the other hand some permutation ϕ map $\mathcal{D}(\pi_1)$ onto $\mathcal{D}(\pi_2)$ and fix $v+1$. Since $\mathcal{D}_1$ is obtained from $\mathcal{D}(\pi_1)$ and from $\mathcal{D}(\pi_2)$ by forming the derived design at $v+1$, ϕ restricted to V is an automorphism α_1 of $\mathcal{D}_1$. Also the residual designs at $v+1$ are mapped one onto the other such that $\pi_1 \phi \pi_2^{-1}$ restricted to V is an automorphism α_2 of $\mathcal{D}_2$. So, $\alpha_1 = \pi_1 \alpha_2 \pi_2^{-1}$ such that

$$\alpha_2^{-1} \pi_1^{-1} \alpha_1 = \pi_2^{-1}$$

and

$$\alpha_1 \pi_1 \alpha_2^{-1} = \pi_2.$$

This means that

$$A_1 \pi_1 A_2 = A_1 \pi_2 A_2.$$

$\square$

Corollary 4.2: *Let A_1 be the automorphism group of a_1 isomorphism types of t-$(v, k-1, \lambda)$ designs and A_2 be the automorphism group of a_2 isomorphism types of t-$(v, k, \lambda\frac{(v-k+1)}{(k-t)})$ designs. Then there exist at least*

$$\frac{a_1 \cdot a_2}{v+1} |A_1 \backslash S_{v+1} / A_2|$$

isomorphism types of extensions to t-$(v+1, k, \lambda\frac{(v-t+1)}{(k-t)})$ *designs.*

Proof Assume that all given designs are defined on the point set V. Each orbit of the automorphism group of an extension corresponds to one extension with the added new point $v+1$. So, at most $v+1$ of the extensions with fixed added point $v+1$ may be isomorphic to the same design where the added point may be different from $v+1$. Thus, among all

$$a_1 \cdot a_2 |A_1 \backslash S_{v+1} / A_2|$$

isomorphism classes with fixed added point $v+1$ there are at least

$$\frac{a_1 \cdot a_2}{v+1} |A_1 \backslash S_{v+1} / A_2|$$

general isomorphism types. □

The numbers of double cosets appearing in Corollary 4.2 actually are rather large. So, we list some examples for automorphism groups where many designs had been found already, see [16].

Double Cosets	Number
$PSL(2,23) \backslash S_{24} / PSL(2,23)$	16,828,376,982,435,832
$PSL(2,23) \backslash S_{24} / PGL(2,23)$	8,414,188,491,217,916
$PGL(2,23) \backslash S_{24} / PGL(2,23)$	4,207,094,330,061,055
$PGL(2,25) \backslash S_{26} / PGL(2,25)$	1,657,180,580,754,274,540
$P\Gamma L(2,25) \backslash S_{26} / P\Gamma L(2,25)$	414,295,145,235,066,413
$PGL(2,25) \backslash S_{26} / P\Gamma L(2,25)$	828,590,290,377,152,694
$AGL(1,25p) \backslash S_{26} / P\Gamma L(2,25)$	10,771,673,642,332,865,588
$AGL(3,3) \backslash S_{27} / AGL(3,3)$	118,397,102,441,920,363
$ASL(3,3) \backslash S_{27} / AGL(3,3)$	236,794,204,702,349,473
$ASL(3,3) \backslash S_{27} / ASL(3,3)$	473,588,409,404,698,946
$AGL(3,3) \backslash S_{27} / P\Gamma L(2,25)+)$	1,150,819,833,931,867,436
$Sp(6,2)_{28} \backslash S_{28} / Sp(6,2)_{28}$	144,708,746,195,525,184

The numbers of double cosets are computed by SYMMETRICA, using Redfield's cap-product and cycle indicator polynomials, see Kerber's book [27].

As an application we give some figures on the number of isomorphism types of 8-designs on 28 points. By prescribing $ASL(3,3)$ we obtained from DISCRETA 8-$(27,13,\lambda)$ designs. The numbers of isomorphism types are collected in the following table. Actually, the number of solutions is twice as big because $AGL(3,3)$ is the normalizer in S_{27} and it has orbits of length 2 on the set of solutions.

Each 8-$(27, 13, \lambda)$ design with automorphism group $ASL(3,3)$ extended together with its supplementary 8-$(27, 14, \lambda')$ design yields
$$|SL(3,3)\backslash S_{27}/ASL(3,3)|/28 \geq 16913871764453533$$

isomorphism types of 8-$(28, 14, \lambda + \lambda')$ designs with various groups of automorphisms. Exactly one of them is a 9-$(28, 14, \lambda)$ design by Alltop's construction. This one has automorphism group $ASL(3,3)+$.

Parameter set	Constructed by	No. of isomorphism types
8-$(27, 11, 432)$	$ASL(3,3)$	1
8-$(27, 12, 1296)$	$ASL(3,3)$	4336
8-$(27, 12, 1932)$	$ASL(3,3)$	2110899
8-$(27, 13, 3204)$	$ASL(3,3)$	538218
8-$(27, 13, 3240)$	$ASL(3,3)$	618421
8-$(27, 13, 4608)$	$ASL(3,3)$	235678089
8-$(27, 13, 5076)$	$ASL(3,3)$	many
8-$(27, 13, 5148)$	$ASL(3,3)$	many
8-$(28, 13, 5832)$	$ASL(3,3)+$	$\geq 5 \times 10^9$
8-$(28, 13, 7080)$	$ASL(3,3)+$	many
8-$(28, 13, 7128)$	$ASL(3,3)+$	many
8-$(28, 14, 10680)$	8-$(27, 13, 3204)$	$\geq 4.6 \times 10^{27}$
8-$(28, 14, 10800)$	8-$(27, 13, 3240)$	$\geq 6.1 \times 10^{27}$
8-$(28, 14, 14040)$	$ASL(3,3)+$	≥ 1
8-$(28, 14, 15360)$	8-$(27, 13, 4608)$	$\geq 8.4 \times 10^{32}$
8-$(28, 14, 16920)$	8-$(27, 13, 5076)$	$\geq 1.6 \times 10^{16}$
8-$(28, 14, 17160)$	8-$(27, 13, 5148)$	$\geq 1.6 \times 10^{16}$
8-$(28, 14, 18600)$	$ASL(3,3)+$	≥ 1
9-$(28, 14, 3204)$	$ASL(3,3)+$	538218
9-$(28, 14, 3240)$	$ASL(3,3)+$	618421
9-$(28, 14, 4608)$	$ASL(3,3)+$	235678089
9-$(28, 14, 5076)$	$ASL(3,3)+$	many
9-$(28, 14, 5148)$	$ASL(3,3)+$	many

These designs might be contained in designs with an even larger value of t.

- 9-$(28, 14, 3204)$ could be residual of 10-$(29, 14, 1068)$ with derived 9-$(28, 13, 3204)$ and which again could be residual of 11-$(30, 14, 267)$

- 9-$(28, 14, 3240)$ could be residual of 10-$(29, 14, 1080)$

- 9-$(28, 14, 4608)$ could be residual of 10-$(29, 14, 1536)$

- 9-$(28, 14, 5076)$ could be residual of 10-$(29, 14, 1692)$

- 9-$(28, 14, 5148)$ could be residual of 10-$(29, 14, 1716)$

No design with any of these parameters has been found yet. So, 9 is the largest value of t for which a simple t-design could be found on up to 40 points.

References

[1] S. AJOODANI-NAMINI : Extending large sets of t-designs. *J. Combin. Theory A* **76**(1996), 139-144.

[2] W. O. ALLTOP: Extending t-designs. *J. Combin. Theory A* **18** (1975), 177-186.

[3] M. ASCHBACHER : On collineation groups of symmetric block designs. *J. Combin. Theory A* **11**(1971),272-281.

[4] A. BETTEN: Schnittzahlen von Designs. *Bayreuther Math. Schr.* **58** (2000).

[5] A. BETTEN, A. KERBER, A. KOHNERT, R. LAUE, A. WASSERMANN: The discovery of simple 7-designs with automorphism group $P\Gamma L(2,32)$. *Proc of AAECC 11, Springer* LN in Computer Science **948** (1995), 131-145.

[6] A. BETTEN, R. LAUE, A. WASSERMANN: Simple 7-Designs With Small Parameters. *J. Combinatorial Designs* **7** (1999), 79-94.

[7] A. BETTEN, R. LAUE, A. WASSERMANN: Some simple 7-designs. *Geometry, Combinatorial Designs and Related Structures, Proceedings of the First Pythagorean Conference.* J. W. P. Hirschfeld, S. S. Magliveras, M. J. de Resmini eds. Cambridge University Press, LMS Lecture Notes **245** (1997), 15-25.

[8] A. BETTEN, R. LAUE, A. WASSERMANN: Simple 6- and 7-designs on 19 to 33 points. *Congressus Numerantium* **123** (1997), 149-160.

[9] A. BETTEN, E. HABERBERGER, R. LAUE: Genealogy of t-designs I: ancestors and families. Preprint.

[10] A. BETTEN, M. C. KLIN, R. LAUE, A. WASSERMANN: Graphical t-Designs via polynomial Kramer-Mesner matrices. *Discrete Mathematics* **197/198** (1999), 83-109.

[11] A. BETTEN, A. KERBER, R. LAUE, A. WASSERMANN: Simple 8-designs with small parameters. *Designs, Codes and Cryptography* **15** (1998), 5-27.

[12] A. BETTEN, R. LAUE, A. WASSERMANN: Simple 8-(40,11,1440) designs. *Discrete Applied Mathematics* **95** (1999), 109-114.

[13] A. BETTEN, R. LAUE, A. WASSERMANN: A Steiner 5-Design on 36 Points. *Designs, Codes and Cryptography* **17** (1999), 181-186.

[14] A. BETTEN, S. MOLODTSOV, R. LAUE, A. WASSERMANN: Steiner systems with automorphism group $PSL(2,71)$, $PSL(2,83)$ and $P\Sigma L(2,3^5)$. *J. of Geometry* **67** (2000), 35–41.

[15] W. BURNSIDE: Theory of groups of finite order. Dover Publ. , New York, 1955, reprint of 2nd. edition 1911.

[16] http://www.mathe2.uni-bayreuth.de/ discreta/

[17] C. J. COLBOURN,J. H. DINITZ ED.: *The CRC Handbook of Combinatorial Designs.* CRC Press, 1996.

[18] Z. ESLAMI, G. B. KHOSROVSHAHI: Some New 6-$(14,7,4)$ Designs. *J. Combin. Theory A* **93** (2001), 141-152.

[19] THE GAP-TEAM Groups, algorithms, and programming. Version 4. *Lehrstuhl D für Mathematik, RWTH Aachen, Germany and School of Mathematical and Computational Sciences, U. St.Andrews, Scotland, 1997.*

[20] M. J. GRANNELL, T. S. GRIGGS: Some applications of computers in design theory. *Computers in math. research* Inst. Math. Appl. Conf. Ser. **Ser 14** Oxford Univ. Press (1988), 135-148.

[21] M. J. GRANNELL, T. S. GRIGGS, R. A. MATHON: On Steiner systems $S(5,6,48)$. *J. Comb. Math. Comb. Comput.* **12** (1992), 77-96.

[22] M. J. GRANNELL, T. S. GRIGGS, R. A. MATHON: Some Steiner 5-designs with 108 and 132 points. *JCD* **1** No. 3, (1993), 213-238.

[23] M. J. GRANNELL, T. S. GRIGGS: A Steiner system $S(5,6,108)$. *Discrete Mathematics* **125** (1995), 183-186.

[24] E. HABERBERGER, A. BETTEN, R. LAUE: Isomorphism classification of t-designs with group theoretical localization techniques applied to some Steiner quadruple systems on 20 points. *Congressus Numerantium* **142** (2000), 75-96.

[25] B. HUPPERT: *Endliche Gruppen I.* Springer Grundlehren der mathematischen Wissenschaften Bd **134**, (1967).

[26] C. JORDAN: Sur la limite de transitivite des groupes non alternes. *Bull. Soc. Math. France* **1** (1973), 40-71.

[27] A. KERBER: *Algebraic combinatorics via finite group actions.* BI-Wissenschaftsverlag Mannheim, 1991, 2.nd ed. Springer 1999.

[28] G. B. KHOSROVSHAHI, S. AJOODANI-NAMINI : Combining t-designs. *J. Combin. Theory A* **58**(1991), 26-34.

[29] G. B. KHOSROVSHAHI, M. MOHAMMAD-NOORI, B. TAYFEH-REZAIE: Classification of 6-(14,7,4) designs with nontrivial automorphism groups. preprint.

[30] E. S. KRAMER, D. M. MESNER: t-designs on hypergraphs. *Discrete Math.* **15** (1976), 263–296.

[31] D. L. KREHER, S. P. RADZISZOWSKI: The existence of simple 6−(14,7,4) designs. *J. Combin. Theor. A* **43** (1986), 237-243.

[32] R. LAUE: Eine konstruktive Version des Lemmas von Burnside. *Bayreuther Math. Schr.* **28** (1989), 111-125.

[33] R. LAUE Constructing objects up to isomorphism, simple 9-designs. Proceedings of ALCOMA99, Springer 2001, 232-260.

[34] R. LAUE, A. BETTEN, E. HABERBERGER: A new smallest simple 6-design with automorphism group A_4. *Congressus Numerantium* **150** (2001), 145-153.

[35] R. LAUE, S. S. MAGLIVERAS, A. WASSERMANN: New large sets of t-designs. *J. Combinatorial Designs* **9** (2001), 40-59.

[36] M. W. LIEBECK, C. E. PRAEGER, J. SAXL: A classification of the maximal subgroups of the finite alternating and symmetric groups. *J. Algebra* **111** (1987), 365-383.

[37] R. MATHON: Searching for spreads and packings. *Geometry, Combinatorial Designs and Related Structures, Proceedings of the First Pythagorean Conference.* J. W. P. Hirschfeld, S. S. Magliveras, M. J. de Resmini eds. Cambridge University Press, LMS Lecture Notes **245** (1997), 161-176.

[38] G.-C. ROTA, D. A. SMITH: Enumeration under group action. *Annali Scuola Normale Superiore-Pica. Classe de Scienze* **(4)4**, (1977), 637-646.

[39] B. SCHMALZ: The t-Designs with prescribed automorphism group, new simple 6-designs. *J. Combinatorial Designs* **1** (1993),125-170.

[40] TRAN VAN TRUNG: On the construction of t-designs and the existence of some new infinite series of simple 5-designs. *Arch. Math.* **47** (1986), 187-192.

[41] D. C. VAN LEIJENHORST: Orbits on the projective line. *J. Combin. Theory A* **31** (1981), 146–154.

[42] A. WASSERMANN: Finding simple *t*-designs with enumeration techniques. *J. Combinatorial Designs* **6** (1998), 79-90.

Chapter 12

FINDING DOUBLE YOUDEN RECTANGLES

N. C. K. Phillips

Department of Computer Science,
Southern Illinois University,
Carbondale, Illinois 62901, USA

nick@cs.siu.edu

D. A. Preece

School of Mathematical Sciences,
Queen Mary, University of London,
Mile End Road, London E1 4NS
and
Institute of Mathematics and Statistics,
University of Kent at Canterbury,
Canterbury, Kent CT2 7NF, UK

D.A.Preece@qmul.ac.uk, D.A.Preece@ukc.ac.uk

1.   Introduction

The role of computers in searches for Graeco-Latin squares (and indeed for hyper-Graeco-Latin squares) is well known. We now explain how we have used the computer to find examples of a less well-known type of Graeco-Latin design: the *double Youden rectangle*. We report the discovery of the first known double Youden rectangle of size 13×40; it is 13-cyclic.

2.   Definitions and Literature

Let k and v be integers with $0 < k < v$, let σ be the integer part of v/k, and let μ be the corresponding remainder, *i.e.* $\mu = v - \sigma k$. Let $\mathcal{X}$ be a set of v elements, and $\mathcal{Y}$ a set of k elements. Let $\mathbf{I}_k$ denote a $k \times k$ identity matrix, and $\mathbf{J}_k$ a $k \times k$ matrix all of whose elements are 1.

A *double Youden rectangle* (DYR) of size $k \times v$ is [1, p. 40] a $k \times v$ arrangement of the kv distinct ordered pairs x, y ($x \in \mathcal{X}$, $y \in \mathcal{Y}$) that satisfies the following conditions:

(i) each element of $\mathcal{X}$ appears exactly once per row,

(ii) each element of $\mathcal{Y}$ appears exactly once per column,

(iii) the sets of members of $\mathcal{X}$ in the columns are the blocks of a symmetric balanced incomplete block design (SBIBD, or symmetric 2-design), and

(iv) each element of $\mathcal{Y}$ appears either σ or $\sigma + 1$ times in each row, and either (a) $\mu = 1$ or (b) the sets of elements of $\mathcal{Y}$ in the rows become the blocks of size μ of an SBIBD when σ occurrences of each element from $\mathcal{Y}$ are removed from each set.

If the elements of $\mathcal{Y}$ are omitted from a DYR, the surviving rectangular arrangement of elements of $\mathcal{X}$ is what is unhappily known as a *Youden square* [6, 10].

For all $k > 3$, DYRs with $k = v - 1$ exist [9]. Otherwise, few DYRs are known. Vowden [14, 15] gave two infinite series, but other known constructions for particular sizes lack generality [5, 7, 8, 9, 11, 12, 13]. These *ad hoc* constructions were achieved by restricting attention to design-structures that involve some form of cyclic generation, *i.e.* by searching for designs whose automorphism groups would have some inherent cyclic structure.

3.   Examples

Consider the following 7×15 DYR with $\mathcal{X} = \{*, A, B, \ldots, G, a, b, \ldots, g\}$ and $\mathcal{Y} = \{1, 2, \ldots, 7\}$, and with $(\sigma, \mu) = (2, 1)$:

	∞	P	Q	R	S	T	U	V	p	q	r	s	t	u	v
1	A1	*1	c4	e7	E3	b6	C2	B5	a1	g6	f4	F3	d7	G2	D5
2	B2	C6	*2	d5	f1	F4	c7	D3	E6	b2	a7	g5	G4	e1	A3
3	C3	E4	D7	*3	e6	g2	G5	d1	B4	F7	c3	b1	a6	A5	f2
4	D4	e2	F5	E1	*4	f7	a3	A6	g3	C5	G1	d4	c2	b7	B6
5	E5	B7	f3	G6	F2	*5	g1	b4	C7	a4	D6	A2	e5	d3	c1
6	F6	c5	C1	g4	A7	G3	*6	a2	d2	D1	b5	E7	B3	f6	e4
7	G7	b3	d6	D2	a5	B1	A4	*7	f5	e3	E2	c6	F1	C4	g7

For convenience we take row-labels from the set $\mathcal{R} = \{1, 2, \ldots, 7\}$ and column-labels from the set $\mathcal{C} = \{\infty, P, Q, \ldots, V, p, q, \ldots, v\}$. This design is 7-cyclic in the sense that it can, in an obvious way, be generated cyclically from the first element of column ∞ and from columns P and p, by using the permutations

$$(ABCDEFG)(abcdefg), \quad (PQRSTUV)(pqrstuv), \quad (1234567),$$

which are of order 7. The DYR can be written alternatively as the following 7-cyclic *incidence array*, where the rows now correspond to the elements of $\mathcal{X}$, and the elements ρ in the ordered pairs ρ, y correspond to the elements of $\mathcal{R}$:

	∞	P	Q	R	S	T	U	V	p	q	r	s	t	u	v
*		11	22	33	44	55	66	77							
A	11				67		74	46				52		35	23
B	22	57				71		15	34				63		46
C	33	26	61				12		57	45				74	
D	44		37	72				23		61	56				15
E	55	34		41	13				26		72	67			
F	66		45		52	24				37		13	71		
G	77			56		63	35				41		24	12	
a					75		43	62	11	54	27		36		
b		73				16		54		22	65	31		47	
c	65	14					27				33	76	42		51
d			76	25				31	62			44	17	53	
e	42		17	36						73			55	21	64
f		53			21	47			75		14			66	32
g			64		32	51			43	16		25			77

Now, each element from $\mathcal{R}$ occurs exactly once in each row and exactly once in each column, as does each element of $\mathcal{Y}$. Also, the labellings of the elements of $\mathcal{R}$ and $\mathcal{Y}$ have been synchronised so that each pair α, α ($\alpha = 1, 2, \ldots, 7$) occurs 3 times overall ($\sigma + 1 = 3$), whereas each pair α, β ($\alpha \neq \beta$; $\alpha, \beta = 1, 2, \ldots, 7$) occurs twice ($\sigma = 2$).

If each ordered pair in the above incidence array is replaced by 1, and 0 is entered in all other positions in the array, we have the 7-cyclic incidence matrix of the SBIBD on which the DYR is founded. If the first entry in each ordered pair is removed, we are left with a 7-cyclic incidence array for a Youden square, as we are if we remove the second entry throughout, instead of the first. This observation is the basis of our method for finding incidence arrays for DYRs. Using this method in our 7×15 situation, we first convert the 7-cyclic incidence matrix for the SBIBD into 7-cyclic incidence arrays for Youden squares; each such incidence array is obtained by overwriting the entries 1 in the matrix with integers from $\{1, 2, \ldots, 7\}$ in such a way that none of the integers $1, 2, \ldots, 7$ is repeated in any row or column. The incidence arrays so obtained become "candidate arrays" in the construction process. The candidate arrays are then paired successively. Pairings that do not satisfy the conditions for a DYR are rejected; the search continues until a satisfactory pairing is obtained, or until "sufficiently many" have been found.

In the 7×15 example, the sets $\mathcal{C}$ and $\mathcal{X}$ contain elements (respectively ∞ and $*$) that are invariant under the 7-cyclic permutations, but there are no invariant members of $\mathcal{R}$ and $\mathcal{Y}$. A different situation is seen in the following incidence array for a 4×7 DYR:

	∞	P	Q	R	p	q	r
$*$	00	13	21	32			
A	11	20				03	32
B	22		30		13		01
C	33			10	02	21	
a			12	01		30	23
b		02			31		10
c		31	03		20	12	

Here, the zero member of the set $\{0, 1, 2, 3\}$ is invariant under the 3-cyclic permutation (123) that is used in conjunction with the permutations

$(ABC)(abc)$ and $(PQR)(pqr)$. However, our method of construction, using 3-cyclic candidate arrays, is still available.

An even more complex situation is seen in the following 3-cyclic incidence array for a 5×11 DYR:

	P	Q	R	S	T	U	V	W	X	Y	Z
p		44				23	35	52		11	
q			44	31			53		15	22	
r	44					12	25	51		33	
s			32	55				43	24		11
t	13					55	34		41		22
u		21		55			42	14			33
v		53	25		34	42	11				
w	35		51	43		14	22				
x	52	15		24	41			33			
y	21	32	13							45	54
z				12	23	31				54	45

Here, the elements 4 and 5 from the set $\{1,2,3,4,5\}$ are invariant in the 3-cyclic generation of the array, as are the elements Y and Z from $\{P, Q, \ldots, Z\}$ and the elements y and z from $\{p, q, \ldots, z\}$. The permutations corresponding to the 3-cyclic structure are $(PQR)(STU)(VWX)$, $(pqr)(stu)(vwx)$, and (123). However, 3-cyclic candidate arrays can still be constructed and paired until DYRs are found.

This last example was found by hand, without aid from a computer. Larger similar examples would have to be found using a computer.

This paper illustrates our general computer approach, used to produce DYRs given in [5, 12, 13], by taking an example not previously covered in the literature. The example, whilst having a simple cyclic structure, is not trivial computationally.

4.   Searching for a 13-cyclic 13 × 40 DYR

At least 389 non-isomorphic SBIBDs are available for a search for 13 × 40 DYRs [2, p. 16]. We use a particular SBIBD whose incidence matrix can be written either with a very basic 40-cyclic structure (from a difference set) or with a simple 13-cyclic structure. With **0** and **1** denoting column vectors all of whose elements are, respectively, 0 and 1, and **0'** and **1'** denoting the corresponding row vectors, the 40 × 40 incidence matrix can be written in 13-cyclic form as

$$
\mathcal{N} = \begin{bmatrix} 0 & \mathbf{1'} & \mathbf{0'} & \mathbf{0'} \\ \mathbf{1} & \mathbf{L} & \mathbf{L} & \mathbf{L} \\ \mathbf{0} & \mathbf{L} & \mathbf{M} & \mathbf{N} \\ \mathbf{0} & \mathbf{L} & \mathbf{N} & \mathbf{M} \end{bmatrix}
$$

where **L**, **M** and **N** are cyclically generated 13 × 13 matrices satisfying

$$\mathbf{L} + \mathbf{M} + \mathbf{N} = \mathbf{J} ,$$

with

$$
\mathbf{L} = \begin{bmatrix} 0 & 1 & \cdots \\ 0 & 0 \\ 0 & 0 \\ 0 & 0 \\ 0 & 0 \\ 0 & 0 \\ 1 & 0 \\ 0 & 1 \\ 1 & 0 \\ 0 & 1 \\ 0 & 0 \\ 1 & 0 \\ 1 & 1 \end{bmatrix} , \quad
\mathbf{M} = \begin{bmatrix} 0 & 0 & \cdots \\ 0 & 0 \\ 1 & 0 \\ 0 & 1 \\ 0 & 0 \\ 0 & 0 \\ 0 & 0 \\ 1 & 0 \\ 0 & 1 \\ 1 & 0 \\ 0 & 1 \\ 0 & 0 \\ 0 & 0 \end{bmatrix} , \quad
\mathbf{N} = \begin{bmatrix} 1 & 0 & \cdots \\ 1 & 1 \\ 0 & 1 \\ 1 & 0 \\ 1 & 1 \\ 1 & 1 \\ 0 & 1 \\ 0 & 0 \\ 0 & 0 \\ 0 & 0 \\ 1 & 0 \\ 0 & 1 \\ 0 & 0 \end{bmatrix} ,
$$

where each successive column of each of **L**, **M** and **N** is obtained from the previous column by letting each entry drop to the row below, except of course that the bottom entry moves to the top. As the 13-cyclic representation was found by hand, and there were risks of transcriptions errors, the first computing tasks after input were to check that

$$\mathbf{L} + \mathbf{M} + \mathbf{N} = \mathbf{J} ,$$

and

$$\mathcal{N}\mathcal{N}' = 4\mathbf{I} + 9\mathbf{J} .$$

To implement our method, we treat $\mathcal{N}$ as being blocked, by rows, as follows:

$$\begin{bmatrix} \text{row 1} \\ \text{slab 1 (rows 2 thro 14)} \\ \text{slab 2 (rows 15 thro 27)} \\ \text{slab 3 (rows 28 thro 40)} \end{bmatrix}$$

where row 1 is the row

$$\begin{bmatrix} 0 & \mathbf{1'} & \mathbf{0'} & \mathbf{0'} \end{bmatrix},$$

slab 1 is the blocked matrix

$$\begin{bmatrix} \mathbf{1} & \mathbf{L} & \mathbf{L} & \mathbf{L} \end{bmatrix},$$

etc. Then, without any loss of generality, we start to obtain a candidate array by replacing the component $\mathbf{1'}$ of row 1 by

$$1 \quad 2 \quad 3 \quad 4 \quad 5 \quad 6 \quad 7 \quad 8 \quad 9 \quad 10 \quad 11 \quad 12 \quad 13 .$$

In principle, proceeeding with our construction beyond this point is easy:

repeat
　replace the 1's in the first row of each slab by a permutation of $1, \ldots, 13$;
　generate the rest of each slab cyclically
until each column has no repeated element.
But this is computationally hard. If we simply choose each of the three permutations at random, the above loop runs "for ever".

Generating slab 1 cyclically from a permutation may or may not yield a "good" slab 1, *i.e.* one such that no column in

$$\begin{bmatrix} \text{row 1} \\ \text{slab 1} \end{bmatrix} \qquad\qquad (4.1)$$

has repeated elements. However, a good slab 1 is found very quickly by trying random permutations for its first row, generating the slab cyclically, and testing for repeated elements in columns of (1).

Having found a good slab 1, we use it to find similarly a "good" slab 2, *i.e.* one such that no column in

$$\begin{bmatrix} \text{row 1} \\ \text{slab 1} \\ \text{slab 2} \end{bmatrix} \qquad (4.2)$$

has repeated elements. A fast PC took on average about 1 second to put good slabs 1 and 2 into place.

Having found good slabs 1 and 2, we use them to find a "good" slab 3, *i.e.* one such that no column in

$$\begin{bmatrix} \text{row 1} \\ \text{slab 1} \\ \text{slab 2} \\ \text{slab 3} \end{bmatrix} \qquad (4.3)$$

has repeated elements; we then have a candidate array. However, if we now proceed, as we did previously, by testing random permutations for the first row of slab 3, the process will take "for ever", as 13! is vastly larger than the number of permutations that might lead to a good slab 3. So another approach is needed here.

The structure of a good slab 3 is

$$\begin{bmatrix} \mathbf{0} & \mathbf{L} & \mathbf{N} & \mathbf{M} \end{bmatrix} ,$$

and it is completely determined by the entries in the first columns of $\mathbf{L}$, $\mathbf{M}$ and $\mathbf{N}$. These entries are in turn determined up to permutation by the entries in the corresponding columns of (2). Thus there are $4! \times 6! \times 3!$ possibilities for slab 3. For each of these, the cyclic structure enables the slab's first row to be obtained easily. As soon as we here have a first row where all the non-zero elements are distinct, we have a good slab 3 and therefore a candidate array.

Thus, in summary, our algorithm for finding candidate arrays is this:

repeat
    construct a good slab 1 as above;
    construct a good slab 2 as above;
    determine the sets of non-zero elements that would appear
        in the first columns of $\mathbf{L}$, $\mathbf{M}$ and $\mathbf{N}$ in a good slab 3;
    in each of the $4! \times 6! \times 3!$ possible ways:
        "back calculate" the first row of the slab 3
            that gives the current first columns of $\mathbf{L}$, $\mathbf{M}$ and $\mathbf{N}$;
        if the non-zero elements in this row are distinct,
            complete the corresponding good slab 3

and output the candidate array;
until sufficient candidate arrays have been found.

We could only guess at the number to be regarded as "sufficient". We chose 2000, and on a fast PC we found them at the rate of about 20 per minute.

The candidate arrays then had to be paired until an incidence array for a DYR was found. In such an array, with an ordered pair of symbols in each position where the corresponding SBIBD incidence matrix contains a 1, the possible distinct ordered pairs of symbols must occur as nearly as possible the same number of times. Here this means that each symbol from either one of the candidate arrays occurs exactly $4 (= \sigma + 1)$ times with just $1 (= \mu)$ of the symbols from the other candidate array, and exactly $3 (= \sigma)$ times with each of the other 12. We allowed our program to run until some 50 pairings had been found that gave incidence arrays for DYRs; over that period, we found that about 4 million pairings were needed on average to produce a successful outcome.

Converting the incidence array for a DYR to the corresponding DYR is straightforward.

The first incidence array that we obtained for a 13×40 DYR is in Table 1, where $10, 11, 12, 13$ are denoted respectively as t, e, f, g. The corresponding 13×40 DYR is in Table 2, where $\mathcal{X} = \{A, B, \ldots, Z, 1, 2, \ldots, 9, t, u, v, w, x\}$ and $\mathcal{Y} = \{a, b, \ldots, m\}$.

5.   Isomorphism

Even if two different successful pairings of candidate arrays are obtained from our construction-process as described above, the two corresponding DYRs could well be isomorphic to one another. This is but one reason for enhancing the programming by incorporating use of the computer package **nauty** [3] so that appropriate automorphism groups, and their orders, can readily be obtained. Although we have thought it right to defer detailed isomorphism studies of our 13×40 DYRs until further thought has been given to certain theoretical matters (see §6 below), we explored the possibility of implementing calls on **nauty** as follows, and we recommend making such calls in general.

If $\mathcal{N}$ is the incidence matrix of an SBIBD, then so is $\mathcal{N}'$. If the two corresponding SBIBDs are isomorphic to one another, then each is said to be *self-dual*. We therefore first used **nauty** to obtain the

Table 1. Incidence array for 13 × 40 double Youden rectangle

automorphism group of the incidence matrix $\mathcal{N}$ for the SBIBD that we were using. We found the order of this group to be 12130560, namely $40 \times 13 \times 2^5 \times 3^6$; that this is a multiple of 40 agrees with our assertion that the SBIBD could be obtained from a difference set, whereas the factor 13 agrees with the 13-cyclic structure that we have described. We then modified the nauty input to allow automorphisms where the rows of $\mathcal{N}$ map onto columns and the columns onto rows; we then found the order of the automorphism group to be twice what it had been previously, confirming that the rows can indeed be mapped onto columns and *vice versa*, thus indicating that the SBIBD is self-dual.

(When nauty is used to study a combinatorial design of any sort, it operates on a graph corresponding to that design. We here omit details

Left part:

Ha	Aa	2k	Pg	De	Lj	Bl	Fi	4b	Rh	9m	5h	Xj	Th	Qm	Kc	Ck	td	3i	Ml
Ib	Ui	Ab	3l	Qh	Ef	Mk	Cm	Gj	5c	Si	ta	6i	Yk	Vk	Ra	Ld	Dl	ue	4j
Jc	Zl	Vj	Ac	4m	Ri	Fg	Nl	Da	Hk	6d	Tj	ub	7j	Xf	Wl	Sb	Me	Em	vf
Kd	8k	1m	Wk	Ad	5a	Sj	Gh	Bm	Eb	Il	7e	Uk	vc	Md	Yg	Xm	Tc	Nf	Fa
Le	wd	9l	Oa	Xl	Ae	6b	Tk	Hi	Ca	Fc	Jm	8f	Vl	5j	Ne	Zh	Ya	Ud	Bg
Mf	Wm	xe	tm	Pb	Ym	Af	7c	Ul	Ij	Db	Gd	Ka	9g	3c	6k	Bf	1i	Zb	Ve
Ng	th	Xa	2f	ua	Qc	Za	Ag	8d	Vm	Jk	Ec	He	Lb	6h	4d	7l	Cg	Oj	1c
Bh	Mc	ui	Yb	3g	vb	Rd	1b	Ah	9e	Wa	Kl	Fd	If	2l	7i	5e	8m	Dh	Pk
Ci	Jg	Nd	vj	Zc	4h	wc	Se	Oc	Ai	tf	Xb	Lm	Ge	Hg	3m	8j	6f	9a	Ei
Dj	Hf	Kh	Be	wk	1d	5i	xd	Tf	Pd	Aj	ug	Yc	Ma	ve	Ih	4a	9k	7g	tb
Ek	Nb	Ig	Li	Cf	xl	Oe	6j	2e	Ug	Qe	Ak	vh	Zd	7a	wf	Ji	5b	tl	8h
Fl	1e	Bc	Jh	Mj	Dg	2m	Pf	7k	3f	Vh	Rf	Al	wi	Ni	8b	xg	Kj	6c	um
Gm	xj	Of	Cd	Ki	Nk	Eh	3a	Qg	8l	4g	Wi	Sg	Am	Jb	Bj	9c	2h	Lk	7d

Right part:

8e	wb	uk	xf	Ja	Vd	Uj	Oi	Zf	6e	vm	Nc	Wg	Yd	Gg	Ic	7b	Ea	1l	Sf
Nm	9f	xc	vl	2g	Kb	We	Tg	Pj	1g	7f	wa	Bd	Xh	Ze	Hh	Jd	8c	Fb	Om
5k	Ba	tg	2d	wm	3h	Lc	Pa	Uh	Qk	Oh	8g	xb	Ce	Yi	1f	Ii	Ke	9d	Gc
wg	6l	Cb	uh	3e	xa	4i	Hd	Qb	Vi	Rl	Pi	9h	2c	Df	Zj	Og	Jj	Lf	te
Gb	xh	7m	Dc	vi	4f	2b	uf	Ie	Rc	Wj	Sm	Qj	ti	3d	Eg	1k	Ph	Kk	Mg
Ch	Hc	2i	8a	Ed	wj	5g	Nh	vg	Jf	Sd	Xk	Ta	Rk	uj	4e	Fh	Ol	Qi	Ll
Wf	Di	Id	3j	9b	Fe	xk	Mm	Bi	wh	Kg	Te	Yl	Ub	Sl	vk	5f	Gi	Pm	Rj
Od	Xg	Ej	Je	4k	tc	Gf	Sk	Na	Cj	xi	Lh	Uf	Zm	Vc	Tm	wl	6g	Hj	Qa
Ql	Pe	Yh	Fk	Kf	5l	ud	Rb	Tl	Bb	Dk	2j	Mi	Vg	1a	Wd	Ua	xm	7h	Ik
Fj	Rm	Qf	Zi	Gl	Lg	6m	Jl	Sc	Um	Cc	El	3k	Nj	Wh	Ob	Xe	Vb	2a	8i
uc	Gk	Sa	Rg	1j	Hm	Mh	9j	Km	Td	Va	Dd	Fm	4l	Bk	Xi	Pc	Yf	Wc	3b
9i	vd	Hl	Tb	Sh	Ok	Ia	4c	tk	La	Ue	Wb	Ee	Ga	5m	Cl	Yj	Qd	Zg	Xd
va	tj	we	Im	Uc	Ti	Pl	Ye	5d	ul	Mb	Vf	Xc	Ff	Hb	6a	Dm	Zk	Re	1h

Table 2. 13 × 40 double Youden rectangle corresponding to the incidence array in Table 1

of how such a graph should be set up. But mention must be made of a "trick" needed when, for example, an SBIBD is tested for self-duality. The nauty graph will have one node for each row of $\mathcal{N}$ and one node for each column. We then ask whether *all* the row-nodes can be mapped 1–1 onto the column nodes and *vice versa*. Mapping just *some* row-nodes onto column-nodes and the others onto row-nodes is not acceptable. To avoid any possibility of this, two additional special nodes, here called *factor-nodes*, are needed: a *row-factor node* and a *column-factor node*. In the graph, each row node must be joined to the row-factor node, and each column-node must be joined to the column-factor node. If the SBIBD is self-dual, the row-nodes will map 1–1 onto the column-nodes and *vice versa*, and the row-factor and coumn-factor nodes will map onto one another.)

If the roles of $\mathcal{C}$ and $\mathcal{X}$ are interchanged in a Youden square with column labels drawn from $\mathcal{C}$ and entries drawn from $\mathcal{X}$, we then still have a Youden square. This result can be expressed in another way by saying that the transpose of the incidence array for a Youden square is also an incidence array for a Youden square. If these two incidence arrays are isomorphic to one another, we say that the Youden square is *self-dual* (which can happen only if the inherent SBIBD is self-dual). So a candidate array in our construction process might *prima facie* be (a) isomorphic to another candidate array, (b) isomorphic to the transpose of another candidate array, or (c) isomorphic to its own transpose. Each of these possibilities can be explored using nauty. In our quest for 13×40 DYRs, we offered each of our candidate arrays to nauty, to obtain a report on its automorphism group. A possibility that we anticipated was that some candidates would have larger automorphism groups than others (all the orders being, of course, multiples of 13). In the belief (partly based on previous experience) that candidates with larger automorphism groups would be the more likely to be enhanceable into incidence arrays for DYRs, we envisaged ranking the candidates by the orders of the groups. However, we found in every instance that the order was 13, so the proposed ranking was not open to us. We could of course have used nauty to detect whether any of the candidates corresponded to self-dual Youden squares, and then ranked the candidates on this alternative criterion; this would have involved modifying the nauty input to allow automorphisms where the rows of a candidate map onto columns and the columns onto rows, possibly in conjunction with a permutation of the elements $1, 2, \ldots, 13$ in the candidate. (Two factor-nodes would again have been needed.) However, intuition and experience suggested no great gain from ranking self-dual candidates ahead of non-self-dual ones, so we did not pursue this approach.

Now let us consider a DYR with row labels drawn from $\mathcal{R}$ and column labels drawn from $\mathcal{C}$. If (a) the roles of $\mathcal{C}$ and $\mathcal{X}$ are interchanged and/or (b) the roles of $\mathcal{R}$ and $\mathcal{Y}$ are interchanged, we then still have a DYR. Thus a DYR can be said to have three duals. The four DYRs made up from a single DYR and its three duals may be (i) all isomorphic to one another, or (ii) isomorphic to one another in pairs, or (iii) all non-isomorphic to one another. Which of these possibilities holds for a particular DYR can readily be determined by nauty by using simple extensions of the procedures already described. (Four factor-nodes are needed.) The 13×40 DYR in Table 2 is not isomorphic to any of its duals.

6. Further Properties of our new DYRs

In the incidence array that Table 1 gives for a DYR, each of the two candidate arrays that have been paired contains the entries $8, 9, t, e, f, g, 1, 2, \ldots, 7$ in positions $2, 3, \ldots, 14$ respectively of column 1. Furthermore, we found that *all* our 2000 candidate arrays (see §4 above) had this property. We do not know the reason for this, although we have seen something similar for other sizes of DYR. The property suggests that $\mathcal{N}$ could have been more elegantly presented in a "canonical" form obtainable by taking the rows of each of **L**, **M** and **N** in the order $7, 8, \ldots, 13, 1, 2, \ldots, 6$, so that the first column of the incidence array for any DYR formed from the candidate arrays would contain the entries $11, 22, \ldots, gg$ in that order.

Having noted this property, we decided to take account of it by modifying our algorithm for finding candidate arrays. Any random permutation for the first row of slab 1 was now rejected immediately if it did not begin with an 8. The rate at which candidate arrays were produced was thereby improved from about 20 per minute to about 300 per minute.

A further and very convenient property for which we have no explanation, although a theoretical reason for it must presumably exist, is that the incidence arrays for *all* our 13×40 DYRs have the pairs α, α ($\alpha = 1, 2, \ldots, g$) each occurring 4 times overall ($\sigma + 1 = 4$), and the pairs α, β ($\alpha \neq \beta$; $\alpha, \beta = 1, 2, \ldots, g$) each occurring 3 times overall ($\sigma = 3$).

We judge that an understanding of the properties described in this Section might best be saught by seeing to what extent analogous properties hold for DYRs of smaller size.

7.   Check for Balance

A DYR is a design having "general balance" in the sense recognised by the statistical computer package Genstat [4]. This means that the ANOVA directive of Genstat can be used to check the correctness of a design believed to be a DYR. We have ourselves repeatedly used such a check in our quests for DYRs of different sizes. Two restrictions apply to such checks. First, the size of the purported DYR must not lead the relevant storage limits of the program to be exceeded; what these limits are depends on the implementation of Genstat that is being used. Second, because of the arithmetic way in which Genstat does its test for general balance, an occasional "false negative" outcome may be obtained, rejecting a correct DYR; this problem can be overcome by use of the TOLERANCES option of the ANOVA directive [4, §9.7.5].

References

[1] R. A. Bailey. Designs: mappings between structured sets. In J. Siemons, editor, *Surveys in Combinatorics*, pages 22–51. Cambridge Univ. Press, Cambridge, 1989.

[2] R. Mathon and A. Rosa. 2-(v, k, λ) designs of small order. In C. J. Colbourn and J. H. Dinitz, editors, *The CRC Handbook of Combinatorial Designs*, pages 3–41. CRC Press, Boca Raton, Florida, 1996.

[3] B. D. McKay. *nauty* User's Guide (Version 1.5), Technical Report TR-CS-90-02. Computer Science Department, Australian National University, Canberra, 1990.

[4] Genstat 5 Committee of the Statistics Department, Rothamsted Experimental Station. *Genstat 5 Release 3 Reference Manual*. Oxford Science Publications, Clarendon Press, Oxford, 1993.

[5] N. C. K. Phillips, D. A. Preece, and D. H. Rees. Double Youden rectangles for the four biplanes with $k = 9$. *Journal of Combinatorial Mathematics and Combinatorial Computing*, to appear.

[6] D. A. Preece. Fifty years of Youden squares: a review. *Bull. Inst. Math. Applic.*, 26:65–75, 1990.

[7] D. A. Preece. Double Youden rectangles of size 6×11. *Mathematical Scientist*, 16:41–45, 1991.

[8] D. A. Preece. A set of double Youden rectangles of size 8×15. *Ars Combinatoria*, 36:215–219, 1993.

[9] D. A. Preece. Double Youden rectangles — an update with examples of size 5×11. *Discrete Math.*, 125:309–317, 1994.

[10] D. A. Preece. Youden squares. In C. J. Colbourn and J. H. Dinitz, editors, *The CRC Handbook of Combinatorial Designs*, pages 511–514. CRC Press, Boca Raton, Florida, 1996.

[11] D. A. Preece. Some 6×11 Youden squares and double Youden rectangles. *Discrete Math.*, 167/168:527–541, 1997.

[12] D. A. Preece, B. J. Vowden, and N. C. K. Phillips. Double Youden rectangles of sizes $p \times (2p+1)$ and $(p+1) \times (2p+1)$. *Ars Combinatoria*, 51:161–171, 1999.

[13] D. A. Preece, B. J. Vowden, and N. C. K. Phillips. Double Youden rectangles of sizes $(p + 1) \times (p^2 + p + 1)$. *Utilitas Mathematica*, 59:139–154, 2001.

[14] B. J. Vowden. Infinite series of double Youden rectangles. *Discrete Math.*, 125:385–391, 1994.

[15] B. J. Vowden. A new infinite series of double Youden rectangles. *Ars Combinatoria*, 56:133–145, 2000.

Chapter 13

KIRKMAN TRIPLE SYSTEMS AND THEIR GENERALIZATIONS: A SURVEY

Rolf S. Rees

Department of Mathematics and Statistics
Memorial University of Newfoundland

W. D. Wallis

Department of Mathematics
Southern Illinois University, Carbondale

Abstract

Over 150 years ago, the Reverend T.P. Kirkman posed the following problem as Query 6 in the Lady's and Gentlemen's Diary [26].

Fifteen young ladies in a school walk out three abreast for seven days in succession: it is required to arrange them daily, so that no two will walk twice abreast.

Both Kirkman [27] and Cayley [13] published solutions to this problem in that same year. The search for a solution to the general problem of arranging $v \equiv 3 \pmod 6$ ladies into groups of three over a series of $(v-1)/2$ days became known as the *Kirkman Schoolgirl Problem*, one of the most famous problems in combinatorial design theory. It took well over a century before the first published proof of the existence of these designs for all $v \equiv 3 \pmod 6$ was given by D.K. Ray-Chaudhuri and R.M. Wilson [38].

In this paper we update and expand upon the survey on Kirkman Triple Systems and related designs published by D.R. Stinson a decade ago [51]. The study of various generalizations of the Kirkman Schoolgirl Problem remains a very active area of research, and much work has been done in the past ten years. This paper does not cover all of the topics discussed in Stinson's survey; on the other hand we discuss several topics not covered by that survey. Thus our paper is not meant so much to replace Stinson's survey as it is to complement it.

1.   Introduction

1.1   Historical background

Our aim is to survey some of the generalizations of a famous problem in block designs, *Kirkman's schoolgirl problem.*

Although most readers will be familiar with the basic definitions, we include some for completeness. A *balanced incomplete block design* or *BIBD* is a way of selecting subsets of size k, called *blocks*, from a set S of objects called *treatments*, in such a way that every member of S occurs in r blocks and any two members of S occur together in precisely λ blocks. The usual notations for the number of blocks and the cardinality of S are b and v respectively, and the design is often denoted a (v, b, r, k, λ)-*design*. The parameters v, k, λ determine b and r, so we often write (v, k, λ)-*BIBD*. In particular, a $(v, 3, 1)$-*BIBD* is called a *Steiner triple system.*

In 1850, the Reverend T. P. Kirkman proposed the following problem [26]: A schoolmistress has 15 girl pupils and she wishes to take them on daily walks for a week. On each day the girls are to walk in five rows of three girls each. It is required that no two girls should walk in the same row more than once over the week. (Kirkman's exact wording was given in the *Abstract*). Can this be done? This problem is generally known as "Kirkman's schoolgirl problem." There are seven days in a week, and every girl will walk in the company of two others each day. As no repetitions are allowed, a girl has 14 companions over the week; since there are only 15 in the class, it follows that every pair of girls must walk together in a row at least once; in view of the requirements, every pair of girls walk together *precisely* once. Every girl walks on each day of the week. So, if we treat the girls as "treatments" and the rows in which they walk as "blocks," we are required to select 35 blocks of size 3 from a set of 15 treatments, in such a way that every treatment occurs in 7 blocks and every pair of treatments occur together in precisely one block — that is, the schoolmistress requires a balanced incomplete block design with parameters $(15, 35, 7, 3, 1)$.

But there is a further constraint — since all the girls must walk every day, it is necessary that the block set be partitioned into seven groups of triples, so that every girl appears once in each group — these will correspond to the seven days. Borrowing from geometry, a *parallel* class of blocks in a design is defined to be a set of blocks that between them contain every element precisely once. (A parallel class consisting entirely of triples is called a *triangle-factor*, and one consisting entirely of pairs is a *one-factor*.) A design is called *resolvable* when its set of blocks admits a partition into parallel classes. So Kirkman was in fact asking: Is there a resolvable Steiner triple system on 15 treatments?

Kirkman's problem has a solution: for example, we may use the schedule

Monday:	$1, 2, 3$	$4, 5, 6$	$7, 8, 9$	$10, 11, 12,$	$13, 14, 15$
Tuesday:	$1, 4, 11$	$2, 5, 10$	$3, 8, 13$	$6, 7, 14$	$9, 12, 15$
Wednesday:	$1, 5, 9$	$2, 6, 8$	$3, 11, 15$	$4, 12, 14$	$7, 10, 13$
Thursday:	$1, 6, 13$	$2, 4, 15$	$3, 9, 10$	$5, 7, 12$	$8, 11, 14$
Friday:	$1, 7, 15$	$2, 9, 14$	$3, 6, 12$	$4, 8, 10$	$5, 11, 13$
Saturday:	$1, 8, 12$	$2, 7, 11$	$3, 5, 14$	$4, 9, 13$	$6, 10, 15$
Sunday:	$1, 10, 14$	$2, 12, 13$	$3, 4, 7$	$5, 8, 15$	$6, 9, 11$

More generally, Kirkman asked: for how many girls can a complete set of daily walks be devised in which every girl walks in a row with each classmate exactly once? In other words, for which v does there exist a resolvable Steiner triple system on v treatments? Since every girl will have $2r$ companions, we have $v = 2r + 1$, and so v is odd. As every day's parade is to be made up of complete rows, v must be a multiple of 3. So a necessary condition is $v \equiv 3 \pmod 6$. A solution is called a *Kirkman Triple System*, and a Kirkman Triple System on v points will be denoted $KTS(v)$. Writing $v = 6m + 3$ for some integer m, the parameters of a $KTS(v)$ are

$$(6m + 3, (2m + 1)(3m + 1), 3m + 1, 3, 1).$$

There are $2m + 1$ blocks in each parallel class, so there are $3m + 1$ parallel classes.

The smallest case, a $KTS(3)$, is trivial, containing only one block. Two further examples — systems on 9 and 21 points — are shown in Table 1.1.

$KTS(9):$

1	2	3	1	4	7	1	5	9	1	6	8
4	5	6	2	5	8	2	6	7	2	4	9
7	8	9	3	6	9	3	4	8	3	5	7

$KTS(21):$

1	2	3	4	5	6	7	8	9	10	11	12	13	14	15
4	7	13	7	10	16	10	13	19	1	13	16	4	16	19
5	8	14	8	11	17	11	14	20	2	14	17	5	17	20
6	9	15	9	12	18	12	15	21	3	15	18	6	18	21
10	17	21	3	13	20	2	6	16	5	9	19	1	8	12
11	18	19	1	14	21	3	4	17	6	7	20	2	9	10
12	16	20	2	15	19	1	5	18	4	8	21	3	7	11

16	17	18	19	20	21	10	18	20	11	16	21	12	17	19
1	7	19	1	4	10	2	13	21	3	14	19	1	15	20
2	8	20	2	5	11	3	5	16	1	6	17	2	4	18
3	9	21	3	6	12	6	8	19	4	9	20	5	7	21
4	11	15	7	14	18	1	9	11	2	7	12	3	8	10
5	12	13	8	15	16	4	12	14	5	10	15	6	11	13
6	10	14	9	13	17	7	15	17	8	13	18	9	14	16

Table 13.1.   Small Kirkman Triple Systems

There is an interesting construction for a $KTS(15)$ using a structure that was not formally defined until a century after Kirkman's original query, that structure being a *Room square*. This is an $r \times r$ array whose cells are either empty or contain an unordered pair from $\{0, 1, \ldots, r\}$, with the property that every pair occurs exactly once and every symbol appears once in each row and once in each column. If $\mathcal{R}$ is a room square, we define its *incidence matrix* to be the $r \times r$ matrix with 1 in the (i, j) position if the (i, j) cell of $\mathcal{R}$ is occupied, and 0 otherwise. In the case $r = 7$, the parameters are such that the incidence matrix might be the complement of the incidence matrix of a *Fano plane*, or $(7, 3, 1)$-*BIBD*, and in that case the Room square is *balanced*. Table

13.2 shows a balanced room square of side 7, with its columns labeled A, B, C, D, E, F, G. The square has obviously been constructed in a cyclic fashion, but this is not essential to the construction: only the room square properties, and the balance, are essential.

A	B	C	D	E	F	G
01			36		27	45
56	02			47		31
42	67	03			51	
	53	71	04			62
73		64	12	05		
	14		75	23	06	
		25		16	34	07

Table 13.2.   A balanced Room square of side 7.

We construct a $KTS(15)$ as follows. The fifteen schoolgirls are the symbols $\{0, 1. \ldots, 7, A, B, \ldots, G\}$. The seven days of the week correspond to the seven rows of the square. For a given day, four triples are constructed from the four pairs in the row: if $\{x, y\}$ appears in row Z, then the corresponding triple is $\{x, y, Z\}$. The fifth triple consists of the labels of the three columns that are unoccupied in that row. The result is a $KTS(15)$.

Another interesting small case is $v = 39$. The treatments are represented by the integers modulo 39. Thirteen parallel classes are obtained by adding, in turn, $i = 0, 3, 6, 9, 12, 15, 18, 21, 24, 27, 30, 33, 36$ to each element in the blocks of the following parallel class.

$$\{5, 6, 25\}, \quad \{9, 14, 31\}, \quad \{2, 10, 18\}, \quad \{20, 22, 24\},$$
$$\{27, 28, 29\}, \quad \{30, 34, 38\}, \quad \{1, 8, 33\}, \quad \{0, 3, 12\},$$
$$\{15, 21, 36\}, \quad \{13, 19, 37\}, \quad \{4, 7, 16\}, \quad \{23, 26, 35\},$$
$$\{11, 17, 32\}.$$

Each of the remaining six parallel classes consists of the 13 blocks

$$\{\{x + i, y + i, z + i\} : i = 0, 3, 6, 9, 12, 15, 18, 21, 24, 27, 30, 33, 36\},$$

where $\{x, y, z\}$ is the *starter block* for the parallel class. The six starter blocks are

$\{0, 13, 26\}, \{3, 19, 35\}, \{12, 23, 37\}, \{4, 15, 32\}, \{11, 16, 21\}, \{7, 17, 36\}.$

We shall see that the condition $v \equiv 3(\bmod 6)$ is in fact sufficient — a fact that was not proven for over a century after Kirkman originally posed the problem. Lu Jiaxi [33] finally solved the problem in 1965, but his solution was not published at that time (due to the Great Cultural Revolution in China), so the first known solution was published by Ray-Chaudhuri and Wilson in 1971 [38].

A natural question to ask is: what if the class size is not an odd multiple of 3? In that case one can ask for a set of daily walks in which as many pairs walk together as is possible without repetitions (called a *Kirkman packing design*), or that every pair walk together and the number of repetitions is minimized (a *Kirkman covering design*). When the number of schoolgirls is not divisible by 3, one row of 2 or 4 girls is allowed. Another solution in the case of $v \equiv 1(\bmod 3)$ is to allow one girl to walk alone; each day's walk is called a *near-triangle-factor*, and one can discuss packings and coverings with these factors. Similarly, in the case of $v \equiv 2(\bmod 3)$, packings with one row of five girls has been considered.

In those cases where a Kirkman Triple System cannot exist, we would like to find the maximum number of rounds in a Kirkman packing. Writing $KP(v, r)$ for a Kirkman packing with r rounds (or *of length r*), we would like to determine $r(v)$, the largest value of r such that a $KP(v, r)$ exists. If $b(v)$ denotes the number of unordered pairs of objects covered by a round of a $KP(v, r)$ then the number of pairs covered by the design is $rb(v)$; clearly, this cannot exceed $\frac{1}{2}v(v-1)$, so

$$r(v) \le M(v) = \lfloor \frac{v(v-1)}{2b(v)} \rfloor.$$

A Kirkman packing that attains this upper bound $M(v)$ will be called *optimal*. Our question is: for what values of v does an optimal Kirkman packing exist?

Suppose the number of schoolgirls is an even multiple of 3. Then the nearest one can come to the exact solution is a design in which each pair will occur exactly once, except that, given a schoolgirl x, there will exist a schoolgirl x' such that the pair xx' never occurs.

Resolvable packing designs called *Nearly Kirkman Triple Systems* were introduced by Kotzig and Rosa [28]. The same idea was invented independently by Irving [24]. A *Nearly Kirkman Triple System* $NKTS(v)$ is a way of selecting from a v-set S one one-factor and $\frac{1}{2}v - 1$ triangle-factors which together contain every pair of elements precisely once. If the one-factor is deleted, the triangle factors form a $KP(v, \frac{1}{2}v - 1)$. When $v = 6m$, $b(v) = 6m$ also (there are $2m$ triples, and each covers 3 pairs). So

$$M(v) = \lfloor \frac{v(v-1)}{2v} \rfloor = \frac{1}{2}v - 1.$$

Thus Nearly Kirkman Triple Systems provide optimal Kirkman packing designs for the case $v \equiv 0 (\bmod 6)$.

No $NKTS(6)$ or $NKTS(12)$ exists [28]. When $v = 6$, obviously only one round is possible, and $r(6) = 1$. For $v = 12$, a hand computation shows that five rounds cannot be constructed. On the other hand, a set of 4 rounds can be constructed from the affine plane on 16 points. Select one line; delete all points on this line and all lines in its parallel class. The remaining blocks form the design, with the rounds corresponding to the original parallel classes (see Table 13.3). So $r(12) = 4$. It is well known that an $NKTS(v)$ exists whenever v is a multiple of 6 greater than 12, see for example [46]. We will give a new proof of this result in Section 2.2.

1	2	3	4
5	6	7	8
9	10	11	12
13	14	15	16

1	5	9	13		1	7	12	14		1	5	9		1	7	12
2	6	10	14		2	8	11	13		2	6	10		2	8	11
3	7	11	15		3	5	10	16		3	7	11		3	5	10
4	8	12	16		4	6	9	15		4	8	12		4	6	9

$\Rightarrow$

1	6	11	16		1	8	10	15		1	6	11		1	8	10
2	5	12	15		2	7	9	16		2	5	12		2	7	9
3	8	9	14		3	6	12	13		3	8	9		3	6	12
4	7	10	13		4	5	11	14		4	7	10		4	5	11

Table 13.3. Construction of $KP(12, 4)$.

If the number of schoolgirls is not a multiple of 3, the situation is more complicated. A Kirkman packing design (or *project design* — for the history of this name, see [14]) of order v and length r, or $KP(v, r)$, is a way of choosing sets called blocks from a set of v objects, and of partitioning the set of blocks into r subsets called rounds, so that: each object occurs exactly once per round; all blocks in each round are triples except for at most one, and that one can contain 2 or 4 objects;

and each object-pair occurs in at most one block in the design. Again, the existence of optimal packings will be discussed.

All of these packing problems have analogous covering problems, and we shall discuss those also.

1.2   Outline of the paper

In the remainder of Section 1 we introduce some combinatorial designs as well as several constructions that will be used in the sequel. Section 2.1 contains a proof that Kirkman Triple Systems exist for all suitable v. We then consider Nearly Kirkman Triple Systems (the earliest generalization of Kirkman Triple Systems that was considered) in Section 2.2, proving their existence for all $v \equiv 0(\bmod 6), v \geq 18$.

In Section 3, we discuss more of the earlier generalizations of Kirkman Triple Systems that were considered. Specifically, we discuss Minimum resolvable coverings of v points by triangle-factors in Section 3.1, the Chromatic Index problem for Steiner Triple Systems of order $v \equiv 1(\bmod 6)$ in Section 3.2 and the spectrum for Resolvable group-divisible designs with block size three in Section 3.3; we will see that all of these problems have now been completely solved. (This was not the case at the time of the Stinson survey.)

In Section 4 we discuss generalizations of Kirkman Triple Systems to packings and coverings of v points where v is not a multiple of 3. These resolvable configurations will have the property that all but one block in each parallel class has size three; the remaining block may take sizes one or four (when $v \equiv 1(\bmod 3)$) or two or five (when $v \equiv 2(\bmod 3)$), depending on the application. In Section 4.1 we give several examples of these designs; then in Sections 4.2 and 4.3 we discuss respectively the cases $v \equiv 2(\bmod 3)$ and $v \equiv 1(\bmod 3)$. Some interesting open problems are posed.

Finally, in Section 5 we briefly discuss four further generalizations of Kirkman Triple Systems that have been researched recently. These are Kirkman Triple Systems and Nearly Kirkman Triple Systems with Subsystems (Section 5.1), Kirkman Squares (Section 5.2), Maximal Sets of Triangle-Factors (Section 5.3) and Large Sets of Kirkman Triple Systems (Section 5.4). We will see that in each case (except for Kirkman Triple Systems with Subsystems, whose spectrum had been completely determined at the time of the Stinson survey) it remains a challenging open problem to complete their spectrum.

Kirkman Triple Systems can be generalized in many other ways. Not only would discussion of them make this paper unwieldy, but other generalizations and extensions are discussed in the excellent survey paper by Stinson [51], which complements the present paper. We have confined our bibliography to references to results actually given in this paper; for a more extensive bibliography of articles written prior to 1991 we refer the reader to the Stinson survey.

1.3 Some combinatorial designs — definitions, notation and some basic constructions

Latin Squares A *Latin square* of side n is an $n \times n$ array based on some set S of n symbols (treatments), with the property that every row and every column contains every symbol exactly once. In other words, every row and every column is a permutation of S.

Two Latin squares A and B of the same side n are called *orthogonal* if the n^2 ordered pairs (a_{ij}, b_{ij}) — the pairs formed by superimposing one square on the other — are all different. More generally, one can speak of a set of k *mutually orthogonal* Latin squares, that being a set of squares $A_1, A_2, \ldots, A_k$ such that A_i is orthogonal to A_j whenever $i \neq j$. If $N(n)$ represents the size of the largest possible set of mutually orthogonal Latin squares of side n, then $N(n) \leq n - 1$, with equality when n is a prime power [4].

Tarry [53] proved that $N(6) = 1$ (see also Stinson [49]). Subsequently Bose, Parker and Shrikhande [5] proved

Theorem 1 $N(n) \geq 2$ *if and only if $n > 0$, except for $n = 2$ or 6.*

Various researchers worked together to show that $N(n) \geq 3$ when $n > 14$ (the results are surveyed in [55]). Then Todorov [54] proved that $N(14) \geq 3$. It was already known that $N(12) \geq 5$ [25], so we have the following.

Theorem 2 $N(n) \geq 3$ *if and only if $n > 0$, except for $n = 2, 3$ or 6 and except possibly for $n = 10$.*

The existence or otherwise of three orthogonal Latin squares of side 10 is a famous open problem.

Pairwise Balanced Designs, Transversal Designs   A *pairwise balanced design* of index λ is a way of selecting blocks from a set of treatments such that any two treatments occur together in λ blocks. If there are v treatments and if every block size is a member of some set K of positive integers, the design is designated a $(v, K; \lambda)$-*PBD*. It is not required that every member of K be a block size. So any $(v, K; \lambda)$-*PBD* is also a $(v, L; \lambda)$-*PBD* whenever L contains K. If $\lambda = 1$ we will write simply (v, K)-*PBD*.

A *simple group divisible design* or *SGDD* is constructed from a pairwise balanced design with $\lambda = 1$ by selecting a parallel class and deleting those blocks (but not the points contained therein). (Of course, not every pairwise balanced design with $\lambda = 1$ contains a parallel class, so they cannot all be used to construct SGDDs.) The deleted blocks are called "groups"; thus two treatments either determine a common block or a common group, but not both. An *SGDD* whose sets of treatments, groups, and blocks are X, $\mathcal{G}$, and $\mathcal{A}$, respectively, will sometimes simply be denoted by the triple $(X, \mathcal{G}, \mathcal{A})$. We shall sometimes refer to the *SGDD* as a $GD(v; G; A)$, where G and A are the sets of cardinalities of members of $\mathcal{G}$ and $\mathcal{A}$, respectively. (If G or A has only one element, we may sometimes write that number, omitting set brackets.) If $\mathcal{G} \cup \mathcal{A}$ is interpreted as a set of blocks, we have a pairwise balanced design: if there exists a $GD(v; G; A)$, then there exists a $(v, G \cup A)$-*PBD*. By $TD(k, n)$ we denote a *transversal design* or *transversal system*, which is an *SGDD* with uniform block size k, with uniform group size n, and with k groups. It follows that each block is a *transversal* of the groups: each block contains precisely one element from each group. The number of blocks is n^2, and there are kn treatments.

The relationship between transversal designs and Latin squares is very important:

Theorem 3 *The existence of a set of $k - 2$ mutually orthogonal Latin squares of order n is equivalent to the existence of a $TD(k, n)$.*

Proof. Suppose a $TD(k, n)$ with groups $G_1, G_2, \ldots, G_k$ is given. Relabel the elements so that

$$G_h = \{x_{h1}, x_{h2}, \ldots, x_{hn}\}.$$

Then for every h, i, and j with $1 \leq h \leq k - 2$ and $1 \leq i, j \leq n$, define

$$a_{ij}^h = m,$$

where m is the integer such that x_{hm} is the (unique) member of G_h in the (unique) block of the $TD(k, n)$ that contains both $x_{k-1,i}$ and $x_{k,j}$. It is easy to verify that the array A_h with (i, j) element a_{ij}^h is a Latin square of side n, and that $A_1, A_2, \ldots, A_{k-2}$ are orthogonal. The construction may be reversed. $\qquad\qquad\qquad\qquad\qquad\qquad\qquad\qquad\qquad\qquad\qquad\square$

Assume a $TD(k, n)$ exists. Then it is easy to see that a $(kn, \{k, n\})$-PBD exists: one simply takes a pairwise balanced design whose blocks are the blocks and the groups of the transversal design. By adding a common further element to all the groups of size n, one obtains a $(kn + 1, \{k, n + 1\})$-PBD. So we have:

Corollary 3.1 *If there exists a set of $k - 2$ pairwise orthogonal Latin squares of side n, then there exist a $(kn, \{k, n\})$-PBD and a $(kn + 1, \{k, n + 1\})$-PBD.*

In particular, the existence of a Latin square of side n for every positive integer n proves that there is a $(3n, \{3, n\})$-PBD and a $(3n + 1, \{3, n + 1\})$-PBD for every n.

Transversal designs (and hence sets of mutually orthogonal Latin squares) play an invaluable role in the construction and consequent uses of pairwise balanced designs, see for example Section 2.1 in this survey. A more general discussion of these applications is given by Wilson in [56].

Group Divisible Designs and the Wilson Fundamental Construction Let $(X, \mathcal{G}, \mathcal{A})$ be a group divisible design, or GDD (we drop the adjective *simple* for brevity). The *type* (or *group-type*) of the GDD is the multiset $S = \{|G| : G \in \mathcal{G}\}$. A K-GDD of type S then is a GDD in which every block has size from the set K and in which S is the multiset of group sizes. We also use an exponential notation: thus a K-GDD of type $g_1^{u_1} g_2^{u_2} \ldots g_s^{u_s}$ has all of its block sizes in the set K and has u_i groups of size g_i for $i = 1, 2, \ldots, s$. A GDD is called resolvable if its block set admits a partition into parallel classes. Thus for example a Nearly Kirkman Triple System $NKTS(v)$ is equivalent to a resolvable $\{3\}$-GDD of type $2^{v/2}$ (the groups are the pairs of elements not covered by any of the triangle-factors). It is not difficult to see that in any resolvable group divisible design with constant block size, the group size must also be constant.

Example 4 *A resolvable* $\{3\}$*-GDD of type* 6^4

Point Set: $\mathbb{Z}_8 \times \{1, 2, 3\}$

Groups: $\{\{0 + i, 4 + i\} \times \{1, 2, 3\} : i = 0, 1, 2, 3\}$

Parallel classes: We obtain four classes by developing the following parallel class mod 8 (adding 4 to every element leaves this class fixed, and so we get an orbit of length four).

$\{(2, 1), (4, 1), (5, 1)\}, \{(4, 2), (6, 2), (7, 2)\},$

$\{(5, 3), (7, 3), (0, 3)\}, \{(7, 1), (1, 2), (2, 3)\},$

$\{(6, 1), (0, 1), (1, 1)\}, \{(0, 2), (2, 2), (3, 2)\},$

$\{(1, 3), (3, 3), (4, 3)\}, \{(3, 1), (5, 2), (6, 3)\}$

Five more classes are obtained by developing each of the following five starter blocks mod 8.

$\{(0, 1), (1, 2), (7, 3)\}, \{(0, 1), (6, 2), (1, 3)\}, \{(0, 1), (3, 2), (5, 3)\},$

$\{(0, 1), (5, 2), (2, 3)\}, \{(0, 1), (7, 2), (6, 3)\}$

A *double* group divisible design is a quadruple $(X, \mathcal{H}, \mathcal{G}, \mathcal{A})$ where X is a set of points, $\mathcal{H}$ and $\mathcal{G}$ are partitions of X (into holes and groups, respectively) and $\mathcal{A}$ is a collection of subsets of X (blocks) such that

(i) for each block $b \in \mathcal{A}$ and each hole $Y \in \mathcal{H}, |b \cap Y| \leq 1$, and

(ii) any pair of points from X which are not in the same hole occur together in some group or in exactly one block, but not both.

A K-$DGDD$ of type $(g, h^v)^u$ is a double group divisible design in which every block has size from the set K and in which there are u groups of size g, each of which intersects each of the v holes in h points. (Thus, $g = hv$. Obviously not every $DGDD$ can be expressed this way, but this is the most general type that we will require.) A K-$DGDD$ is called resolvable if its block set admits a partition into parallel classes, each parallel class being a partition of the point set. Note that, in particular, a resolvable 3-$DGDD$ of type $(g, h^v)^u$ contains $h(v-1)(u-1)/2$ parallel classes. Now by Theorem 1 there exists a resolvable $TD(3, n)$ for every

$n > 0, n \neq 2, 6$. That is, for these values of n there exists a resolvable 3-$DGDD$ of type $(n, 1^n)^3$ (having n-1 parallel classes of triples). It follows easily that if there exists a resolvable 3-GDD of type g^u and $n > 0, n \neq 2, 6$ then we can replace each point in the GDD by n new points (a so-called *weighting* operation) and each block in the GDD by a resolvable 3-$DGDD$ of type $(n, 1^n)^3$ to obtain a resolvable 3-$DGDD$ of type $(ng, g^n)^u$ (see Construction 6). We will use these designs in our proof of the existence of Nearly Kirkman Triple Systems in Section 2.2.

We conclude this subsection with the following well known construction (see, e.g. [15]). In what follows, the weight function w need not be a constant function.

Construction 5 (Wilson Fundamental Construction for GDDs)
Let $(X, \mathcal{G}, \mathcal{A})$ be a GDD and let $w : X \to \mathbb{Z}^+ \cup \{0\}$ be a weight function on X. Suppose that for each block $b \in \mathcal{A}$ there is a K-GDD of type $\{w(x) : x \in b\}$. Then there is a K-GDD of type $\{\sum_{x \in G} w(x) : G \in \mathcal{G}\}$.

1.4 Transversal design inflation and the doubling construction for triangle-factors

If G is a graph, we denote by $G \otimes I_n$ the graph whose vertex-set is formed by replacing each vertex x of G by n vertices $x_1, x_2, \ldots, x_n$, with x_i adjacent to y_j if and only if x is adjacent to y in G. A *triangle-factor* in G is a vertex-disjoint union of triangles (K_3s) in G which covers all the vertices. In this terminology, a Nearly Kirkman Triple System of order $6n$ is a decomposition of $K_{3n} \otimes I_2$ into edge-disjoint triangle-factors.

As noted in Section 1.3, the existence of a resolvable $TD(3, n)$ for every $n > 0$ except $n = 2$ or 6 yields the following simple construction.

Construction 6 (TD Inflation) *If the graph G admits an edge-decomposition into triangle-factors and $n > 0, n \neq 2, 6$, then the graph $G \otimes I_n$ admits an edge-decomposition into triangle-factors. In particular, if we start with a resolvable 3-GDD of type g^u we can produce a resolvable 3-DGDD of type $(ng, g^n)^u$.*

Rees [43] gives a condition under which one can apply "weight" 2 (and therefore "weight" 6) to a collection of triangle-factors:

Theorem 7 (Doubling Construction) *If the graph G admits an edge-decomposition into an even number of triangle-factors then the graph $G \otimes I_2$ admits an edge-decomposition into triangle-factors.*

Proof. It suffices to prove the assertion for two triangle-factors T and T^*. Let H be the graph whose vertices are the triangles of $T \cup T^*$, where two vertices in H are adjacent if and only if the corresponding triangles intersect in a vertex of G. Then H is a 3-regular bipartite graph and therefore has a one-factorization. This in turn induces a strong 3-coloring of the vertices of G; that is, G admits a 3-coloring $V(G) = C_1 \cup C_2 \cup C_3$ such that if $x_1 x_2 x_3 \in T$ and $x_1^* x_2^* x_3^* \in T^*$ then we can assume $x_i, x_i^* \in C_i, i = 1, 2, 3$. The four triangle-factors on $G \otimes I_2$ are then given by

$$\{x_{11}, x_{21}, x_{31} : x_1 x_2 x_3 \in T\} \quad \cup \quad \{x_{12}^*, x_{22}^*, x_{32}^* : x_1^* x_2^* x_3^* \in T^*\},$$
$$\{x_{11}, x_{22}, x_{32} : x_1 x_2 x_3 \in T\} \quad \cup \quad \{x_{12}^*, x_{21}^*, x_{31}^* : x_1^* x_2^* x_3^* \in T^*\},$$
$$\{x_{12}, x_{22}, x_{31} : x_1 x_2 x_3 \in T\} \quad \cup \quad \{x_{11}^*, x_{21}^*, x_{32}^* : x_1^* x_2^* x_3^* \in T^*\},$$
$$\{x_{12}, x_{21}, x_{32} : x_1 x_2 x_3 \in T\} \quad \cup \quad \{x_{11}^*, x_{22}^*, x_{31}^* : x_1^* x_2^* x_3^* \in T^*\}.$$

$\square$

As a direct consequence of the foregoing we have:

Construction 8 *If the graph G admits an edge-decomposition into an even number of triangle-factors and $n > 0$ then the graph $G \otimes I_n$ admits an edge-decomposition into triangle-factors.*

Example 9 *We illustrate Theorem 7 by constructing an $NKTS(18)$ from a $KTS(9)$.*

We use the following $KTS(9)$ (see Table 1.1), which we interpret as an edge-decomposition of K_9 into triangle-factors:

$$
\begin{array}{ccc}
147 & 258 & 369 \\
159 & 267 & 348 \\
\\
123 & 456 & 789 \\
186 & 429 & 753
\end{array}
$$

The doubling construction gives the following:

```
1 4 7   2 5 8   3 6 9      1*5*9*  2*6*7*  3*4*8*
1 4*7*  2 5*8*  3 6*9*     1*5 9   2*6 7   3*4 8
1*4*7   2*5*8   3*6*9      1 5 9*  2 6 7*  3 4 8*
1*4 7*  2*5 8*  3*6 9*     1 5*9   2 6*7   3 4*8

1 2 3   4 5 6   7 8 9      1*8*6*  4*2*9*  7*5*3*
1 2*3*  4 5*6*  7 8*9*     1*8 6   4*2 9   7*5 3
1*2*3   4*5*6   7*8*9      1 8 6*  4 2 9*  7 5 3*
1*2 3*  4*5 6*  7*8 9*     1 8*6   4 2*9   7 5*3
```

We get an edge-decomposition of $K_9 \otimes I_2$ into triangle-factors (that is, an $NKTS(18)$). Note that the uncovered pairs are $\{i, i^*\}$ for $i = 1, 2, 3, 4, 5, 6, 7, 8, 9$.

1.5 Kirkman Frames and frame constructions

A group-divisible design $(X, \mathcal{G}, \mathcal{A})$ is called *frame-resolvable* if its block-set $\mathcal{A}$ admits a partition into holey parallel classes, each holey parallel class being a partition of $X \backslash G$ for some group $G \in \mathcal{G}$. The groups in a frame-resolvable GDD are usually referred to as holes. A frame-resolvable GDD in which all blocks have size three is called a *Kirkman Frame*; in such a structure the number of holey parallel classes corresponding to a given hole G is $|G|/2$, from which it follows that all holes in a Kirkman Frame must have an even number of points (see Stinson [50]).

Example 10 *Kirkman Frames of type* 2^4 *and* 2^7:

2^4	Hole	Holey parallel class
	1, 5	3, 4, 6 7, 8, 2
	2, 6	4, 5, 7 8, 1, 3
	3, 7	5, 6, 8 1, 2, 4
	4, 8	6, 7, 1 2, 3, 5

2^7	Hole	Holey parallel class			
	1, 8	2, 4, 10	3, 7, 12	5, 6, 9	11, 13, 14
	2, 9	3, 5, 11	4, 1, 13	6, 7, 10	12, 14.8
	3, 10	4, 6, 12	5, 2, 14	7, 1, 11	13, 8, 9
	4, 11	5, 7, 13	6, 3, 8	1, 2, 12	14, 9, 10
	5, 12	6, 1, 14	7, 4, 9	2, 3, 13	8, 10, 11
	6, 13	7, 2, 8	1, 5, 10	3, 4, 14	9, 11, 12
	7, 14	1, 3, 9	2, 6, 11	4, 5, 8	10, 12, 13

Observe that if we add the symbol 0 to each hole in the frame of type 2^4 we obtain the four parallel classes of a $KTS(9)$; similarly, if we add the symbol 0 to each hole in the frame of type 2^7, we obtain the seven parallel classes of a $KTS(15)$. It is easy to see that in general, adding a common point to the holes in a frame of type 2^u gives rise to a $KTS(2u + 1)$. Conversely, if we start with a $KTS(2u + 1)$ and delete a point x we obtain a Kirkman Frame of type 2^u; the holes in the frame are just the pairs of elements that formed triples with x in the original KTS.

Example 11 *A Kirkman Frame of type 2^{10}.*

Delete any point from the $KTS(21)$ given in Table 1.1.

Kirkman Frames were first formally introduced by Stinson [50], who determined their spectrum in the case where all holes have the same size:

Theorem 12 *A Kirkman Frame of type g^u exists if and only if g is even, $u \geq 4$ and $g(u - 1) \equiv 0(\bmod 3)$.*

Conceptually, the idea of Kirkman Frames had been around for some time; both Lu Jiaxi [33] and Ray-Chaudhuri and Wilson [38] exploited them in their original proofs of the existence of Kirkman Triple Systems. Since their introduction by Stinson however, Kirkman Frames have proven to be the single most valuable tool for the construction of the various generalizations of KTSs that will be discussed in this survey. A particularly powerful use of frames occurs in the following construction. (Compare to the Wilson Fundamental Construction.)

Construction 13 (Fundamental Frame Construction) [50]   Let $(X, \mathcal{G}, \mathcal{A})$ be a GDD, and let $w : X \to \mathbb{Z}^+ \cup \{0\}$ be a weight function on X. Suppose that for each block $b \in \mathcal{A}$ there is a Kirkman

Frame of type $\{w(x) : x \in b\}$. *Then there is a Kirkman Frame of type* $\{\sum_{x \in G} w(x) : G \in \mathcal{G}\}$.

A proof of Construction 13 can be found in the Stinson survey [51]. A simple consequence of the Fundamental Frame Construction is that, for each fixed g, the set of values u for which there is a Kirkman Frame of type g^u is PBD-closed.

Theorem 14 *For each (even) $g > 0$, $\{u : \exists$ a Kirkman Frame of type $g^u\}$ is PBD-closed.*

Proof. Write $\mathcal{U} = \{u : \exists$ a Kirkman Frame of type $g^u\}$ and suppose that $v \in B(\mathcal{U})$. Delete a point from a $(v, \mathcal{U})$-PBD to yield a $GDD(X, \mathcal{G}, \mathcal{A})$ with $|b| \in \mathcal{U}$ for every block $b \in \mathcal{A}$, and $|G| \in \mathcal{U} - 1 = \{u - 1 : u \in \mathcal{U}\}$ for every group $G \in \mathcal{G}$. Apply the Fundamental Frame Construction with $w(x) = g$ for every $x \in X$ to yield a Kirkman Frame of type $\{\sum_{x \in G} g : G \in \mathcal{G}\}$, that is, a frame in which the size of each hole belongs to the set $\{g(u - 1) : u \in \mathcal{U}\}$. Now adjoin g ideal points to the frame. On each hole of size $g(u-1)$ in the frame together with the ideal points construct a Kirkman Frame of type g^u, aligning one of its holes H_∞ on the ideal points. Now pair off each of the $g(u-1)/2$ holey parallel classes corresponding to the hole of size $g(u - 1)$ with the holey parallel classes in the frame of type g^u used to 'fill' that hole, *not* using the $g/2$ holey parallel classes corresponding to H_∞. (Thus we have $g(u - 1)/2$ such holey parallel classes, as required.) Finally, there remain on each hole of size $g(u - 1)$ the $g/2$ holey parallel classes corresponding to H_∞; taking unions yields the final $g/2$ holey parallel classes corresponding to H_∞. The result is a Kirkman Frame of type g^v, so that $v \in \mathcal{U}$. Hence $\mathcal{U}$ is PBD-closed. $\square$

The particular case $g = 2$ of Theorem 14 is of historical interest, for it is equivalent to the assertion that the set of replication numbers for Kirkman Triple Systems is PBD-closed. Wilson [56] exploited this fact in his proof of the existence of KTSs by showing that $B(\{4, 7, 10, 19\}) = \{v : v \equiv 1 \pmod 3\}$, whereupon constructions for $KTS(v)$ were required only for $v \in \{9, 15, 21, 39\}$ (see Section 1.1).

Theorem 14 provides an example of what is generally called a 'filling in holes' construction. In this case the product is again a Kirkman Frame, but there are equally effective analogues for filling in holes in frames to produce, for example, resolvable group-divisible designs and resolvable designs having (or missing) subdesigns (see Stinson [51]). We shall see applications of all of these constructions throughout the survey, in addition to applications of Kirkman Frames to produce other combinatorial objects, from packings and coverings to maximal sets of triangle-factors. In particular, we shall provide a new proof of the existence of Nearly Kirkman Triple Systems, which uses as its main ingredient Kirkman Frames with hole size six — see Section 2.

2.   Constructions for Kirkman Triple Systems and Nearly Kirkman Triple Systems for all admissible orders

2.1   Construction for Kirkman Frames with hole size 6 and the Stinson Construction for Kirkman Triple Systems

We begin by determining the spectrum for Kirkman Frames with hole size 6, following Stinson [51]. We shall use the fact that $\{u \geq 4 : \exists$ a Kirkman Frame of type $6^u\}$ is PBD-closed (see Theorem 14).

Lemma 15 (Brouwer, Hanani and Schrijver) [8]
$$B(\{4, 5, 6, 7, 8, 9, 10, 11, 12, 14, 15, 18, 19, 23\}) = \{v : v \geq 4\}.$$

Proof. By deleting some number of points in projective planes of orders 3, 4 and 5 we get $13 \in B(\{4\})$, $16, 17, 20, 21 \in B(\{4, 5\})$, and $v \in B(\{4, 5, 6\})$ for $24 \leq v \leq 31$. By adjoining 7 infinite points to the parallel classes of a $KTS(15)$ we have $22 \in B(\{4, 7\})$. Now suppose $44 \leq v \leq 47$. Adjoin $v - 40$ infinite points to a resolvable $(40, 4, 1)$-$BIBD$ to get $v \in B(\{4, 5, 6, 7\})$.

We now proceed by induction on v. Write $v = 4t + 4, 4t + 5, 4t + 6$ or $4t + 7$ where $t \geq 7, t \neq 10$. Take a $TD(5, t)$ (which is equivalent to a set of three mutually orthogonal Latin squares of order t) and truncate a group to 4, 5, 6 or 7 points to get $v \in B(\{4, 5, 6, 7, t\})$; as $t < v$ the result now follows by induction.   □

Remark 16 As observed in [8], for each $v \in S = \{4, 5, 6, 7, 8, 9, 10, 11, 12, 14, 15, 18, 19, 23\}$, there does not exist a PBD on v points with block sizes in $S\backslash\{v\}$; thus every element of S is essential.

Theorem 17 *There exists a Kirkman Frame of type 6^u for every $u \geq 4$.*

Proof. From Lemma 15 it suffices to prove the Theorem for $u \in S = \{4, 5, 6, 7, 8, 9, 10, 11, 12, 14, 15, 18, 19, 23\}$.

The following is a Kirkman Frame of type 6^6:

Points: $(((\mathbb{Z}_5 \times \{1, 2\}) \cup \{\infty\}) \times \mathbb{Z}_3) \cup \{\alpha_1, \alpha_2, \alpha_3\}$

Holes:

$$\{\{(0, 1, 0), (0, 1, 1), (0, 1, 2), (0, 2, 0), (0, 2, 1), (0, 2, 2)\} \bmod (5, -; -)\}$$

$$\cup\{\{(\infty, 1), (\infty, 2), (\infty, 0), \alpha_1, \alpha_2, \alpha_3\}\}$$

Holey Parallel Classes:

$$\left.\begin{array}{ll}
\{(\infty, 1), (1, 1, 0), (3, 1, 2)\} & \{\alpha_3, (4, 1, 1), (2, 2, 0)\} \\
\{(\infty, 2), (4, 2, 1), (3, 2, 2)\} & \{(2, 1, 1), (3, 1, 1), (4, 1, 2)\} \\
\{(\infty, 0), (3, 1, 0), (2, 2, 1)\} & \{(1, 2, 2), (3, 2, 1), (4, 2, 2)\} \\
\{\alpha_1, (1, 1, 1), (2, 2, 2)\} & \{(4, 1, 0), (2, 1, 0), (1, 2, 0)\} \\
\{\alpha_2, (2, 1, 2), (1, 2, 1)\} & \{(3, 2, 0), (4, 2, 0), (1, 1, 2)\}
\end{array}\right\} \bmod (5, -; 3)$$

together with

$$\{(4, 1, 2), (3, 1, 0), (0, 2, 2)\}\{(2, 2, 0), (4, 2, 1), (1, 1, 1)\} \bmod (5, -; 3).$$

To obtain Kirkman Frames of type 6^u for $u \in S\backslash\{6\}$ we construct a $(3u + 1, \{4, 7, 10\})$-*PBD* with a 4-head (i.e. a point which is contained exclusively in blocks of size 4). Removing the 4-head yields a $\{4, 7, 10\}$-*GDD* of type 3^u; applying the Fundamental Frame Construction with weight 2, using frames of types 2^4, 2^7 and 2^{10} as input designs then yields the result (see Section 1.5).

For $u \in \{4, 5, 8, 9, 12\}$ we take a $(3u + 1, 4, 1)$-*BIBD*. For $u \in \{7, 10\}$ we take a $KTS(2u+1)$ and adjoin u infinite points to its parallel classes, while for $u = 11$ we take the resolvable 3-*GDD* of type 4^6 given in [51] and adjoin 10 infinite points to its parallel classes. Constructions for $u = 14, 15, 18$ and 19 are from Brouwer [6]: for $u = 15$ and 19 he gives direct constructions for $(3u + 1, \{4, 7^*\})$-*PBD*s, while for $u = 14$ and 18 he gives simple recursive constructions for $(3u + 1, \{4, 7^*\})$-*PBD*s starting with a $TD(4, 10)$ and a $TD(4, 13)$. Finally, for $u = 23$ we start with the following 4-*GDD* of type $2^9 5^1$, also from Brouwer [6]:

Points: $(\mathbb{Z}_2 \times \mathbb{Z}_3 \times \mathbb{Z}_3) \cup \{\infty_0, \infty_1, \infty_2, \infty_3, \infty_4\}$

Groups: $\{\{(0, 0, 0), (1, 0, 0)\} \bmod (-, 3, 3)\} \cup \{\{\infty_0, \infty_1, \infty_2, \infty_3, \infty_4\}\}$

Blocks:
$$\{\{(0,0,0),(0,1,0),(0,2,0),\infty_4\} \bmod (2,-,3)\}$$
$$\cup \quad \{\{(0,0,2),(0,1,1),(0,2,0),\infty_3\},\{(1,0,0),(1,0,1),(1,0,2),\infty_3\}$$
$$\bmod (-,3,-)\}$$
$$\cup \quad \{\{(0,0,0),(1,0,1),(1,1,2),\infty_0\},\{(0,0,1),(1,0,2),(1,1,0),\infty_1\},$$
$$\{(0,0,2),(1,0,0),(1,1,1),\infty_2\} \bmod (2,3,-)\}$$
$$\cup \quad \{\{(0,0,0),(0,0,1),(1,1,1),(1,2,0)\} \bmod (-,3,3)\}$$

Apply the Wilson Fundamental Construction (see Section 1.3) to this GDD with weight 3, using $TD(4,3)$s as input designs, to obtain a 4-GDD of type $6^9 15^1$. Adjoining one infinite point to the groups and constructing a $(16,4,1)$-$BIBD$ on the group of size 15 together with the infinite point yields the desired PBD.

This completes the proof. $\square$

Theorem 18 *There exists a Kirkman Triple System $KTS(v)$ if and only if $v \equiv 3(\bmod 6)$.*

Proof. A $KTS(3)$ consists of a single triple, while $KTS(v)$ for $v = 9$, 15 and 21 were constructed in Section 1.1. Assume $v \geq 27$ and write $v = 6u + 3, u \geq 4$. By Theorem 17 there is a Kirkman frame of type 6^u. Adjoin three ideal points to this frame and on each hole in the frame together with the three ideal points construct a $KTS(9)$ in which the ideal points form a block. For each hole in the frame we pair off its three holey parallel classes with the three parallel classes in the $KTS(9)$ used to fill that hole that do *not* contain the block formed by the ideal points. This yields $3u$ parallel classes. One more parallel class is obtained by taking the union of the remaining parallel classes in each of the $KTS(9)$s. The result is a $KTS(v)$, as desired (see Section 1.5). $\square$

2.2   A Unified Construction for Nearly Kirkman Triple Systems

In this section we present a new proof for the existence of $NKTS(v)$ for $v \equiv 0(\bmod 6), v \geq 18$, relying heavily on the Kirkman Frames of type 6^u that were constructed in Theorem 17.

We begin by indicating constructions for a number of small designs.

Lemma 19 *There exist $NKTS(v)$ for $v = 30, 36$ and 48.*

Proof. These designs are all directly constructed by Baker and Wilson in [3]. There is a minor error in the NKTS(36), which we fix presently.

$NKTS(36)$

Point set: $(\mathbb{Z}_{17} \times \{1, 2\}) \cup \{\infty_1, \infty_2\}$

Groups: $\{\{(i, 1), (i, 2)\} : i \in \mathbb{Z}_{17}\} \cup \{\{\infty_1, \infty_2\}\}$

Parallel Classes:

$$
\left.
\begin{array}{ccccccccc}
(2,1) & (7,2) & (8,2) & (12,1) & (10,2) & (13,2) & (3,1) & (12,2) & (0,2) \\
(11,1) & (14,2) & (4,2) & (7,1) & (6,2) & (15,2) & (0,1) & (9,1) & (11,2) \\
(13,1) & (8,1) & (3,2) & (10,1) & (14,1) & (4,1) & (15,1) & (16,1) & (1,1) \\
(16,2) & (1,2) & (5,2) & \infty_1 & (5,1) & (9,2) & (6,1) & (2,2) & \infty_2
\end{array}
\right\} \text{ mod } 17.
$$

$\square$

Lemma 20 *There exist* $NKTS(v)$ *for* $v = 24, 54, 60, 72, 78$ *and* 144.

Proof. The following self-explanatory constructions are from Kotzig and Rosa [28] (both constructions use resolvable $TD(3, n)$s).

(i) If there exists a $KTS(v), v \geq 9$, then there exists an $NKTS(3(v-1))$.

Apply this construction to $v = 24 = 3 \cdot 8$, $v = 60 = 3 \cdot 20$ and $v = 78 = 3 \cdot 26$. The required Kirkman Triple Systems exist by Theorem 18.

(ii) If there exists an $NKTS(v)$ then there exists an $NKTS(3v)$.

Apply this construction to $v = 54 = 3 \cdot 18$, $v = 72 = 3 \cdot 24$ and $v = 144 = 3 \cdot 48$. An $NKTS(18)$ was constructed in Example 9 and an $NKTS(48)$ exists by Lemma 19. $\square$

Lemma 21 *There exists an* $NKTS(84)$.

Proof. This design is constructed by Rees and Stinson [46]; they construct a 4-GDD of type $6^5 9^1$ and apply the Fundamental Frame Construction with weight 2, using frames of type 2^4, to produce a Kirkman Frame of type $12^5 18^1$. Adjoin six ideal points; on each hole of size 12 together with the ideal points construct a copy of the following $INKTS(18, 6)$ (this design was constructed by Brouwer [7]; for the definition of an $INKTS$ see Section 5.1), aligning the points $\{a, b, c, d, e, f\}$ on the ideal points. Then construct an $NKTS(24)$ (Lemma 20) on the hole of size 18 together with the ideal points. The construction works as follows. For each hole of size 12 in the frame we pair off its 6 holey parallel classes with the six parallel classes in the $INKTS(18, 6)$ used to fill that hole. Then the 9 holey parallel classes in the frame corresponding to the hole of size 18 are paired with nine of the parallel classes in the $NKTS(24)$. The last two parallel classes in the $NKTS(24)$ are paired with the holey parallel classes in the $INKTS(18, 6)$s for two more parallel classes. This yields in all a total of 41 parallel classes of triples, as desired.

$INKTS(18, 6)$

Points: $\{0, 1, ..., 11\} \cup \{a, b, c, d, e, f\}$

Hole: $\{a, b, c, d, e, f\}$

Groups: $\{\{i, i + 6\} : i = 0, 1, \ldots, 5\}$

Holey Parallel Classes:

$0, 1, 2$	$3, 4, 5$	$6, 7, 8$	$9, 10, 11$
$0, 3, 7$	$1, 4, 11$	$2, 6, 9$	$5, 8, 10$

Parallel Classes:

$a, 0, 11$	$b, 1, 3$	$c, 4, 8$	$d, 6, 10$	$e, 2, 7$	$f, 5, 9$
$a, 7, 9$	$b, 0, 10$	$c, 1, 5$	$d, 2, 4$	$e, 8, 11$	$f, 3, 6$
$a, 4, 6$	$b, 2, 11$	$c, 0, 9$	$d, 5, 7$	$e, 3, 10$	$f, 1, 8$
$a, 1, 10$	$b, 5, 6$	$c, 2, 3$	$d, 0, 8$	$e, 4, 9$	$f, 7, 11$
$a, 3, 8$	$b, 4, 7$	$c, 6, 11$	$d, 1, 9$	$e, 0, 5$	$f, 2, 10$
$a, 2, 5$	$b, 8, 9$	$c, 7, 10$	$d, 3, 11$	$e, 1, 6$	$f, 0, 4$

$\square$

Remark 22 The idea of using six ideal points in the type of construction given in Lemma 21 was introduced by Brouwer [7] to construct $NKTS(v)$ for all $v \equiv 6(\bmod 12), v \geq 66$. His constructions rely on the existence of 4-GDDs of type $6^u, u \geq 5$.

We are now ready to give the main constructions for $NKTS(v)$.

Theorem 23 *There exists an $NKTS(v)$ for all $v \equiv 18(\bmod 24)$.*

Proof. Write $v = 24u + 18$, where $u \geq 0$. Take a $KTS(12u + 9)$ (Theorem 18); this KTS has an even number $6u + 4$ of parallel classes.

Apply the Doubling Construction (Theorem 7) to obtain an $NKTS(v)$, as desired. (See Example 9.) □

Theorem 24 *There exists an $NKTS(v)$ for all $v \equiv 0(\bmod 24)$.*

Proof. The cases $v = 24,48,72$ and 144 have all been handled in Lemmas 19 and 20. Now assume $v = 24u, u \geq 4, u \neq 6$. Take a resolvable 3-$GDD$ of type 6^4 (see Example 4) and apply Construction 6 with $g = 6, u = 4$ and $n = u$ to obtain a resolvable 3-$DGDD$ of type $(6u, 6^u)^4$. On each group construct a Kirkman Frame of type 6^u (Theorem 17) aligning the holes in the frame on the intersections of the holes in the $DGDD$ with that particular group, and on each hole construct an $NKTS(24)$ (Lemma 20). Now the $DGDD$ itself has $9u - 9$ parallel classes of triples. We obtain $3u$ further parallel classes by taking 3 parallel classes in each of the u $NKTS(24)$s and forming unions with the 3 holey parallel classes in each of the four frames. There then remain 8 parallel classes in each of the $NKTS(24)$s; take the unions over these to form the 8 final parallel classes of the design. We have in all $(9u - 9) + 3u + 8 = 12u - 1$ parallel classes of triples, forming an $NKTS(24u)$, as desired. □

Remark 25 Note that, as an intermediary design in the proof of Theorem 24, we produce a $\{3\}$-GDD of type 24^u whose set of triples can be partitioned into $9u - 9$ parallel classes and $3u$ holey parallel classes. This design is an example of what is called a *semiframe*; for more on these designs we refer the reader to [40] and to the Stinson survey [51].

Theorem 26 *There exist $NKTS(v)$ for all $v \equiv 6$ or $12(\bmod 24)$ with $v \geq 30$.*

Proof. The cases $v = 30, 36, 54, 60, 78$ and 84 were handled in Lemmas 19, 20 and 21.

Assume $u \geq 4$ and take a Kirkman frame of type 6^u (Theorem 17). Apply Construction 6 with $n = 4$ to obtain a Kirkman frame of type 24^u. We now consider two cases:

(i). $v \equiv 6(\bmod 24), v \geq 102$.

Write $v = 24u + 6, u \geq 4$. Take a Kirkman Frame of type 24^u and adjoin 6 ideal points. On all but one of the holes of size 24 together with the ideal points construct a copy of the following $INKTS(30, 6)$ (this design was constructed by Rees [41]), aligning the points $\{a, b, c, d, e, f\}$

on the ideal points. Then construct an $NKTS(30)$ on the last hole of size 24 together with the ideal points. (Compare with the proof of Lemma 21.)

$INKTS(30,6)$

Points: $(\mathbb{Z}_{12} \times \{0,1\}) \cup \{a,b,c,d,e,f\}$

Hole: $\{a,b,c,d,e,f\}$

Groups: $\{\{(i,0),(6+i,0)\},\{(i,1),(6+i,1)\} : i = 0,1,\ldots,5\}$

Holey Parallel Classes:

Develop the triples $(1,0),(5,0),(9,0)$ and $(0,0),(1,1),(2,1)$ mod 12 for two classes.

Parallel Classes:

$$\left.\begin{array}{ll}
(0,0),(1,0),(3,0) & (6,0),(3,1),b \\
(0,1),(2,1),(5,1) & (7,0),(1,1),c \\
(5,0),(10,0),(8,1) & (8,0),(7,1),d \\
(2,0),(6,1),(10,1) & (9,0),(4,1),e \\
(4,0),(9,1),a & (11,0),(11,1),f
\end{array}\right\} \text{ mod } 12.$$

(ii). $v \equiv 12(\bmod 24), v \geq 108$.

We proceed as in Case (i), writing $v = 24u + 12, u \geq 4$ and adjoining 12 ideal points to a Kirkman Frame of type 24^u. We require the following $INKTS(36,12)$ (this design can be found in Deng, Rees and Shen [21]) and an $NKTS(36)$.

$INKTS(36,12)$

Points: $(\mathbb{Z}_{12} \times \{1,2\}) \cup (\mathbb{Z}_3 \times \{3\}) \cup (\mathbb{Z}_2 \times \{4,5\}) \cup \{a,b,c,d,e\}$

Hole: $(\mathbb{Z}_3 \times \{3\}) \cup (\mathbb{Z}_2 \times \{4,5\}) \cup \{a,b,c,d,e\}$

Groups: $\{\{(i,1),(6+i,1)\},\{(i,2),(6+i,2)\} : i = 0,1,\ldots,5\}$

Holey Parallel Classes:

Develop the triples $(0,1),(7,2),(5,2)$ and $(9,2),(8,1),(10,1)$ mod 12 for three classes;

develop the triples $(0,1)$, $(4,1)$, $(8,1)$ and $(1,1)$, $(9,2)$, $(10,2)$ mod 12 for two further classes.

Parallel Classes:

$$
\left.
\begin{array}{lll}
(0,1),(1,1),(0,3) & (6,2),(9,2),(1,4) & (6,1),(8,2),b \\
(0,2),(4,2),(2,3) & (4,1),(9,1),(0,5) & (10,1),(1,2),c \\
(3,1),(7,2),(1,3) & (10,2),(3,2),(1,5) & (8,1),(2,2),d \\
(2,1),(5,1),(0,4) & (11,1),(11,2),a & (7,1),(5,2),e
\end{array}
\right\} \text{ mod } 12.
$$

Elements with second coordinate 3 are to be developed mod 3 while elements with second coordinate 4 or 5 are to be developed mod 2. □

Collecting Theorems 23, 24 and 26 now yields the spectrum for Nearly Kirkman Triple Systems:

Theorem 27 *There exists a Nearly Kirkman Triple System $NKTS(v)$ if and only if $v \equiv 0(\text{ mod } 6)$, $v \geq 18$.*

Remark 28 It is easy to see that no $NKTS(6)$ can exist, and Kotzig and Rosa [28] eliminated the possible existence of an $NKTS(12)$ by direct computation. It seems to the authors that there should be a theoretical proof for the non-existence of an $NKTS(12)$ but we are not aware of any at this time.

We conclude this section by noting that determining the spectrum for Nearly Kirkman Triple Systems solves a special case (namely, $a = 1$) of the following general problem. For which non-negative integers a, b and n with $a + 2b = 6n - 1$ does the complete graph K_{6n} admit an edge-decomposition into a one-factors and b triangle factors? A complete solution for $a > 1$ was first given by Rees [39] in a lengthy article in 1987; a much shorter proof (using the Doubling Construction, see Theorem 7) appears in Rees [42]:

Theorem 29 *The complete graph K_{6n} admits an edge-decomposition into a one-factors and b triangle factors for all non-negative integers a, b and n with $a > 1$ and $a + 2b = 6n - 1$.*

3.   Early Generalizations

3.1   Minimum Resolvable Coverings by Triangle-Factors

Recall the exact wording of the Kirkman Schoolgirl Problem:

> Fifteen young ladies in a school walk out three abreast for seven days
> in succession: it is required to arrange them daily, so that no two will
> walk twice abreast.

The condition that no two walk twice abreast implies that every pair walks abreast exactly once.

The earliest generalization of Kirkman Triple Systems, namely Nearly Kirkman Triple Systems, were motivated by considering Kirkman's problem for 18 girls instead of 15. The solution is no longer exact; if no two are to walk twice abreast, the maximum number of days is 8, and some pairs of girls never walk together. But suppose that Kirkman had posed his original problem as follows.

> Fifteen young ladies in a school walk out three abreast for seven days
> in succession: it is required to arrange them daily, so that each pair will
> walk at least once abreast.

The solution of course is still a $KTS(15)$, but what if we consider this problem for 18 girls? It is now necessary to walk for at least 9 days, and some pairs will walk more than once. More generally we may ask, what is the smallest number of triangle-factors required to cover the pairs of a v-set where $v \equiv 0 \pmod 6$? Clearly this number must be at least $v/2$, and when achieved the excess of the cover will be a one-factor. It turns out that this bound cannot be achieved for $v = 6$ or 12 (one needs, respectively, 4 or 7 triangle-factors). Due in large part to Theorem 12, a complete answer to the foregoing question is now known (see Assaf, Mendelsohn and Stinson [2] and Lamken and Mills [30]):

Theorem 30 *There exists a resolvable covering of v points by $v/2$ triangle-factors if and only if $v \equiv 0 \pmod 6, v \geq 18$.*

Remark 31 It is easy to see that for $v \equiv 0 \pmod 6, v \geq 18$ we can take the triangle-factors of a Nearly Kirkman Triple System and the triangle-factors of a minimum resolvable covering on the same v-set of points so that their union forms the resolution classes of a resolvable

$(v, 3, 2)$-$BIBD$. For $v = 12$, there is a unique maximum packing of 4 triangle-factors on 12 points, constructed in Section 1.1 (see Table 13.3), and it follows from that construction that the leave of the packing is three disjoint copies of K_4. It is not difficult to construct a resolvable covering of 12 points by 7 triangle-factors so that its excess consists of three disjoint K_4s (see Table 13.4). These two designs together constitute a resolvable $(12, 3, 2)$-$BIBD$.

Theorem 32 *For every $v \equiv 0(\bmod\, 3), v \neq 6$, there exists a resolvable $(v, 3, 2)$-$BIBD$ whose triangle-factors admit a partition into a maximum packing and a minimum covering of v points by triangle-factors.*

Proof. No resolvable $(6, 3, 2)$-$BIBD$ exists. For $v \equiv 0(\bmod\, 6), v \geq 12$, see Remark 31. For $v \equiv 3(\bmod\, 6)$ we take two copies of a $KTS(v)$. $\square$

A resolvable $(v, 3, 2)$-$BIBD$ of the type described in Theorem 32 will be called *separable*; in Section 4.3 we shall consider the analogous problem for near-resolvable $(v, 3, 2)$-$BIBDs$ — that is, we shall discuss the existence of near-resolvable $(v, 3, 2)$-$BIBDs$ whose near-triangle-factors admit a partition into a maximum packing and a minimum covering. The Hanani Triple Systems considered in the next subsection will play a critical rôle in the solution to this problem.

3.2 Hanani Triple Systems (The Chromatic Index Problem)

The *chromatic index* of a Steiner Triple System $(X, \mathcal{B})$, denoted $\chi'(X, \mathcal{B})$, is the smallest number of colors required to color the blocks of $\mathcal{B}$ in such a way that intersecting blocks do not receive the same color; more formally, it is the smallest size of a set C for which there exists a

packing				covering			
1, 5, 9	2, 6, 10	3, 7, 11	4, 8, 12	1, 5, 9	2, 6, 10	3, 7, 11	4, 8, 12
1, 6, 11	2, 5, 12	3, 8, 9	4, 7, 10	1, 6, 11	2, 5, 12	3, 8, 9	4, 7, 10
1, 7, 12	2, 8, 11	3, 5, 10	4, 6, 9	1, 7, 12	2, 8, 11	3, 5, 10	4, 6, 9
1, 8, 10	2, 7, 9	3, 6, 12	4, 5, 11	1, 8, 10	2, 3, 4	5, 6, 7	9, 11, 12
				4, 5, 11	1, 2, 3	6, 7, 8	9, 10, 12
				2, 7, 9	1, 3, 4	5, 6, 8	10, 11, 12
				3, 6, 12	1, 2, 4	5, 7, 8	9, 10, 11

Table 13.4. Resolvable packing and covering forming a resolvable $(12, 3, 2)$-$BIBD$

function $f : \mathcal{B} \rightarrow C$ with the property that $f^{-1}(c)$ is a disjoint set of blocks for every $c \in C$. It is easy to see that in any STS of order v, the chromatic index is at least $(v-1)/2$; moreover a coloring with $(v-1)/2$ colors necessarily has the property that each color class is in fact a parallel class of triples, and the color classes between them form a Kirkman triple system $KTS(v)$. It follows immediately that if $v \equiv 1(\bmod 6)$ then the chromatic index of an $STS(v)$ is at least $(v+1)/2$. For $v = 7$ the unique $STS(v)$ has the property that every pair of blocks intersects, so seven colors are required. Neither of the two $STS(13)$s has chromatic index 7. The following is an $STS(19)$ with chromatic index 10:

Example 33 *Point Set:* $(\mathbb{Z}_9 \times \{1,2\}) \cup \{\infty\}$

Nine Color Classes:

$$
\left.
\begin{array}{ccc}
\infty, & (0,1), & (0,2) \\
(1,1), & (3,2), & (8,2) \\
(2,1), & (3,1), & (8,1)
\end{array}
\qquad
\begin{array}{ccc}
(4,1), & (5,2), & (7,2) \\
(5,1), & (7,1), & (4,2) \\
(6,1), & (1,2), & (2,2)
\end{array}
\right\} \text{ mod } 9
$$

Tenth Color Class:

$$(0,2), (3,2), (6,2) \quad (1,2), (4,2), (7,2) \quad (2,2), (5,2), (8,2)$$

This block coloring has a particularly nice form. Nine of its color classes are in fact near-parallel classes, while the tenth 'short' class covers precisely the nine points that are not covered by all of the near-parallel classes. A *Hanani Triple System* $HTS(v)$ is a Steiner Triple System $STS(v)$, $v \equiv 1(\bmod 6)$, with chromatic index $(v+1)/2$, whose block-set admits a partition into $(v-1)/2$ near-parallel classes together with one further class of $(v-1)/6$ disjoint triples. Thus Example 33 illustrates an $HTS(19)$. The spectrum for Hanani Triple Systems has been determined (see Carter *et al* [12]):

Theorem 34 *There exists a Hanani Triple System $HTS(v)$ if and only if $v \equiv 1(\bmod 6)$ and $v \geq 19$.*

It is not difficult to see how Kirkman Frames come to be so useful in the proof of Theorem 34. For example, for $v \equiv 1(\bmod 18), v \geq 73$ we can construct an $HTS(v)$ by starting with a frame of type $18^{(v-1)/18}$ (see Theorem 12) and adjoining one ideal point; on each hole in the frame together with the ideal point construct a copy of the $HTS(19)$ given in Example 33, taking ∞ to be the ideal point. Each of the nine holey parallel classes with respect to a given hole in the frame is paired with a near-parallel class in the $HTS(19)$ used to fill that hole; finally, the union of the short classes in the $HTS(19)s$ forms the short class in the resulting $HTS(v)$.

3.3 Resolvable Group-Divisible Designs with Block Size Three and the Rees Construction

The spectrum for resolvable 3-$GDDs$ of type g^u has been completely settled (see Mendelsohn and Shen [35], Assaf and Hartman [1], Rees and Stinson [46] and Rees [42]):

Theorem 35 *There exists a resolvable* 3-GDD *of type* g^u *if and only if* $u \geq 3$, $gu \equiv 0 (\mathrm{mod}\, 3)$ *and* $g(u - 1)$ *is even, except for* $(g, u) \in \{(2, 3), (2, 6), (6, 3)\}$.

For the most part the proof of this theorem involved the use of Kirkman frames and TD inflation; the only unsettled case at the time of the Stinson survey was that where $u = 6$ and $g \equiv 2$ or $10 (\mathrm{mod}\, 12)$, $g \geq 22$ (Assaf and Hartman [1] had constructed resolvable 3-$GDDs$ of types 10^6 and 14^6). The problem here was the non-existence of a resolvable 3-GDD of type 2^6 (i.e. an $NKTS(12)$), whereupon one could not use TD inflation.

Rees [42] completed the spectrum with the following result.

Theorem 36 *A resolvable* 3-GDD *of type* g^6 *exists for every* $g \equiv 2$ *or* $10 (\mathrm{mod}\, 12)$ *with* $g > 2$.

The proof uses the following construction, which is a special case of [42, Construction 1]. In what follows, a *set system* $(X, \mathcal{B})$ consists simply of a set X together with a collection $\mathcal{B}$ of subsets of X. A set system is called resolvable if $\mathcal{B}$ admits a partition into parallel classes, each parallel class being a partition of X. An α-parallel class of blocks in a GDD $(X, \mathcal{G}, \mathcal{B})$ is a subset $B \subseteq \mathcal{B}$ with the property that each point $x \in X$ is contained in exactly α blocks in B. A GDD is called A-resolvable if its block set $\mathcal{B}$ admits a partition into subsets $B_1, B_2, ..., B_r$ where for each $i = 1, 2, ..., r$, B_i is an α_i-parallel class of blocks, $\alpha_i \in A$.

Construction 37 *Let* $(X, \mathcal{G}, \mathcal{B})$ *be an* A-resolvable 3-GDD *of type* m^u *in which for each* $\alpha_i \in A$ *there are* r_i α_i-parallel classes of triples. Sup-pose that there is a difference matrix* $DM(h, u, 1)$ *over some group* $\mathcal{H}$,

and let $\{H_j\}$ be a collection of subsets of $\mathcal{H}$, there being r_i such subsets of size α_i for each $\alpha_i \in A$, where $\bigcap_j H_j \neq \emptyset$. If the set system
$(\mathcal{H}, \{H_j * \tau : \tau \in \mathcal{H}, \ j = 1, 2, \ldots, \ \sum_i r_i\})$ is resolvable then there is a
resolvable 3-GDD of type $(hm)^u$.

We illustrate Construction 37 by proving Theorem 36, using $g = 22$ as
an example, i.e. constructing a resolvable 3-GDD of type 22^6. We start
with the following 3-GDD of type 2^6 whose block set can be partitioned
into a 2-parallel class and a 3-parallel class.

$$
\begin{aligned}
X = \ & \mathbb{Z}_{13} - \{12\} \\
\mathcal{G} = \ & \{\{0,3\}, \{1,7\}, \{2,11\}, \{4,6\}, \{5,10\}, \{8,9\}\} \\
\mathcal{B} = \ & \{\{0,1,4\}, \{0,2,8\}, \{1,3,9\}, \{2,4,10\}, \{3,5,11\}, \\
& \{5,6,9\}\{6,7,10\}, \{7,8,11\}\} \\
& \bigcup \{\{1,2,5\}, \{2,3,6\}, \{3,4,7\}, \{4,5,8\}, \{5,7,0\}, \{6,8,1\} \\
& \{7,9,2\}, \{8,10,3\}, \{9,10,0\}, \{9,11,4\}\{10,11,1\}, \{11,0,6\}\}
\end{aligned}
$$

We apply Construction 37 with $A = \{2,3\}$ where $\alpha_1 = 2$, $\alpha_2 = 3$,
$r_1 = r_2 = 1$ and $m = 2$, $u = 6$. Because of the previously known
existence of resolvable 3-GDDs of types 10^6 and 14^6 we may assume that
$g \geq 22$ and (by TD inflation) that $\gcd(g, 5) = 1$. Thus we have $g = 2h$
where $h \equiv 1$ or $5(\bmod 6)$, $h \geq 11$ and $\gcd(h, 5) = 1$. Hence there is a
difference matrix $DM(h, 6, 1)$ over $\mathbb{Z}_h$ (take as its (i, j)-entry the product
$(i-1) \cdot j(\bmod h)$ where $i = 1, 2, \ldots, 6$ and $j = 0, 1, \ldots, h-1 \in \mathbb{Z}_h$). Let
$H_1 = \{0, 1\}$ and $H_2 = \{0, 1, 2\}$. It is not difficult to show that the set
system $(\mathbb{Z}_h, \{H_j + \tau : \tau \in \mathbb{Z}_h, \ j = 1, 2\})$ is resolvable. As an example we
have the following resolution for $h = 11$.

$\{9, 10\} \ \{0, 1, 2\} \ \{3, 4, 5\} \ \{6, 7, 8\}; \ \{1, 2\} \ \{3, 4\} \ \{5, 6\} \ \{7, 8\} \ \{9, 10, 0\};$
$\{10, 0\} \ \{1, 2, 3\} \ \{4, 5, 6\} \ \{7, 8, 9\}; \ \{2, 3\} \ \{4, 5\} \ \{6, 7\} \ \{8, 9\} \ \{10, 0, 1\};$
$\{0, 1\} \ \{2, 3, 4\} \ \{5, 6, 7\} \ \{8, 9, 10\}.$

Now the resulting GDD will have point set $X \times \mathbb{Z}_h$, with groups $G_i \times \mathbb{Z}_h$,
$G_i \in \mathcal{G}$. Let P_1 be the 2-parallel class of triples in $(X, \mathcal{G}, \mathcal{B})$ and from its
triples form the following disjoint spanning set of triples on $X \times H_1$ in
the obvious way.

$$
\begin{aligned}
P_1' = \{ & \{(0,0), (1,0), (4,0)\}, \ \{(0,1), (2,0), (8,0)\}, \ \{(1,1), (3,0), (9,0)\}, \\
& \{(2,1), (4,1), (10,0)\}, \ \{(3,1), (5,0), (11,0)\}, \ \{(5,1), (6,0), (9,1)\}, \\
& \{(6,1), (7,0), (10,1)\}, \ \{(7,1), (8,1), (11,1)\}\}
\end{aligned}
$$

In a similar manner construct from the 3-parallel class P_2 of triples in $(X, \mathcal{G}, \mathcal{B})$ the following disjoint spanning set of triples on $X \times H_2$.

$$
\begin{aligned}
P_2' = \{ & \{(1,0),(2,0),(5,0)\}, \{(2,1),(3,0),(6,0)\}, \quad \{(3,1),(4,0),(7,0)\}, \\
& \{(4,1),(5,1),(8,0)\}, \quad \{(5,2),(7,1),(0,0)\}, \quad \{(6,1),(8,1),(1,1)\}, \\
& \{(7,2),(9,0),(2,2)\}, \quad \{(8,2),(10,0),(3,2)\}, \quad \{(9,1),(10,1),(0,1)\}, \\
& \{(9,2),(11,0),(4,2)\}, \{(10,2),(11,1),(1,2)\}, \{(11,2),(0,2),(6,2)\}\}
\end{aligned}
$$

Now from the $DM(h,6,1)$ we get a resolvable $TD(6,h)$ (the base blocks are the columns of the difference matrix, each one generating a parallel class when developed modulo h). Let the points of the TD be labeled (i, τ) where $i = 1, 2, \ldots, 6$ and $\tau \in \mathbb{Z}_h$. Let $b_1' = \{(x_1, h_1), (x_2, h_2), (x_3, h_3)\} \in P_1'$. Then for each block $\{(1, \tau_1), (2, \tau_2), \ldots, (6, \tau_6)\}$ in the TD we construct, from b_1', the block

$$
\{(x_1, h_1 + \tau_{g(x_1)}), (x_2, h_2 + \tau_{g(x_2)}), (x_3, h_3 + \tau_{g(x_3)})\}
$$

where $g(x_i)$ denotes that value j for which $x_i \in G_j$. For example, for $b_1' = \{(0,1),(2,0),(8,0)\}$ we construct for each block in the TD the block $\{(0, 1 + \tau_1), (2, 0 + \tau_3), (8, 0 + \tau_6)\}$. In this way the original block $(x_1, x_2, x_3) \in P_1$ is being replaced by the h^2 triples of a $TD(3, h)$. Perform the analogous procedure for each triple $b_2' \in P_2'$ to again replace each block (x_1, x_2, x_3) in P_2 by the h^2 triples of a $TD(3, h)$. At this stage we have constructed a 3-GDD of type $(2h)^6 = g^6$; it remains to be shown that this GDD is resolvable. We do this by showing that each parallel class Q of sets in $(\mathbb{Z}_h, \{H_j + \tau : \tau \in \mathbb{Z}_h, j = 1, 2\})$ gives rise to h parallel classes of triples in our 3-GDD.

We label the parallel classes in the $TD(6, h)$ as π_j, $j \in \mathbb{Z}_h$; in other words π_j is the parallel class generated by column j in the $DM(h, 6, 1)$. Now let Q be a parallel class of sets from $(\mathbb{Z}_h, \{H_j + \tau : \tau \in \mathbb{Z}_h, j = 1, 2\})$. Let $S = \{\tau \in \mathbb{Z}_h : H_j + \tau \in Q$ for some $j = 1, 2\}$. Now fix $j \in \mathbb{Z}_h$ and let $C = \{b \in \pi_j : (1, s) \in b$ for some $s \in S\}$. For example, with $h = 11$ take $Q = \{9, 10\}, \{0, 1, 2\}, \{3, 4, 5\}, \{6, 7, 8\}$, and take $j = 1$. This yields $S = \{0, 3, 6, 9\}$ and $C = \{\{(1,0),(2,1),(3,2),(4,3),(5,4),(6,5)\}, \{(1,3),(2,4),(3,5),(4,6),(5,7),(6,8)\}, \{(1,6),(2,7),(3,8),(4,9),(5,10), (6,0)\}, \{(1,9),(2,10),(3,0),(4,1),(5,2),(6,3)\}\}$.

Consider now the restriction of the foregoing construction to the action of each block $b \in C$ on the triples of $P_j' = P'(H_j)$ where $H_j + s \in Q$, $(1, s) \in b$. Thus in our example with $h = 11$, where $C = \{c_0, c_3, c_6, c_9\}$, we are considering the action of c_0 on P_2', c_3 on P_2', c_6 on P_2' and c_9 on

P_1'. The action of c_9 on P_1' yields the disjoint triples

$$(0, 9), (1, 10), (4, 1) \quad (0, 10), (2, 0), (8, 3) \quad (1, 0), (3, 9), (9, 3)$$
$$(2, 1), (4, 2), (10, 2) \quad (3, 10), (5, 2), (11, 0) \quad (5, 3), (6, 1), (9, 4)$$
$$(6, 2), (7, 10), (10, 3) \quad (7, 0), (8, 4), (11, 1)$$

while the action of c_{3i} on P_2', $i = 0, 1, 2$ yields the disjoint triples

$$(1, 3i + 1), (2, 3i + 2), (5, 3i + 4) \quad (2, 3i + 3), (3, 3i), (6, 3i + 3)$$
$$(3, 3i + 1), (4, 3i + 3), (7, 3i + 1) \quad (4, 3i + 4), (5, 3i + 5), (8, 3i + 5)$$
$$(5, 3i + 6), (7, 3i + 2), (0, 3i) \quad (6, 3i + 4), (8, 3i + 6), (1, 3i + 2)$$
$$(7, 3i + 3), (9, 3i + 5), (2, 3i + 4) \quad (8, 3i + 7), (10, 3i + 4), (3, 3i + 2)$$
$$(9, 3i + 6), (10, 3i + 5), (0, 3i + 1) \quad (9, 3i + 7), (11, 3i + 2), (4, 3i + 5)$$
$$(10, 3i + 6), (11, 3i + 3), (1, 3i + 3) \quad (11, 3i + 4), (0, 3i + 2), (6, 3i + 5).$$

In all, the foregoing 44 triples form a parallel class on $X \times \mathbb{Z}_{11}$.

In this way we get a parallel class of blocks on our 3-GDD of type g^6 for every pair (Q, π_j) where Q is a parallel class of sets from $(\mathbb{Z}_h, \{H_j + \tau : \tau \in \mathbb{Z}_h, j = 1, 2\})$ and $j \in \mathbb{Z}_h$, for a total of $5h$ parallel classes, as required. To see that each triple in our GDD occurs in exactly one of these $5h$ parallel classes we observe that the triple c is obtained by considering the action of some block b in the $TD(6, h)$ on some set $P_j' = P'(H_j)$ of triples on $X \times H_j$. If b contains $(1, s)$ then the triple c will be picked up when (and only when) considering the parallel class of sets Q containing $H_j + s$. This completes the proof of Theorem 36.

4.   Resolvable Packings and Coverings of v points where v ≢ 0 (mod 3)

4.1   Introduction and Examples

A few years ago the second author received a request for help to solve the following problem.

A high school teacher wants to break his class into groups of three students each (or, if $v \equiv 1 \pmod{3}$ exactly one group of size four is allowed while if $v \equiv 2 \pmod{3}$ exactly one group of size two is allowed) during the semester to work on various small group projects. He wants to find the maximum number r of projects possible without having any pair of students work together twice.

Now of course if $v \equiv 0 \pmod{3}$ then we have a complete solution from Section 2:

Theorem 38 (i) *For* $v \equiv 3(\bmod 6)$, $r(v) = (v-1)/2$;
(ii) $r(6) = 1$ *and* $r(12) = 4$; *for* $v \equiv 0(\bmod 6)$ *with* $v \geq 18$, $r(v) = \frac{1}{2}v - 1$.

Proof. Use a Kirkman Triple System $KTS(v)$ or a Nearly Kirkman Triple System $NKTS(v)$. It is obvious that $r(6) = 1$, while Table 1.3 illustrates the (unique) solution showing $r(12) = 4$. $\qquad\square$

Example 39 (i) $r(17) = 8$
Here each project involves breaking the class up into five triples and one pair, for a total of 16 pairs covered. Hence

$$r(17) \leq \left\lfloor \binom{17}{2} / 16 \right\rfloor = 8.$$

We can attain 8 projects by deleting the symbol $9'$ (say) from the $NKTS(18)$ in Example 9:

Students $\{1, 2, \ldots, 9, 1', 2', \ldots, 8'\}$

Project Assignments

1	4	7	2	5	8	3	6	9	2'	6'	7'	3'	4'	8'	1'	5'	
1	4'	7'	2	5'	8'	1'	5	9	2'	6	7	3'	4	8	3	6'	
1'	4'	7	2'	5'	8	3'	6'	9	2	6	7'	3	4	8'	1	5	
1'	4	7'	2'	5	8'	1	5'	9	2	6'	7	3	4'	8	3'	6	
1	2	3	4	5	6	7	8	9	1'	8'	6'	7'	5'	3'	4'	2'	
1	2'	3'	4	5'	6'	1'	8	6	4'	2	9	7'	5	3	7	8'	
1'	2'	3	4'	5'	6	7'	8'	9	1	8	6'	7	5	3'	4	2	
1'	2	3'	4'	5	6'	1	8'	6	4	2'	9	7	5'	3	7'	8	

The uncovered pairs (i.e. students who do not work together on any project) are $\{i, i'\}$, $i = 1, 2, \ldots, 8$. In particular, student 9 works with all of her classmates, exactly one time each.

(ii) $r(22) = 9$
In this case each project involves breaking the class up into six triples and one quadruple, for a total of 24 pairs covered. Hence

$$r(22) \leq \left\lfloor \binom{22}{2} / 24 \right\rfloor = 9.$$

We can attain 9 projects as follows.

Students $(\mathbb{Z}_9 \times \{1, 2\}) \cup \{\infty_1, \infty_2, \infty_3, \infty_4\}$

Project Assignments

$$
\left.
\begin{array}{ll}
(4,1),(8,1),(5,2),(7,2) & \infty_1,(1,1),(6,2) \\
(0,1),(2,1),(3,1) & \infty_2,(5,1),(3,2) \\
(0,2),(1,2),(4,2) & \infty_3,(6,1),(8,2) \\
 & \infty_4,(7,1),(2,2)
\end{array}
\right\} \text{ mod } 9
$$

Here the uncovered pairs are $\{\infty_i, \infty_j\}$ for $1 \le i < j \le 4$ and $\{(i,1),$ $(i,2)\}$ for $i = 0, 1, \ldots, 8$. This is the vertex-disjoint union of a K_4 with nine $K_2 s$, i.e. the project assignments form the parallel classes of a resolvable $\{3,4\}$-GDD of type $2^9 4^1$ with replication number 9.

An equally legitimate request would be for the covering analogue to the high school teacher's problem, i.e. find the minimum number $\bar{r}$ of projects possible so that each pair of students works together at least once.

Again, if $v \equiv 0(\bmod 3)$ we get a complete solution from Section 2.1 and Section 3.1:

Theorem 40   (i) *For $v \equiv 3(\bmod 6)$, $\bar{r}(v) = (v-1)/2$;*

(ii) $\bar{r}(6) = 4$, $\bar{r}(12) = 7$ *and for $v \equiv 0(\bmod 6)$ with $v \ge 18$, $\bar{r}(v) = \frac{1}{2}v$.*

Proof. Use a Kirkman Triple System $KTS(v)$ or a resolvable covering of v points by $v/2$ triangle-factors (Theorem 30). Table 1.4 illustrates a covering of 12 points by 7 triangle-factors, while the following cover shows $\bar{r}(6) = 4$.

$$
\begin{array}{ccc ccc ccc ccc}
1 & 2 & 3 & \quad 1 & 5 & 6 & \quad 1 & 2 & 4 & \quad 1 & 3 & 4 \\
4 & 5 & 6 & \quad 2 & 3 & 4 & \quad 3 & 5 & 6 & \quad 2 & 5 & 6
\end{array}
$$

$\square$

Example 41 *(i)* $\bar{r}(17) = 9$

Each project involves breaking the class up into five triples and one pair, for a total of 16 pairs covered. Hence

$$
\bar{r}(17) \ge \left\lceil \binom{17}{2} / 16 \right\rceil = 9.
$$

We can attain nine projects by deleting a point from a resolvable covering of 18 points by nine triangle-factors (Theorem 30):

Students $\{1, 2, \ldots, 9, 1', 2', \ldots, 8'\}$

Project Assignments

1 2 1'	3 5 8	4 5' 7'	6 4' 8'	7 2' 3'	9 6'
2 3 2'	4 6 9	5 6' 8'	8 3' 4'	1 7' 1'	7 5'
3 4 3'	5 7 1	9 4' 5'	2 8' 2'	8 6' 1'	6 7'
4 5 4'	6 8 2	1 5' 6'	9 7' 2'	7 8' 1'	3 3'
5 6 5'	7 9 3	2 6' 7'	1 8' 3'	4 4' 1'	8 2'
6 7 6'	8 1 4	3 7' 8'	5 5' 2'	9 3' 1'	2 4'
7 8 7'	9 2 5	6 6' 3'	1 4' 2'	3 5' 1'	4 8'
8 9 8'	1 3 6	7 7' 4'	2 5' 3'	4 6' 2'	5 1'
2 4 7	8 8' 5'	3 6' 4'	5 7' 3'	6 2' 1'	9 1

The doubly covered pairs (i.e. students who work together twice) are $\{i, i'\}$, $i = 1, 2, \ldots, 8$. In particular, student 9 works exactly once with each of his classmates.

(ii) $\bar{r}(22) = 10$

Each project involves breaking the class up into six triples and one quadruple, for a total of 24 pairs covered. Hence $\bar{r}(22) \geq \left\lceil \binom{22}{2} / 24 \right\rceil = 10$. We can attain 10 projects as follows.

Students $(\mathbb{Z}_9 \times \{1, 2\}) \cup (\mathbb{Z}_3 \times \{3\}) \cup \{\infty\}$

Project Assignments
Nine projects:

$$
\left.
\begin{array}{ll}
(0,1), (1,1), (1,2), (5,2) & \infty, (8,1), (7,2) \\
(5,1), (2,2), (3,2) & (3,1), (7,1), (0,3) \\
(2,1), (4,1), (4,2) & (6,1), (0,2), (1,3) \\
& (6,2), (8,2), (2,3)
\end{array}
\right\} \text{ mod } 9
$$

(Elements with second coordinate 3 are to be developed mod 3.)

The tenth project assignment is

$(0,3), (1,3), (2,3), \infty$
$(0,1), (3,1), (6,1)$ $\quad (1,1), (4,1), (7,1)$ $\quad (2,1), (5,1), (8,1)$
$(0,2), (3,2), (6,2)$ $\quad (1,2), (4,2), (7,2)$ $\quad (2,2), (5,2), (8,2)$.

Here the doubly covered pairs are $\{(i,1), (i,2)\}$, $i = 0, 1, \ldots, 8$, that is, nine vertex disjoint K_2s. Each of the students in $\{(0,3), (1,3), (2,3), \infty\}$ works exactly once with each of her classmates.

We have seen examples of (optimum) packing and covering designs for $v \equiv 4, 5 \pmod 6$. We will see in the next subsection that dealing with the case $v \equiv 2 \pmod 6$ is a simple consequence of the existence of Kirkman Triple Systems; one merely deletes one point from a $KTS(v+1)$ to get an exact covering of v points by factors of the required type. We finish this subsection by illustrating examples of optimum packing and covering designs for $v \equiv 1 \pmod 6$.

Example 42 *(i)* $\bar{r}(19) = 9$

Students $\{1, 2, \ldots, 18\} \cup \{\infty\}$

Project Assignments

∞ 2 15	4 7 18	5 8 9	6 12 14	11 16 17	1 3 10 13
∞ 3 16	5 8 10	6 9 1	7 13 15	12 17 18	2 4 11 14
∞ 4 17	6 9 11	7 1 2	8 14 16	13 18 10	3 5 12 15
∞ 5 18	7 1 12	8 2 3	9 15 17	14 10 11	4 6 13 16
∞ 6 10	8 2 13	9 3 4	1 16 18	15 11 12	5 7 14 17
∞ 7 11	9 3 14	1 4 5	2 17 10	16 12 13	6 8 15 18
∞ 8 12	1 4 15	2 5 6	3 18 11	17 13 14	7 9 16 10
∞ 9 13	2 5 16	3 6 7	4 10 12	18 14 15	8 1 17 11
∞ 1 14	3 6 17	4 7 8	5 11 13	10 15 16	9 2 18 12

In this assignment the doubly covered pairs form a near-triangle-factor, i.e. $\{1, 4, 7\}$, $\{2, 5, 8\}$, $\{3, 6, 9\}$, $\{10, 13, 16\}$, $\{11, 14, 17\}$, $\{12, 15, 18\}$. Student ∞ works exactly once with each of his classmates.

(ii) $r(25) = 11$

Students $(\mathbb{Z}_{11} \times \{1, 2\}) \cup \{\infty_1, \infty_2, \infty_3\}$

Project Assignments

$$
\left.
\begin{array}{ll}
(3, 1), (7, 1), (5, 2), (6, 2) & (4, 1), (9, 1), (4, 2) \\
(0, 1), (8, 1), (10, 1) & (1, 1), (9, 2), \infty_1 \\
(0, 2), (2, 2), (8, 2) & (5, 1), (1, 2), \infty_2 \\
(2, 1), (3, 2), (7, 2) & (6, 1), (10, 2), \infty_3
\end{array}
\right\} \text{ mod } 11
$$

The uncovered pairs in this case form the single triangle $\{\infty_1, \infty_2, \infty_3\}$. Thus the project assignments form the parallel classes of a resolvable $\{3, 4\}$-GDD of type $1^{22}3^1$ with replication number 11.

4.2 v ≡ 2(mod 3)

Kirkman Packing and Covering Designs Following Phillips, Wallis and Rees [36] we define a Kirkman Packing Design $KPD(v)$ (resp. Kirkman Covering Design $KCD(v)$) to be a resolvable packing (resp. covering) of a v-set by the maximum possible number $r(v)$ (resp. minimum possible number $\bar{r}(v)$) of factors, each of which is composed of $(v-2)/3$ triples and one pair. Thus Example 39(i) is a $KPD(17)$ while Example 41(i) is a $KCD(17)$. For $v \equiv 2(\bmod 3)$ the determination of $r(v)$ and $\bar{r}(v)$ is a simple consequence of Theorems 38 and 40:

Theorem 43 (i) *For every* $v \equiv 2(\bmod 6)$, $r(v) = \bar{r}(v) = v/2$;
(ii) *For every* $v \equiv 5(\bmod 6)$ *with* $v \geq 17$, $r(v) = (v-1)/2$ *and* $\bar{r}(v) = (v+1)/2$.

Proof. Delete one point from an optimum packing or covering on $v+1$ points. □

Now it is easy to see that $r(5) = 1$, and Černý, Horák and Wallis [14] have determined that $r(11) = 4$ (a maximum packing can be obtained by deleting one point from the packing in Table 1.3). On the other hand we have $\bar{r}(5) = 3$:

$$
\begin{array}{ccc}
\begin{array}{cc} 1 & 2 \\ 3 & 4 \ \ 5 \end{array} &
\begin{array}{cc} 2 & 5 \\ 1 & 3 \ \ 4 \end{array} &
\begin{array}{cc} 1 & 5 \\ 2 & 3 \ \ 4 \end{array}
\end{array}
$$

By Theorem 40, $\bar{r}(11) \leq 7$. It remains an open problem to determine whether or not one can construct a resolvable covering of 11 points by six factors of the foregoing type.

Packings by Factors Consisting of (v − 5)/3 Triples and a Quintuple In consideration of applications to cryptography, Cao and Zhu [10] and Cao [9] have established, up to 11 unsettled values of $v \equiv 2(\bmod 3)$, the maximum number $m(v)$ of factors possible in a resolvable packing of v points, where each factor is composed of $(v-5)/3$ triples and one quintuple.

Example 44 $m(20) = 6$
Each factor consists of five triples and a quintuple, for a total of 25 pairs covered. Hence $m(20) \leq \lfloor \binom{20}{2}/25 \rfloor = 7$. Now Cao and Zhu establish that 7 factors cannot be achieved, and provide the following solution with 6 factors.

Points $\{1, 2, \ldots, 18\} \cup \{\infty_1, \infty_2\}$

Factors

5, 6, 13	15, 11, 7	16, 18, 1	3, 12, ∞_1	10, 17, ∞_2	2, 4, 8, 9, 14
6, 1, 14	16, 12, 8	17, 13, 2	4, 7, ∞_1	11, 18, ∞_2	3, 5, 9, 10, 15
1, 2, 15	17, 7, 9	18, 14, 3	5, 8, ∞_1	12, 13, ∞_2	4, 6, 10, 11, 16
2, 3, 16	18, 8, 10	13, 15, 4	6, 9, ∞_1	7, 14, ∞_2	5, 1, 11, 12, 17
3, 4, 17	13, 9, 11	14, 16, 5	1, 10, ∞_1	8, 15, ∞_2	6, 2, 12, 7, 18
4, 5, 18	14, 10, 12	15, 17, 6	2, 11, ∞_1	9, 16, ∞_2	1, 3, 7, 8, 13

Theorem 45 (Cao and Zhu, Cao) *Let* $S = \{23, 29, 59, 83, 107, 155,$ $173, 179, 185, 197\}$. *Then*

(i) $m(v) = 1$ *for* $v \in \{5, 8, 11, 14\}$, *and* $m(17) = 4$;

(ii) $m(v) = (v - 8)/2$ *for* $v \equiv 2(\bmod 6)$ *and* $v \geq 20$, $v \neq 26$;

(iii) $m(v) = (v - 7)/2$ *for every* $v \equiv 5(\bmod 6)$ *with* $v > 17$, *except possibly for* $v \in S$.

The only unsettled value $v \equiv 2(\bmod 6)$ is $v = 26$; in [10] it is indicated that $8 \leq m(26) \leq 10$.

4.3   v ≡ 1(mod 3)

Kirkman Packing and Covering Designs   Again following [36], we define a Kirkman Packing Design $KPD(v)$ (resp. Kirkman Covering Design $KCD(v)$) to be a resolvable packing (resp. covering) of a v-set by the maximum possible number $r(v)$ (resp. minimum possible number $\bar{r}(v)$) of factors, each of which is composed of $(v - 4)/3$ triples and one quadruple. Thus Example 39(ii) is a $KPD(22)$ and Example 41(ii) is a $KCD(22)$, while Example 42 illustrates a $KCD(19)$ and a $KPD(25)$.

Unlike the case $v \equiv 2(\bmod 3)$ where (notwithstanding $v = 5, 11$) the leaves and excesses of $KPD(v)$s and $KCD(v)$s are uniquely determined (i.e. the empty graph when $v \equiv 2(\bmod 6)$, and a near-one-factor when $v \equiv 5(\bmod 6)$), the leave or excess of a $KPD(v)$ or $KCD(v)$ can take many forms when $v \equiv 1(\bmod 3)$.

Example 46 *A* $KCD(19)$

Point Set $\{1, 2, \ldots, 18\} \cup \{\infty\}$

Factors

∞, 7, 14	6, 3, 5	2, 16, 12	8, 17, 11	1, 18, 10	9, 4, 15, 13
∞, 8, 15	7, 4, 6	3, 17, 13	9, 18, 12	2, 10, 11	1, 5, 16, 14
∞, 9, 16	8, 5, 7	4, 18, 14	1, 10, 13	3, 11, 12	2, 6, 17, 15
∞, 1, 17	9, 6, 8	5, 10, 15	2, 11, 14	4, 12, 13	3, 7, 18, 16
∞, 2, 18	1, 7, 9	6, 11, 16	3, 12, 15	5, 13, 14	4, 8, 10, 17
∞, 3, 10	2, 8, 1	7, 12, 17	4, 13, 16	6, 14, 15	5, 9, 11, 18
∞, 4, 11	3, 9, 2	8, 13, 18	5, 14 17	7, 15, 16	6, 1, 12, 10
∞, 5, 12	4, 1, 3	9, 14, 10	6, 15, 18	8, 16, 17	7, 2, 13, 11
∞, 6, 13	5, 2, 4	1, 15, 11	7, 16, 10	9, 17, 18	8, 3, 14, 12

The excess consists of nine disjoint digons (i.e. two copies of $\{i, i+9\}$, $i = 1, 2, \ldots, 9$); compare to Example 42(i).

We define a canonical Kirkman Packing Design $CKPD(v)$ to be a $KPD(v)$ with $\lfloor (v-3)/2 \rfloor$ factors whose leave is a triangle (K_3) if $v \equiv 1 (\bmod 6)$, or the vertex disjoint union of a K_4 with $(v-4)/2$ edges (K_2s) if $v \equiv 4 (\bmod 6)$. Similarly, we define a canonical Kirkman Covering Design $CKCD(v)$ to be a $KCD(v)$ with $\lfloor (v-1)/2 \rfloor$ factors whose excess is the vertex disjoint union of $(v-1)/3$ triangles (K_3s) if $v \equiv 1 (\bmod 6)$, or the vertex disjoint union of $(v-4)/2$ edges (K_2s) if $v \equiv 4 (\bmod 6)$. Thus the packings and coverings given in Examples 39(ii), 41(ii) and 42 are all canonical. The spectrum for $CKPD$s and $CKCD$s has been almost completely determined, see [14], [36], [17] and [9].

Theorem 47 *Let* $v \equiv 1 (\bmod 3)$.

(i) *There exists a canonical Kirkman Packing Design* $CKPD(v)$ *if and only if* $v \geq 22$, *except possibly for* $v \in \{55\}$;

(ii) *There exists a canonical Kirkman Covering Design* $CKCD(v)$ *if and only if* $v > 10$, *except possibly for* $v \in \{13, 16, 67\}$.

Kirkman Frames play an essential role in the proof of Theorem 47. To illustrate how frames come to be used here we construct a class of (non-canonical) Kirkman Covering Designs on $v \equiv 1 (\bmod 6)$ points whose excess consists of the vertex disjoint union of $(v-1)/2$ digons (see [36, Theorem 2.3]):

Theorem 48 *There exists a* $KCD(v)$ *with* $(v-1)/2$ *factors whose excess consists of the vertex disjoint union of* $(v-1)/2$ *digons* ($2K_2$s) *for every* $v \equiv 1 (\bmod 6)$ *except possibly for* $v = 13$.

Proof. For $v = 7$, take the point set $\{\infty, 1, 2, 3, 4, 5, 6\}$ with the three factors $\infty 1 4$, $2 3 5 6$; $\infty 2 5$, $3 1 6 4$; $\infty 3 6$, $1 2 4 5$. For $v = 19$, see Example 46. Now let $v \geq 25$ and write $v = 6u + 1$, $u \geq 4$. By Theorem 17 there is a Kirkman Frame of type 6^u. Adjoin one ideal point to the frame, and on each hole in the frame together with the ideal point construct a copy of the foregoing covering design on 7 points, taking ∞ to be the ideal point. The result follows (compare to the proof of Theorem 18). □

Remark 49 Theorems 47(ii) and 48 raise the following interesting question. Let $v \equiv 1 \pmod 6$ and suppose that k divides $v - 1$. Can one construct a $KCD(v)$ whose excess consists of $(v - 1)/k$ vertex disjoint k-cycles? The foregoing results provide essentially complete solutions for $k = 2$ and 3; a particularly interesting (and undoubtedly difficult) case would be to consider $k = v - 1$, i.e. construct a $KCD(v)$ having a near-hamiltonian excess.

Theorem 50 *Let $v \equiv 1 \pmod 3$.*

(i) $r(4) = r(7) = r(10) = 1$, $r(13) = 4$ *and* $r(v) = \lfloor (v - 3)/2 \rfloor$ *for all* $v \geq 16$ *except possibly for* $v = 19$ *or* $v \in \{55, 61, 67, 73, 85, 109\}$;

(ii) $\bar{r}(v) = \lfloor (v - 1)/2 \rfloor$ *for all* $v \geq 4$.

Proof. (i) From Theorem 47(i) we need consider only $v < 22$. Clearly $r(4) = r(7) = r(10) = 1$. The following $KPD(16)$, having six factors, is from [17] (its leave consists of the three 4-cycles $(1, 2, 3, 4)$, $(5, 6, 7, 8)$, and $(9, 10, 11, 12)$).

6	4	14	5	9	15	11	8	3	2	7	12	1	10	13	16
6	3	13	1	8	15	11	9	7	10	12	4	5	2	14	16
10	7	14	5	12	13	1	3	9	2	4	8	6	11	15	16
10	2	15	7	3	16	1	12	6	5	4	11	9	8	13	14
8	12	16	1	11	14	10	3	5	6	9	2	4	7	13	15
9	4	16	2	11	13	6	8	10	5	7	1	3	12	14	15

Černý, Horák, and Wallis [14] establish that 5 factors on 13 points cannot be achieved; instead $r(13) = 4$ and the unique packing achieving this bound is obtained by adjoining a point ∞ to the packing given in Table 13.3:

1, 5, 9, ∞	2, 6, 10	3, 7, 11	4, 8, 12
1, 6, 11	2, 5, 12	3, 8, 9	4, 7, 10, ∞
1, 7, 12	2, 8, 11, ∞	3, 5, 10	4, 6, 9
1, 8, 10	2, 7, 9	3, 6, 12, ∞	4, 5, 11

(ii) From Theorems 47(ii) and 48 we need consider only $v \in \{4, 10, 13, 16\}$. Clearly $\bar{r}(4) = 1$. Now we can obtain a covering of 10 points with four factors by adjoining a point ∞ to a Kirkman Triple System $KTS(9)$ (see Table 1.1) in the following manner.

∞,	1,	4,	7	2,	5,	8	3,	6,	9
∞,	1,	5,	9	2,	6,	7	3,	4,	8
∞,	1,	2,	3	4,	5,	6	7,	8,	9
∞,	1,	6,	8	2,	4,	9	3,	5,	7

We obtain a covering of 16 points by seven factors similarly, starting with a $KTS(15)$ and adjoining a point ∞ to every triple containing some fixed point x in the KTS. (Clearly this technique will work to produce a $KCD(v)$ for every $v \equiv 4(\mod 6)$, see [36, Theorem 1.10]; the excess will consist of $(v-4)/2$ copies of the edge $\{\infty, x\}$, which may not be desirable for most applications.) Finally, the following covering of 13 points with 6 factors is from [36]; its excess is the graph obtained by splitting one vertex of a cube into three distinct vertices, maintaining 12 edges.

2	7	9	5	6	11	8	12	13	1	3	4	10
2	5	12	4	6	9	1	7	13	3	10	8	11
2	6	10	3	7	12	9	11	13	1	4	5	8
1	2	11	3	6	13	5	7	10	4	8	9	12
2	4	13	6	7	8	10	11	12	1	5	3	9
2	3	8	4	7	11	1	6	12	5	9	10	13

This completes the proof of the Theorem. □

With regards to $v = 19$ in Theorem 50(i), it is indicated in [14] that $6 \leq r(19) \leq 8$.

Separable Exact 2-coverings of v Points by Near-triangle-factors

Phillips, Wallis and Rees [37] consider the following analogue to Theorem 32. For which $v \equiv 1(\mod 3)$ is it possible to construct a near-resolvable $(v, 3, 2)$-$BIBD$ whose near-triangle-factors admit a partition into a maximum packing and a minimum covering of v points by near-triangle-factors?

Example 51 (i) $v = 10$

	Packing				Covering	
4 7 9	3 6 8	2 5 10 (1)		1 3 7	2 4 8	6 9 10 (5)
1 6 9	4 5 8	3 7 10 (2)		1 3 8	2 4 7	5 9 10 (6)
2 8 9	1 5 7	4 6 10 (3)		1 4 9	2 3 10	5 6 8 (7)
3 5 9	2 6 7	1 8 10 (4)		1 4 10	2 3 9	5 6 7 (8)
				1 2 5	3 4 6	7 8 10 (9)
				1 2 6	3 4 5	7 8 9 (10)

Between them these ten near-triangle-factors form a near-resolvable $(10, 3, 2)$-$BIBD$.

(ii) $v = 19$

For the packing we take the near-triangle-factors of a Hanani Triple System $HTS(19)$ (see Example 33):

$$\left.\begin{array}{llllll}
\infty, & (0,1), & (0,2) & (4,1), & (5,2), & (7,2) \\
(1,1), & (3,2), & (8,2) & (5,1), & (7,1), & (4,2) & (6,2) \\
(2,1), & (3,1), & (8,1) & (6,1), & (1,2), & (2,2)
\end{array}\right\} \bmod 9$$

For the covering we then take the near-triangle-factors of a second $HTS(19)$, on the same point set, whose short class is vertex-disjoint from the short class of the first $HTS(19)$:

$$\left.\begin{array}{llllll}
\infty, & (0,2), & (0,1) & (4,2), & (5,1), & (7,1) \\
(1,2), & (3,1), & (8,1) & (5,2), & (7,2), & (4,1) & (6,1) \\
(2,2), & (3,2), & (8,2) & (6,2), & (1,1), & (2,1)
\end{array}\right\} \bmod 9$$

The tenth near-triangle-factor in the covering is the union of the two short classes in the $HTS(19)$s:

$$(0,2), (3,2), (6,2) \quad (1,2), (4,2), (7,2) \quad (2,2), (5,2), (8,2)$$
$$(0,1), (3,1), (6,1) \quad (1,1), (4,1), (7,1) \quad (2,1), (5,1), (8,1) \quad (\infty)$$

Between them these 19 near-triangle-factors form a near-resolvable $(19, 3, 2)$-$BIBD$.

The following result is established in [37].

Theorem 52 *For every $v \equiv 1 \pmod 3$, $v \neq 7$, there exists a near-resolvable $(v, 3, 2)$-$BIBD$ whose near-triangle-factors admit a partition into a maximum packing and a minimum covering of v points by near-triangle-factors.*

In all cases in Theorem 52 a maximum packing of v points consists of $\lfloor(v-1)/2\rfloor$ near-triangle-factors while a minimum covering of v points consists of $\lceil(v+1)/2\rceil$ near-triangle-factors. Colbourn and Zhao [18] have independently determined that for all $v \equiv 1(\bmod 3)$, $v \neq 7$, there exists a balanced packing of $\lfloor(v-1)/2\rfloor$ near-triangle-factors on v points, balanced meaning that each point is not covered by at most one near-triangle-factor (the packings in Theorem 52 necessarily have this property, as do the coverings). With regards the case $v = 7$, there is a near-resolvable $(7,3,2)$-$BIBD$; however a maximum packing consists of just one near-triangle-factor, while a minimum covering consists of five near-triangle-factors (see [37], Proposition 1.6 and Remark 1.7).

5. Other Generalizations

5.1 Subsystems in Kirkman Triple Systems and Nearly Kirkman Triple Systems

If $(X, \mathcal{G}, \mathcal{B})$ is a resolvable 3-GDD of type g^v with a partition P of $\mathcal{B}$ into parallel classes, and $(X', \mathcal{G}', \mathcal{B}')$ is a resolvable 3-GDD of type g^w with a partition P' of $\mathcal{B}'$ into parallel classes, we say that $(X', \mathcal{G}', \mathcal{B}', P')$ is *embedded in* (or *is a subsystem of*) $(X, \mathcal{G}, \mathcal{B}, P)$ if $X' \subseteq X$, $\mathcal{G}' \subseteq \mathcal{G}$, $\mathcal{B}' \subseteq \mathcal{B}$ and if for every $p' \in P'$ there is a $p \in P$ such that $p' \subseteq p$. This last condition implies that if $v > w$ then in fact $v \geq 3w$. If we remove the blocks and groups (but not the points) from the subsystem we obtain an *incomplete* resolvable 3-GDD of type g^v with a hole of size gw, which we denote as 3-$IRGDD(g; gv, gw)$. More generally, we define a 3-$IRGDD(g; gv, gw)$ to be a triple $(X, \mathcal{G}, \mathcal{B})$ where X is a set of gv points, $\mathcal{G}$ is a partition of X into a gw-set H (called the hole) together with $v - w$ g-sets (called groups) and $\mathcal{B}$ is a collection of 3-subsets of X such that

(i) every pair of elements from X is contained either in some member of $\mathcal{G}$ or in exactly one member of $\mathcal{B}$ but not both, and

(ii) $\mathcal{B}$ admits a partition into $(gv - gw)/2$ parallel classes on X and $g(w - 1)/2$ holey parallel classes on $X \backslash H$.

When $g = 1$ a a 3-$IRGDD(1; v, w)$ is an Incomplete Kirkman Triple System, denoted $IKTS(v, w)$. The spectrum for these designs was determined by Rees and Stinson (see [50],[47] and [48]):

Theorem 53 *There exists an $IKTS(v, w)$ if and only if $v \equiv w \equiv 3(\bmod 6)$ and $v \geq 3w$.*

By filling in the hole of size w in Theorem 53 we obtain the following.

Corollary 53.1 *There exists a $KTS(v)$ containing a $KTS(w)$ as a subsystem if and only if $v \equiv w \equiv 3(\bmod 6)$ and $v \geq 3w$.*

More recently, Tang and Shen [52], Deng, Rees and Shen ([21] and [22]) and Deng [20] have considered the case $g = 2$. Here a 3-$IRGDD(2; 2v, 2w)$ is an incomplete Nearly Kirkman Triple System, denoted $INKTS(2v, 2w)$. The following result has been established.

Theorem 54 *There exists an $INKTS(2v, 2w)$ if and only if $v \equiv w \equiv 0(\bmod 3)$ and $v \geq 3w$, except possibly for $3w < v < 3.5w$ where $w \geq 39$.*

Thus the problem of constructing $INKTS$ with 'large' holes remains to be considered. (Deng [20] has reduced the problem of filling in the gap in Theorem 54 to that of determining the existence of just 30 designs.) Note that if $w = 3$ or 6 then the hole in an $INKTS(2v, 2w)$ cannot be filled as there do not exist $NKTS(6)$ or $NKTS(12)$ (for examples of such $INKTS(2v, 2w)$ see the proofs of Lemma 21 and Theorem 26). For $w \geq 9$ we get the following corollary to Theorem 54.

Corollary 54.1 *There exists an $NKTS(v)$ containing an $NKTS(w)$ as a subsystem if and only if $v \equiv w \equiv 0(\bmod 6), w \geq 18$ and $v \geq 3w$, except possibly for $3w < v < 3.5w$ where $w \geq 78$.*

Once again Kirkman Frames play an essential rôle in establishing the foregoing results. Thus for example the Stinson construction for $KTS(v)$s (Theorem 18) actually produces a $KTS(v)$ containing a KTS (9) as a subsystem for all $v \geq 27$. Similarly, the Brouwer construction for $NKTS(v)$s of order $v \equiv 6(\bmod 12), v \geq 66$, produces an $NKTS(v)$ containing an $NKTS(18)$ as a subsystem for those values of v (see [7]).

5.2   Kirkman Squares

Let $r \equiv 1(\bmod 3)$. A Kirkman Square of side r, denoted $KS(r)$, is an $r \times r$ array each of whose cells is either empty or contains an unordered triple of elements from a $(2r + 1)$-set X with the following properties.

(i) The collection of triples in the array forms a Steiner Triple System $STS(2r + 1)$ on the point set X;

(ii) Each row and each column of the array forms a partition of X.

Thus the rows and columns of a $KS(r)$ each form the parallel classes of a Kirkman Triple System $KTS(2r + 1)$, each parallel class of the row-induced system intersecting with each parallel class of the column-induced system in at most one block. Two resolutions of an $STS(2r+1)$ with this property are said to be *orthogonal*.

A considerable amount of effort has been expended, by several researchers, on the existence problem for Kirkman Squares over the last 25 years. At the time of Stinson's survey, only an asymptotic existence result was known. This was because of the scarcity of squares of small side, particularly of sides $r \equiv 4(\bmod 6)$, combined with the known non-existence of squares of sides 4 and 7. Quite recently, Colbourn, Lamken, Ling and Mills [16] constructed squares for several new small sides and have reduced the existence problem for Kirkman Squares to 23 unsettled cases:

Theorem 55 *There exists a $KS(r)$ for every $r \equiv 1(\bmod 3)$ except for $r = 4, 7$ and possibly for $r \in \{10, 28, 34, 46, 49, 52, 58, 70, 73, 76, 82, 88, 91, 94, 100, 115, 118, 124, 130, 133, 142, 175, 178\}$.*

5.3   Maximal Sets of Triangle-Factors

If the graph G admits an edge-decomposition $\mathcal{C}$ into triangle-factors and the complement of G (in $K_{|V(G)|}$) does not contain a triangle-factor then we say that $\mathcal{C}$ is *maximal*.

Example 56 The packing in Table 1.4 constitutes a maximal set of 4 triangle-factors on 12 vertices; consider now the following packing.

1	5	9	1	6	11	1	7	10
2	6	10	2	7	12	2	8	11
3	7	11	3	8	9	3	5	12
4	8	12	4	5	10	4	6	9

It is easy to see that the only triangles in the complement of the graph formed by the pairs in these triangle-factors are contained entirely within the K_4s with vertex-sets $\{1, 2, 3, 4\}$, $\{5, 6, 7, 8\}$ and $\{9, 10, 11, 12\}$. Hence we have a maximal set of 3 triangle-factors on 12 vertices.

Following Rees, Rosa and Wallis [45], we write $F(v)$ to denote the set of all orders k such that there exists a maximal set of k triangle-factors on v vertices. Now a fundamental result of Corrádi and Hajnal [19] states that if G is a graph on $3m$ vertices with minimum degreee $\delta(G) \geq 2m$ then G has a triangle-factor. It follows that $F(3m) \subseteq \{n : m/2 \leq n \leq (3m-1)/2\}$, i.e. $F(v) \subseteq \{n : \lceil v/6 \rceil \leq n \leq \lfloor (v-1)/2 \rfloor\}$.

Obviously $F(3) = F(6) = \{1\}$. Now it is not difficult to see that the only maximal set of triangle-factors on nine vertices forms a $KTS(9)$, i.e. $F(9) = \{4\}$. For $v = 12$ there is only one way to construct two triangle-factors on 12 vertices, that being (up to renumbering) the first two factors in Example 56; this set can be extended (by the third factor therein) and so $F(12)$ does not contain the element 2. On the other hand the non-existence of an $NKTS(12)$ means that $F(12)$ does not contain 5 either. Hence from Example 56 we have $F(12) = \{3, 4\}$.

The spectrum $F(v)$ has been almost completely determined for all but a few values of v (see [45],[43] and [44]):

Theorem 57 (i) *Let* $v \equiv 0(\mathrm{mod}\,6), v \geq 18$. *Then there exists a maximal set of k triangle-factors on v vertices if and only if $v/6 \leq k \leq v/2 - 1$, except for $(k,v) = (3,18)$ and except possibly for $k = v/6$ where $v \equiv 0$ or $12(\bmod 18), v \geq 30$.*

(ii) *Let* $v \equiv 3(\bmod 6), v \geq 15, v \neq 45, 57, 69, 81, 93, 237, 261, 309, 333, 381$. *Then there exists a maximal set of k triangle-factors on v vertices if and only if $(v+3)/6 \leq k \leq (v-1)/2$, except for $(k,v) = (3,15)$ and except possibly for $k = (v+3)/6$ where either $v \equiv 9(\bmod 18)$ or $v = 33$.*

Thus notwithstanding the ten excluded values of v in Theorem 57(ii) (which undoubtedly behave in the same manner as do all other $v \equiv 3(\bmod 6)$) it remains to be determined whether or not $\lceil v/6 \rceil \in F(v)$, where $v \equiv 0, 9$ or $12(\bmod 18), v \geq 27$, or $v = 33$. It is determined in [43] and [44] that if such a maximal set of $k = \lceil v/6 \rceil$ triangle-factors on v vertices exists, then its leave (i.e. its complement in K_v) *cannot* contain an independent set of size greater than $v/3$, and so the methods employed to obtain Theorem 57 (which are very design-theoretical, including the use of resolvable group-divisible designs with block size three and Kirkman Frames) cannot work in these cases. (In all maximal sets constructed to establish Theorem 57 their leaves contain independent sets of size greater than $v/3$, which of course guarantees maximality.) It is our feeling that in fact no maximal sets of triangle-factors with the foregoing parameters exist, but it appears to be a very challenging problem to establish this.

5.4   Large Sets of Kirkman Triple Systems

A *large set* of Kirkman Triple Systems $LKTS(v)$ is a collection of $v-2$ Kirkman Triple Systems of order v, on the same point set X, no two of which have any blocks in common. Equivalently, an $LKTS(v)$ is a partition of the set of 3-element subsets of a v-set X into Kirkman Triple Sysytems of order v. Stinson [51] displays examples of $LKTS(9)$ and $LKTS(15)$ and notes some other small values of v for which $LKTS(v)$ have been constructed, namely $v = 33, 51, 75, 105$ and 129. More recently, Chang and Ge [11] and Ge [23] have given direct constructions for $v = 201, 273$ and 369. Wu ([57] and [58]) has constructed $LKTS(v)$ for $v = 3^n m, n \geq 1$ and $m = 11, 17, 35$ or 41.

One of the most significant recent advances made in the study of large sets is the following tripling construction due to Zhang and Zhu [59], which improves an earlier construction of Denniston by weakening the hypothesis.

Theorem 58 *If there is an $LKTS(v)$ then there is an $LKTS(3v)$.*

From Theorem 58 more general product constructions are developed in Lei ([31] and [32]) and Zhang and Zhu [60]. While these new constructions are useful in generating many infinite families of large sets, determining the full spectrum for $LKTS(v)$ remains a very challenging open problem.

6.   Conclusion and Acknowledgements

We conclude our survey by posing a number of open problems, some of which have already been alluded to in the various sections.

1. Find a theoretical proof for the non-existence of an $NKTS(12)$. (It is determined by computer search in [29] that there does not exist a resolvable {4}-GDD of type 2^{10}. We suspect that these non-exisience results are related.)

2. Determine whether or not there is a resolvable covering of 11 points by 6 factors, each of which is composed of three triples and one pair.

3. Complete the determination of $m(v)$ for all $v \equiv 2(\bmod 3)$ (see Theorem 45, Section 4.2).

4. Complete the spectra for $CKPD$s and $CKCD$s (see Theorem 47, Section 4.3).

5. Determine the packing number $r(19)$ (see Theorem 50, Section 4.3). In particular, if $r(19) = 8$, is there an optimal packing whose leave consists of three vertex-disjoint edges?

6. See Remark 49. In particular, for which $v \equiv 1(\bmod 6)$ does there exist a $KCD(v)$ whose excess is a near-hamiltonian cycle?

7. Complete the spectrum for embeddings of nearly Kirkman triple systems (Section 5.1).

8. Complete the spectrum for Kirkman Squares. In particular, determine whether or not there exists a $KS(10)$ (Section 5.2).

9. Determine whether or not there exists a maximal set of $\lceil v/6 \rceil$ trianglefactors on v vertices where $v \equiv 0, 9$ or $12(\bmod 18), v \geq 27$ or $v = 33$. In particular, is $6 \in F(33)$? (Section 5.3)

10. Complete the spectrum for large sets of Kirkman Triple Systems (Section 5.4).

Much of the research for this paper was conducted while the first author was visiting Southern Illinois University at Carbondale; this author wishes to thank the institution for its hospitality. Research of the first author is supported in part by NSERC grant OGP0107993.

The authors would like to thank Profs. C. Rodger and L. Zhu for many helpful suggestions in the preparation of this article.

References

[1] A. Assaf and A. Hartman, Resolvable group divisible designs with block size 3. *Discrete Math.* **77** (1989), 5–20.

[2] A. Assaf, E. Mendelsohn and D. R. Stinson, On resolvable coverings of pairs by triples. *Utilitas Math.* **32** (1987), 67–74.

[3] R. D. Baker and R. M. Wilson, Nearly Kirkman Triple Systems. *Utilitas Math.* **11** (1977), 289–296.

[4] R. C. Bose, On the application of the properties of Galois fields to the problem of construction of hyper-Graeco-Latin squares. *Sankyha*

3 (1938), 323–338.

[5] R. C. Bose, S. S. Shrikhande and E. T. Parker, Further results on the construction of mutually orthogonal Latin squares and the falsity of Euler's conjecture. *Canad. J. Math.* **12** (1960), 189–203.

[6] A. E. Brouwer, Optimal packings of K_4's into a K_n. *J. Combinatorial Theory* **26A** (1979), 278–297.

[7] A. E. Brouwer, Two new Nearly Kirkman Triple Systems. *Utilitas Math.* **12** (1978), 311–314.

[8] A. E. Brouwer, H. Hanani and A. Schrijver, Group divisible designs with block size four. *Discrete Math.* **20** (1977), 1–10.

[9] H. Cao, Combinatorial constructions for Threshold Schemes, Ph. D. thesis (in Chinese), Suzhou University, March 2002.

[10] H. Cao and L. Zhu, Kirkman Packing Designs KPD($\{3, 5^*\}, v$). *Designs, Codes and Cryptography*, to appear.

[11] Y. Chang and G. Ge, Some new large sets of KTS(v). *Ars. Combin.* **51** (1999), 306–312.

[12] J. Carter, M. Carter, C.J. Colbourn, R. Rees, A. Rosa, P.J. Schellenberg, D.R. Stinson and S.A. Vanstone, Hanani triple systems. *Israel J. Math.* **83** (1993), 305–319.

[13] A. Cayley, On the triadic arrangements of seven and fifteen things. *Philos. Mag.* **37** (1850), 50–53.

[14] A. Černý, P. Horák and W. Wallis, Kirkman's school projects. *Discrete Math.* **167/168** (1997), 189–196.

[15] C. J. Colbourn and J. H. Dinitz (Eds.), *CRC Handbook of Combinatorial Designs* (CRC Press, Boca Raton, FL, 1996).

[16] C. J. Colbourn, E. R. Lamken, A. C. H. Ling and W. H. Mills, The existence of Kirkman squares—doubly resolvable $(v, 3, 1)$-BIBDs. *Designs, Codes and Cryptography*, to appear.

[17] C. J. Colbourn and A. C. H. Ling, Kirkman school project designs. *Discrete Math.* **203** (1999), 49–60.

[18] C. J. Colbourn and S. Zhao, Maximum Kirkman signal sets for synchronous uni-polar multi-user communication systems. *Designs, Codes and Cryptography* **20** (2000), 219–227.

[19] K. Corrádi and A. Hajnal, On the maximal number of independent circuits in a graph. *Acta. Math. Acad. Sci. Hungar.* **14** (1963), 423–439.

[20] D. Deng, The embedding problem for Nearly Kirkman Triple Systems, Ph. D. thesis (in Chinese), Shanghai Jiao Tong University, March 2002.

[21] D. Deng, R. Rees and H. Shen, On the existence and application of incomplete nearly Kirkman triple systems with a hole of size 6 or 12. *Discrete Math.* , to appear.

[22] D. Deng, R. Rees and H. Shen, Further results on embeddings of nearly Kirkman triple systems. *Discrete Math.* , submitted.

[23] G. Ge, More large sets of KTS(v). *J. Comb. Math. Combin. Comp.* , to appear.

[24] R. W. Irving, Generalized Ramsey numbers for small graphs. *Discrete Math.* **3** (1974), 251–264.

[25] D. M. Johnson, A. L. Dulmage and N. S. Mendelsohn, Orthomorphisms of groups and orthogonal Latin squares I. *Canad. J. Math.* **13** (1961), 356–372.

[26] T. P. Kirkman, Query VI. *Lady's and Gentleman's Diary* (1850), 48.

[27] T. P. Kirkman, Note on an unanswered prize question. *Cambridge Dublin Math. J.* **5** (1850), 255–262.

[28] A. Kotzig and A. Rosa, Nearly Kirkman systems. *Congressus Num.* **10** (1974), 607–614.

[29] D. L. Kreher, A. C. H. Ling, R. S. Rees and C. W. H. Lam, A note on {4}-GDDs of type 2^{10}. *Discrete Math.* , to appear.

[30] E. R. Lamken and W. H. Mills, Resolvable Coverings. *Congressus Num.* **96** (1993), 21–26.

[31] J. Lei, On large sets of Kirkman Triple Systems. *Discrete Math.* , to appear.

[32] J. Lei, On large sets of Kirkman Triple Systems with holes, preprint.

[33] J. Lu, *Collected Works on Combinatorial Designs* (Inner Mongolia People's Press, Hinhot, Mongolia, 1990).

[34] R. Mathon and A. Rosa, Nearly Kirkman Triple Systems of order 18. *J. Comb. Math. Combin. Comp.* **39** (2001), 79–91.

[35] E. Mendelsohn and H. Shen, A construction of resolvable group divisible designs with block size three. *Ars. Combin.* **24** (1987), 39–43.

[36] N. C. K. Phillips, W. D. Wallis and R. S. Rees, Kirkman packing and covering designs. *J. Comb. Math. Combin. Comp.* **28** (1998), 299–325.

[37] N. C. K. Phillips, W. D. Wallis and R. S. Rees, Packings and coverings of $v = 3m + 1$ points with near-triangle-factors. *Australas. J. Combin.* **23** (2001), 53–70.

[38] D. K. Ray-Chaudhuri and R. M. Wilson, Solution of Kirkman's schoolgirl problem. *Proc. Symp. Pure Math.* **19** (American Mathematical Society, Providence, RI, 1971), 187–204.

[39] R. Rees, Uniformly resolvable pairwise balanced designs with block sizes two and three. *J. Comb. Theory* **45A** (1987), 207–225.

[40] R. Rees, Semiframes and nearframes. In *Combinatorics 88: Proc. of the Int'l. Conf. on Incidence Geometries and Combinatorial Structures* **2** (Mediterranean Press, Commenda di Rende, Italy, 1991), 359–367.

[41] R. Rees, The existence of restricted resolvable designs I: (1,2)-factorizations of K_{2n}. *Discrete Math.* **81** (1990), 49–80.

[42] R. Rees, Two new direct product-type constructions for resolvable group-divisible designs. *J. Combin. Designs* **1** (1993), 15–26.

[43] R. Rees, Maximal sets of triangle-factors on $v = 6m$ vertices. *J. Combin. Designs* **6** (1998), 235–244.

[44] R. Rees, Maximal sets of triangle-factors on $v = 6m + 3$ vertices. *J. Combin. Designs* **6** (1998), 309–323.

[45] R. Rees, A. Rosa and W. D. Wallis, Maximal Sets of Triangle-Factors. *Australas. J. Combin.* **9** (1994), 67–108.

[46] R. S. Rees and D. R. Stinson, On resolvable group-divisible designs with block size 3. *Ars. Combin.* **23** (1987), 107–120.

[47] R. S. Rees and D. R. Stinson, On the existence of Kirkman triple systems containing Kirkman subsystems. *Ars. Combin.* **26** (1988), 3–16.

[48] R. S. Rees and D. R. Stinson, Kirkman triple systems with maximum subsystems. *Ars. Combin.* **25** (1988), 125–132.

[49] D. R. Stinson, A proof of the non-existence of a pair of orthogonal Latin squares of order 6. *J. Combinatorial Theory* **36A** (1984), 373–376.

[50] D. R. Stinson, Frames for Kirkman Triple Systems. *Discrete Math.* **65** (1987), 289-300.

[51] D. R. Stinson, A survey of Kirkman Triple Systems and related designs. *Discrete Math.* **92** (1991), 371–393.

[52] S. Tang and H. Shen, Embeddings of nearly Kirkman triple systems. *J. Stat. Plan. Infer.* **94** (2001), 327–333.

[53] G. Tarry, Le problème des 36 officiers. *Comptes Rend. Assoc. Fr.* **1** (1900), 122–123; **2** (1901), 170–203.

[54] V. Todorov, Three mutually orthogonal Latin squares of order fourteen. *Ars Combinatoria* **20** (1985), 45–47.

[55] W. D. Wallis, Three orthogonal Latin squares. *Congressus Num.* **42** (1984), 69–86.

[56] R. M. Wilson, Construction and uses of pairwise balanced designs. *Math. Cent. Tracts* **55** (1974), 18–41.

[57] L. Wu, Large sets of KTS(v). In *Combinatorial Designs and Applications, Lecture Notes in Pure and Applied Math.* **126** (Marcel Dekker, New York, 1990), 175–178.

[58] L. Wu, Large Sets of KTS($3^n \cdot 41$). *J. Suzhou Univ.* **14**(2) (1998), 1–2.

[59] S. Zhang and L. Zhu, Transitive resolvable idempotent symmetric quasigroups and large sets of Kirkman Triple Systems. *Discrete Math.* , to appear.

[60] S. Zhang and L. Zhu, An improved product construction for large sets of Kirkman Triple Systems, preprint.